Endorobotics

Endorobotics

Design, R&D and Future Trends

Edited by

Luigi Manfredi
School of Medicine, University of Dundee, UK

ELSEVIER

ACADEMIC PRESS
An imprint of Elsevier

Academic Press is an imprint of Elsevier
125 London Wall, London EC2Y 5AS, United Kingdom
525 B Street, Suite 1650, San Diego, CA 92101, United States
50 Hampshire Street, 5th Floor, Cambridge, MA 02139, United States
The Boulevard, Langford Lane, Kidlington, Oxford OX5 1GB, United Kingdom

Library of Congress Cataloging-in-Publication Data
A catalog record for this book is available from the Library of Congress

British Library Cataloguing-in-Publication Data
A catalogue record for this book is available from the British Library

ISBN 978-0-12-821750-4

For information on all Academic Press publications visit
our website at https://www.elsevier.com/books-and-journals

Publisher: Mara Conner
Acquisitions Editor: Sonnini Yura
Editorial Project Manager: Joshua Mearns
Production Project Manager: Poulouse Joseph
Cover Designer: Miles Hitchen

Typeset by STRAIVE, India

Contents

Contributors

Numbers in paraentheses indicate the pages on which the authors' contributions begin.

Sara-Adela Abad (61), Department of Mechanical Engineering, University College London, London, United Kingdom; Institute for Applied Sustainability Research, Quito, Ecuador

Alberto Arezzo (61), Department of Surgical Sciences, University of Torino, Turin, Italy

Fabio Balducci Romano (297), University of Rome 'Tor Vergata', Roma, Italy

M. Taha Chikhaoui (187), Univ. Grenoble Alpes, CNRS, UMR 5525, VetAgro Sup, Grenoble INP, TIMC, Grenoble, France

Gastone Ciuti (31), The BioRobotics Institute; Department of Excellence in Robotics & AI, Scuola Superiore Sant'Anna, Pisa, Italy

Sandy Cochran (215), University of Glasgow, Glasgow, United Kingdom

Federico Costantini (297, 319), University of Udine, Udine, Italy

Giovanni Dapri (3), International School Reduced Scar Laparoscopy, Minimally Invasive General & Oncologic Surgery Center, Humanitas Gavazzeni University Hospital, Bergamo, Italy

Majdi Elhajji (275), Department of Electrical Engineering, College of Engineering, Shaqra University, Dawadmi, Saudi Arabia

Bingyong Guo (241), College of Engineering, Mathematics and Physical Sciences, University of Exeter, Exeter, United Kingdom; School of Marine Science and Technology, Northwestern Polytechnical University, Xi'an, Shaanxi, China

Shervanthi Homer-Vanniasinkam (61), Department of Mechanical Engineering, University College London, London, United Kingdom

Yu Huan (31), The BioRobotics Institute; Department of Excellence in Robotics & AI, Scuola Superiore Sant'Anna, Pisa, Italy

Nedra Jarray (275), Electronics and Microelectronics Laboratory, Faculty of Sciences of Monastir, University of Monastir, Monastir, Tunisia

Yang Liu (241), College of Engineering, Mathematics and Physical Sciences, University of Exeter, Exeter, United Kingdom

Erwin Loh (23), St Vincent's Health Australia, Melbourne, Australia

Luigi Manfredi (337, 359), Division of Imaging Science and Technology, School of Medicine, University of Dundee, Dundee, United Kingdom

Mario Milazzo (165), The BioRobotics Institute, Scuola Superiore Sant'Anna, Pontedera, PI, Italy

Srinjoy Mitra (215), University of Edinburgh, Edinburgh, United Kingdom

Andrea Moglia (95), EndoCAS, Center for Computer Assisted Surgery, University of Pisa, Pisa, Italy

Alexandru C. Moldovan (215), University of Glasgow, Glasgow, United Kingdom

Paul Motzki (117), Department of Systems Engineering, Saarland University; Center for Mechatronics and Automation Technology (ZeMA gGmbH), Saarbrücken, Germany

Tam Nguyen (23), St Vincent's Hospital Melbourne, Melbourne, Australia

Peter P. Pott (257), Institute of Medical Device Technology, University of Stuttgart, Stuttgart, Germany

Shyam Prasad (241), The Royal Devon and Exeter NHS Foundation Trust, Exeter, United Kingdom

Pietro Raffaini (337), Project Control Management, London, United Kingdom

Gianluca Rizzello (117), Department of Systems Engineering, Saarland University, Saarbrücken, Germany

Fabio Balducci Romano (297), University of Rome 'Tor Vergata', Roma, Italy

Benoît Rosa (187), CNRS, ICube Laboratory, UMR7357 CNRS-Strasbourg University, Illkirch-Graffenstaden, France

Giada Soncini (319), University of Udine, Udine, Italy

Cesare Stefanini (165), The BioRobotics Institute, Scuola Superiore Sant'Anna, Pontedera, PI, Italy; Healthcare Engineering Innovation Center, Department of Biomedical Engineering, Khalifa University of Science and Technology, Abu Dhabi, Unitted Arab Emirates

Jiyuan Tian (241), College of Engineering, Mathematics and Physical Sciences, University of Exeter, Exeter, United Kingdom

Mihnea V. Turcanu (215), University of Glasgow, Glasgow, United Kingdom

Helge A Wurdemann (61), Department of Mechanical Engineering, University College London, London, United Kingdom

Abdelkrim Zitouni (275), Department of Physics, College of Science & Humanities - Jubail, Imam Abdulrahman Bin Faisal University, Jubail Industrial, Saudi Arabia

Preface

Robots are currently being used in several environments and are part of our daily lifes. Advancement in collaborative robots (Cobots) has allowed their direct interaction with humans removing the need for restricted areas. When it comes to medical applications, safety is one of the major concerns, which requires to fulfill strict regulations to ensure that the device is safe. The required performance has been achieved by the progress made in sensors, actuators, fast active control systems, and light and soft materials, opening new design paradigms. Robots for medical applications can overcome limitations of standard diagnostic and therapeutic procedures. They can improve precision, safety, and reliability, enhance the user's dexterity in narrow areas, and provide a wider field for potential interventions with new procedures that are nowadays difficult to predict.

Endoscopy is one of the medical disciplines in which a robot can improve the traditional manual procedure thanks to sophisticated electronics and mechanical components for the implementation of complex tasks. Medicine is moving from endoscopy (endon—internal and skopein—observation, view) to a new discipline, endorobotics (internal—robots). Endorobots represent a new generation of robots that can navigate inside the human body to inspect, localize, treat lesions or diseases, or deliver drugs. A miniaturized high-definition camera allows visualization through natural orifices or small apertures and share high-quality videos not only locally but also remotely. Fast broadband communication has reduced the transmission delay and reduced the time of a control loop with a remote application. Smart instruments can measure the force applied against organs and tissues, they are flexible and soft to move through narrow spaces. Sophisticated consoles reduce stress and fatigue during an operation, provide 3D vision inside the human body, simplify the user's sensory-motor coordination by solving complex kinematic equations. Besides, simulators reduce the long period training of surgeons to become proficient.

The design of endorobots requires an interdisciplinary interface between physics, engineering, life sciences, and medicine. This wide range of expertise needs a common language to ensure an effective and efficient communication process. Their design includes sensors, actuators, mechanical and electronic components to fulfill the medical needs. New materials such as polymers and smart alloys are used to design a new generation of sensors and actuators, being flexible and stretchable or changing from soft to stiff. Robotics technologies have offered many opportunities to push the boundaries of the design.

In the past few years, several companies have brought into the market robots for endoscopic applications with the FDA's approval and CE mark, using technologies that were relegated to research laboratories in the past. The endorobotics global market is increasing exponentially every year. Leading companies are investing in medical robotics: Google, Johnson & Johnson, or Medtronics are just a few examples. Their complexity requires new training for engineers as the designers and for medical doctors as the final users. Biomedical engineering training is now well established. New MSc, such as "technical medicine," is now a new master's degree available at prestigious universities, i.e., TUDelft and Twente University. Worldwide meetings present last R&D technologies in robotics reporting medical applications (e.g., the European Association for Endoscopic Surgery [EAES] or the Society of American Gastrointestinal and Endoscopic Surgeons [SAGES]) and technical developments (e.g., IEEE International Symposium on Medical Robotics, ICRA, IROS).

This book focuses on a new design paradigm for endorobots applications. It addresses the need for endorobots describing the medical requirements and applications. It also presents design tools such as smart and soft materials for sensing and actuation, simulation, and the use of an effective awareness of law regulations processes and project management for product commercialization.

The book is a collection of contributions from the most significant and notable scientists in their field and is divided into three parts.

The first part presents the state of the art of robots for endoscopy (endorobots), including devices already available in the market and those that are still at the R&D stage. These robots will be classified in relation to the applications (e.g., diagnostic, treatment, and theragnostic) and design approach (e.g., master-slave, autonomous, tethered, or wireless). Chapter 1 offers a description of robotic systems for surgery and clinical applications. This covers the evolution of robots in medicine from the first model till the current state of the art of robotics devices. Chapter 2 describes the applications of artificial intelligence (AI) in medicine and how this discipline has transformed the role of doctors in practice medicine. This includes AI in medical robotics for teaching and training, together with ethical and legal issues. In Chapter 3, an evaluation of the different applications in colonoscopy with the current available devices and advancement of AI for imaging analysis is presented. The chapter not only presents the devices that are available at a research level, but also commercial products with the FDA's approval and CE mark. Chapter 4 provides with a deep understanding of how recent advancements on soft materials present advantages and limitations compared to their rigid counterparts. This section will also detail how nature can inspire such design. Virtual reality and different types of simulators for robotic assisted surgery are presented in Chapter 5. Evidence from the published work on validation studies is discussed at the end of the chapter.

The second part focuses on the materials and engineering design. Chapter 6 presents a deep overview on smart materials for the design of miniaturized actuators. Operating principles, major features and challenges, and application-related issues of shape memory alloys (SMAs) and dielectric elastomers are discussed, with a special focus toward robotics and biomedical systems. Chapter 7 presents the fabrication challenges of endoluminal microdevices with details on the physical laws that govern the miniaturization of a device, theoretical mechanical limits of miniaturized structures, and the different fabrication options related to applications. Examples of fabrication processes of rigid and flexible tethered systems for navigation and intervention are presented as well. In the first part of Chapter 8, the modeling approach developed in the last decades is presented. In the second part of the chapter, control strategies adopted for continuum robots are presented. Ultrasound technologies for capsule endoscopy as an alternative imaging modality with the capability to obtain information from beneath the visible surface of the lumen are presented in Chapter 9. Dynamic models of moving capsule endoscopes in the small intestine are debated in Chapter 10. The chapter reviews mathematical and finite element modeling to understand the interaction with the intestine. Haptic interfaces for tele robotic system, including engineering requirements, and technical solutions of existing devices are discussed in Chapter 11. A low-power compression architecture for wireless endoscopy is presented in Chapter 12. Details on the miniaturization approach and coding methods are presented and compared with the well-known approaches, including experimental results.

The third part includes ethics, regulation, and project management. Roboethics and fundamental principles of engineering ethics, including the regulatory framework in the EU, are discussed in Chapter 13. This discussion is extended to Chapter 14, which addresses the EU legal framework analyzing the impact of the field of endorobotics with the most recent regulations, including challenges and opportunities for engineers and designers. Project management of a multidisciplinary team, including Gantt chart, Pert chart, and the Technology Readiness Level approach, are presented in Chapter 15.

Chapter 16 concludes the book by describing the evolution of endorobots and points out the challenges ahead. This covers a description of the organ's anatomy and different design approaches to access difficult-to-reach areas. Challenges ahead include AI, ML (machine learning), miniaturization of smart materials, and ethics and regulation. Commercialization challenges are discussed at the end of the chapter.

Luigi Manfredi
Division of Imaging and Technology, School of Medicine, University of Dundee, Dundee, United Kingdom

Acknowledgments

I would like to start by thanking my awesome Verónica. She has helped me since the beginning, reading the early drafts, advising me on the selection of the covers, and supporting me during this journey. Thank you so much my dear.

Thanks to Alessandro, for giving me so much joy when I see him growing up, learning and taking his first steps in this turbulent, adventurous, and wonderful world. I wish you a bright and joyful future.

Thanks to my family for dealing with my up and down moments and understanding me in every moment of my life. I could not be here without your support.

Thanks to all the contributors to this book, for their commitment and the incredible effort made in writing each of the chapters. Without all of you, this book would not have been possible.

Thanks to Professor Gianfranco Natale for his comments and suggestions on the anatomical description in Chapter 16.

Thanks to the University of Dundee, the School of Medicine, and the Division of Imaging Science and Technology for providing me with a wonderful work environment that has enriched my knowledge, and a laboratory to carry on with my research.

Thanks to the Elsevier team, for the assistance, for the advice, and for giving me this incredible opportunity.

Part 1

State of the art of robots for endoscopy

Chapter 1

Robotics in surgery and clinical application

Giovanni Dapri
International School Reduced Scar Laparoscopy, Minimally Invasive General & Oncologic Surgery Center, Humanitas Gavazzeni University Hospital, Bergamo, Italy

1.1 Introduction

Surgery underwent a radical revolution in the 90s with the advent of minimally invasive surgery (MIS). Thanks to MIS, surgeon started to operate without big scars and without introduction of the hands into the cavities of the body, but watching TV screens and moving the hands together with long instruments through small incisions on the body's surface. MIS was firstly applied in the abdomen, but soon has been implemented in the other parts of the body: chest, neck, and head. The major advantages of MIS, if compared to open surgery, appeared to be a reduced postoperative pain, a reduced hospital length of stay, a reduced incidence of postoperative complications, an increased patient's comfort, and also an improved cosmetic result. Few years later, together with the MIS evolution, robot-assisted MIS (RAMIS) arrived in the surgical field. The main differences of RAMIS, compared to conventional MIS, were the 3D view, the stereoscopic and enhanced vision, the increased degrees of freedom with wristed instruments, the surgeon's tremor cancellation, the improved surgeon ergonomics, and the surgeons' control of a camera lens that locks onto a specific field of view. The evolution of conventional MIS invested the past 30 years and is currently still under further improvement. The evolution of RAMIS was mainly concentrated in the beginning of its arrival, but for 15 years, it was developed by only one surgical company. In the recent 10 years, the research and investment in RAMIS interested other surgical companies, some of them already in the market and some others still under development.

In this chapter, a review of the available surgical robots, of the known companies involved in the development of surgical robots, and also of the robotic tools is reported. A clinical application of the available data is finally inserted at the end of this chapter.

Endorobotics. https://doi.org/10.1016/B978-0-12-821750-4.00001-3

1.2 Robot-assisted minimally invasive surgery (RAMIS)

RAMIS represents a platform with many differences from conventional MIS and open surgery. Current robot systems function through a communication system in which surgical tasks are performed by a platform at the patient bedside, while the surgeon exerts direct control over this platform using a console, removed from direct contact with the patient. The surgeon's console allows the surgeon to control the laparoscopic camera and to "clutch" instruments, making it possible to use their full length, while the console masters are kept at a comfortable distance from the surgeon. The surgeon can employ more than two working arms at a time by swapping control among three engaged instruments. Additionally, the assistant at the patient bedside can change different instruments when necessary.

For more than one decade, the only surgical company available for the robot has been Intuitive Surgical Inc. The company with corporate headquarters in Sunnyvale (CA, United States) was founded in 1995 and pioneered the expanding field of RAMIS (https://intuitive.com). The goal of the company was to provide surgeons with a minimally invasive approach while regaining key benefits of open surgery that were lost with the adoption of laparoscopic surgery: virtual transportation of the surgeon's eyes and hands onto the surgical workspace. Intuitive Surgical Inc. firstly developed two generations of robot prototypes: the Lenny (Fig.1.1) and the Mona (Fig.1.2). The Lenny prototype was taken to animal trials during the summer of 1996. With lessons from the Lenny prototype, the Mona prototype was born with redesign and improvement with the patient-side manipulators, interchangeable architecture, master–slave interface,

FIG. 1.1 Lenny prototype (courtesy of Intuitive Surgical Inc.).

FIG. 1.2 Mona prototype (courtesy of Intuitive Surgical Inc.).

and setup mechanisms [1]. The Mona prototype, named after Leonardo's timeless masterpiece, the Mona Lisa, was the first prototype tested in humans.

In 1997, another company with the name of Computer Motion launched the Zeus Surgical System. The Zeus utilized a voice-controlled endoscopic manipulator aimed at providing laparoscopic surgeons with improved precision and tremor filtration (Fig.1.3). This company seems to have stopped after its origin.

After the Mona prototype, Intuitive Surgical Inc. produced a robot that became until today around the world the most popular and common. This robot was called *da Vinci* System (Fig.1.4) because it was appropriately named during the company's first month of existence for the renowned renaissance polymath, Leonardo da Vinci, given his lasting contributions in the fields of science, art, anatomy, and engineering. Since general surgery was one of the fastest growing specialties using robotic technology, its growth has been seen across most specialties.

The *da Vinci* System was based on three distinct systems: the surgeon-side cart, the vision cart, and the patient-side cart. Six generations of the *da Vinci* System were provided so far. In 2003, a fourth arm was added to the patient-side cart in the creation of the *da Vinci* Standard System (Fig.1.5); the instrumentation expanded from 6 to over 50 units. In 2006, the *da Vinci S* was released focusing on refining the ergonomics of the patient-side cart (Fig.1.6). In 2009, the *da Vinci Si* System was released focusing on the improvement of the surgeon console and vision cart (Fig.1.7). In 2014, the *da Vinci X* System was introduced including redesigned kinematics for patient cart to maximize workspace during an operation (Fig.1.8). In 2017, the *da Vinci Xi* System was introduced providing access to advanced technologies at a more affordable entry point (Fig.1.9).

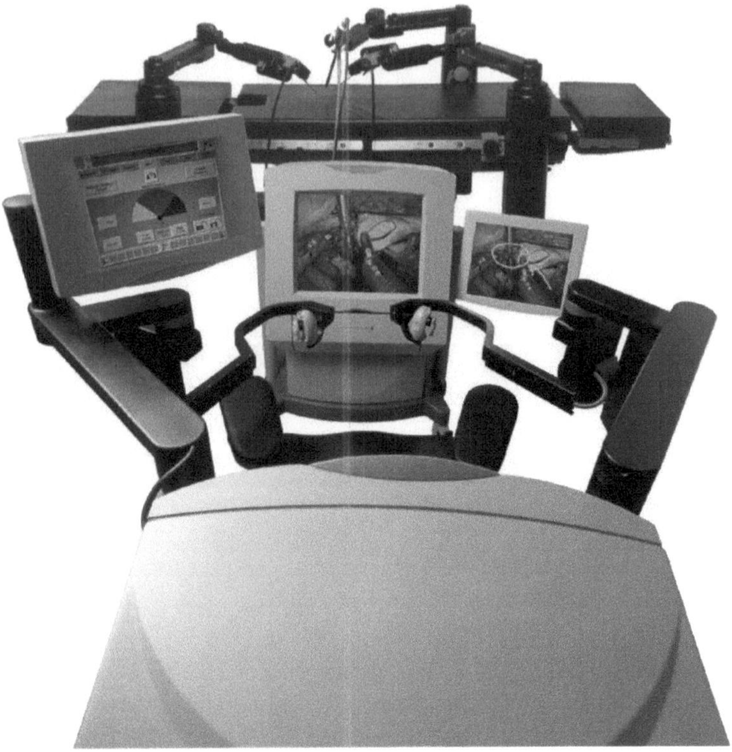

FIG. 1.3 Zeus prototype (courtesy of Intuitive Surgical Inc.).

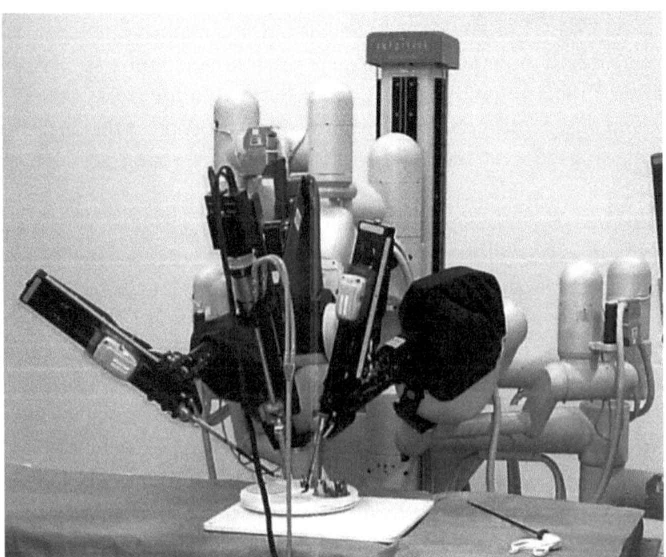

FIG. 1.4 *da Vinci* system—three arms (courtesy of Intuitive Surgical Inc.).

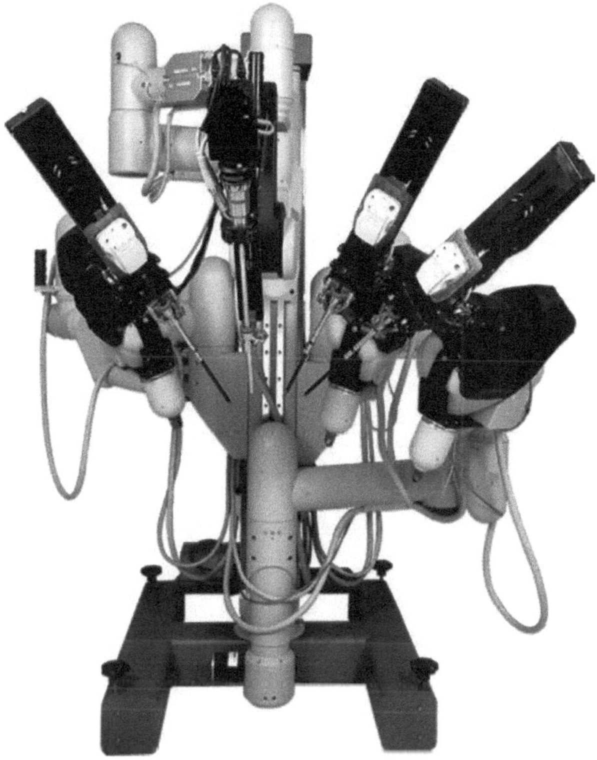

FIG. 1.5 *da Vinci* standard system—four arms (courtesy of Intuitive Surgical Inc.).

Finally, in 2019, the *da Vinci SP* System arrived on the market, introducing the four arms through a single incision (Fig.1.10). In this latter model, the articulated instruments and camera are snake-like robots that can be manipulated independently to provide better surgical access once in the body.

Besides Intuitive Surgical Inc., which has been the dominant company in the field of surgical robots, other companies recently appeared on the market or are still under development and soon will arrive in the surgical theaters. Some of them are here described.

1.2.1 Senhance® Surgical Robotic System

The Senhance® Surgical Robotic System is developed by TransEnterix Inc. (Morrisville, NC, United States) (https://transenterix.com) (Fig.1.11). This robot includes an open console system, glasses for HD 3D display, eye tracking to move the camera, robotic instruments housed in independent arms, and haptic force feedback through controllers that were designed similarly to conventional laparoscopic instrument handles. Most of the instruments do not have increased

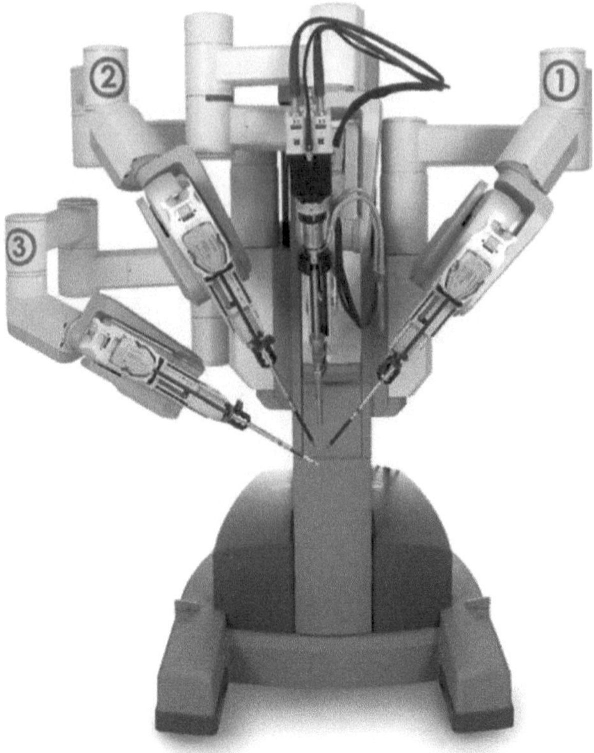

FIG. 1.6 *da Vinci S* (courtesy of Intuitive Surgical Inc.).

degrees of freedom, but a wristed needle driver. Mostly, they are reusable, attached via magnets, facilitating their replacement during surgery [2].

1.2.2 Flex® Robotic System

The Flex® Robotic System is developed by Medrobotics Corp. (Raynham, MA, United States) (htpps://medrobotics.com) (Fig.1.12). This system includes a tower console with a touch screen, magnified HD, two-dimensional visual display, and a joystick controller. It is a robotic platform, which can articulate at nearly 180 degrees, mounted on a flexible endoscope with two articulating 3-mm instruments, like graspers, scissors, monopolar cautery, and needle driver. Control of the instruments is via a handpiece directly connected to the instrument at the patient's bedside. The snake-like flexible scope includes an inner and outer segment with a single point of articulation between adjacent segments. This platform was designed for transoral surgery, particularly oropharyngeal hypopharyngeal, and laryngeal surgery [3]. Recently, another robotic

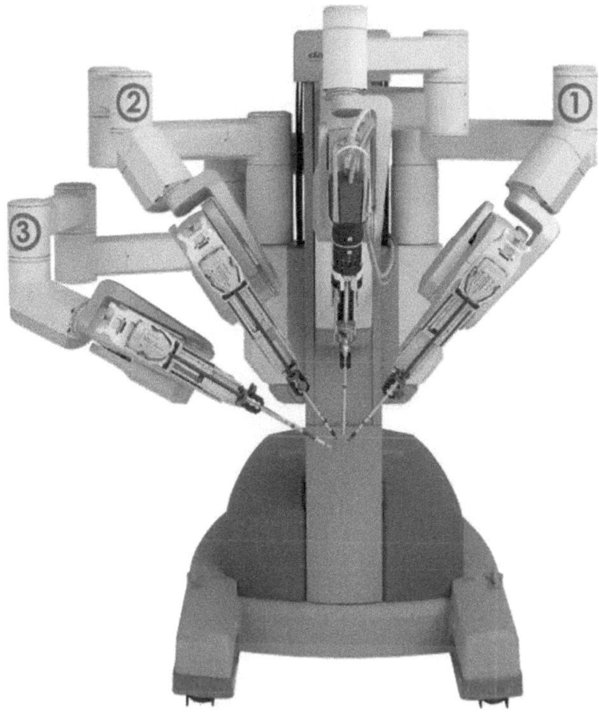

FIG. 1.7 *da Vinci Si* (courtesy of Intuitive Surgical Inc.).

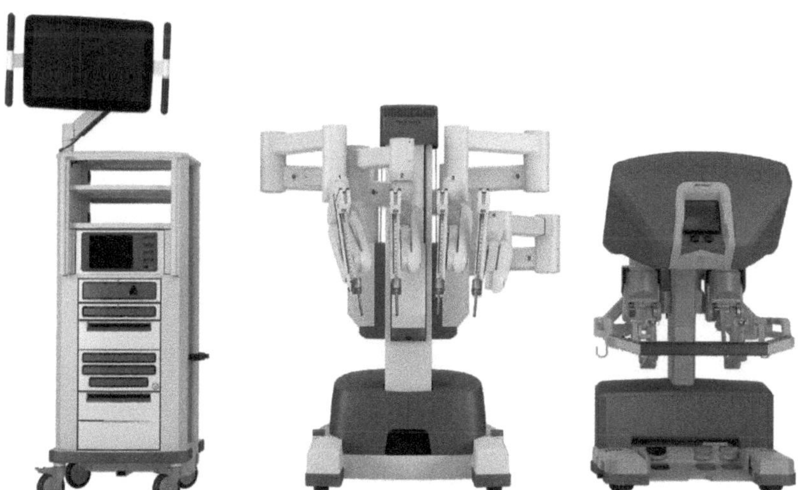
FIG. 1.8 *da Vinci X* (courtesy of Intuitive Surgical Inc.).

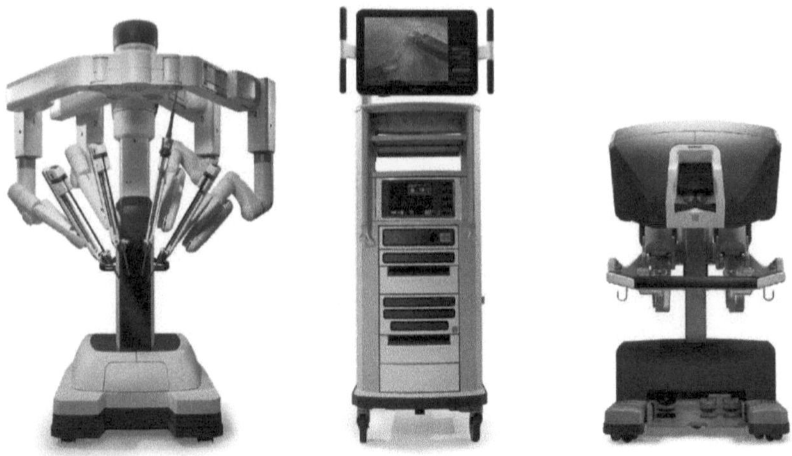

FIG. 1.9 *da Vinci Xi* (courtesy of Intuitive Surgical Inc.).

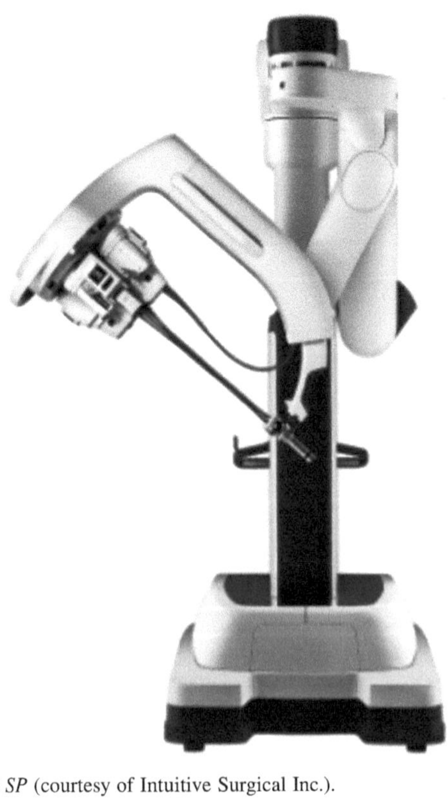

FIG. 1.10 *da Vinci SP* (courtesy of Intuitive Surgical Inc.).

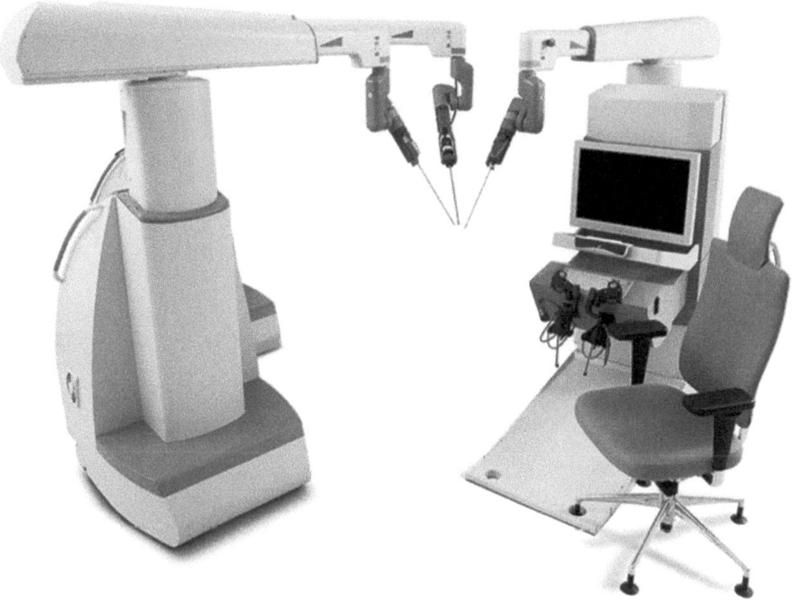

FIG. 1.11 Senhance® Surgical Robotic System (courtesy of TransEnterix Inc.).

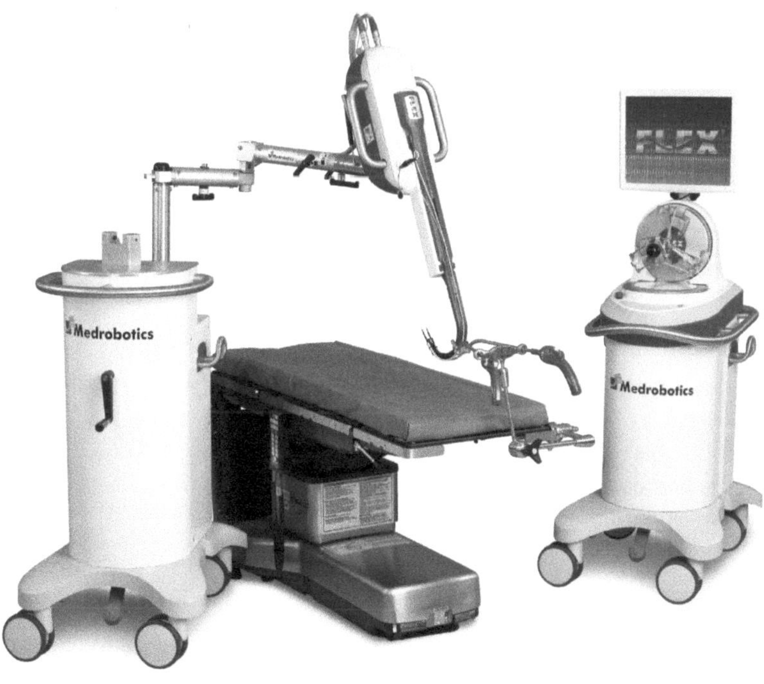

FIG. 1.12 Flex® Robotic System (courtesy of Medrobotics Corp.).

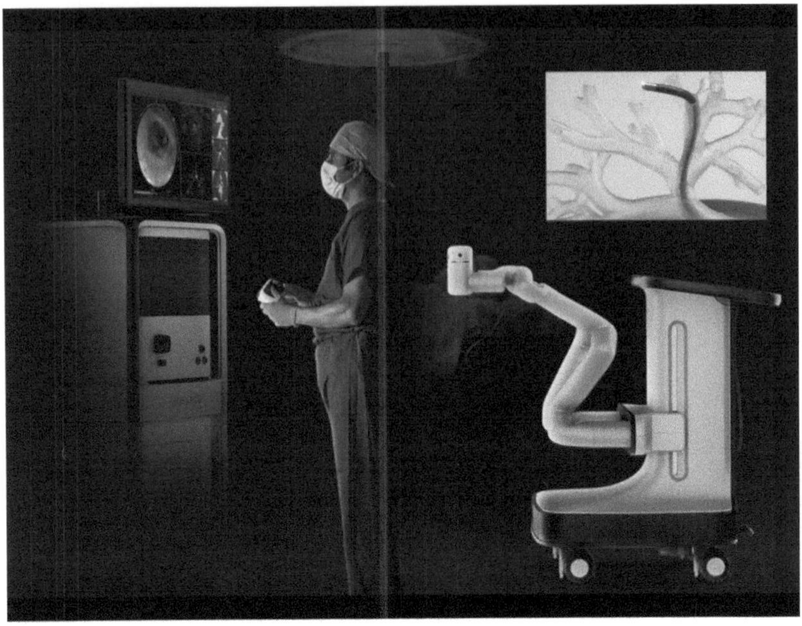

FIG. 1.13 Auris® Robotic Endoscopy System (courtesy of Auris Health Inc.).

system called Flex Colorectal Drive was developed for MIS through the anus as natural orifice surgery.

1.2.3 Auris® Robotic Endoscopy System (ARES)

The Auris® Robotic Endoscopy System (ARES) is the Monarch Platform produced by Auris Health Inc. (San Carlos, CA, United States) (https://aurishealth.com) (Fig.1.13). It is a platform for bronchoscopy, used in diagnostic and therapeutic as well. It consists of a surgeon console, controller cart, patient-side system, and bronchoscope. The patient-side system is made up of a robot cart, robot arms, as well as camera controls and power boxes. The robot has two arms, each with 6 degrees of freedom (DOFs), and an instrument drive with four actuation axes. A flexible bronchoscope can be attached, allowing the endoscopist to bend the bronchoscope in four directions. Moreover, the working channels can be used for standard procedures, like irrigation and aspiration.

1.2.4 University of Nebraska laparoscopic single-incision robot

Virtual Incision Corp. (Lincoln, NE, United States) (https://virtualincision.com) is a company founded by faculty of members at the University of Nebraska-Lincoln and the University of Nebraska Medical Center under the guidance of Prof. D. Oleynikov, who developed the Laparoscopic

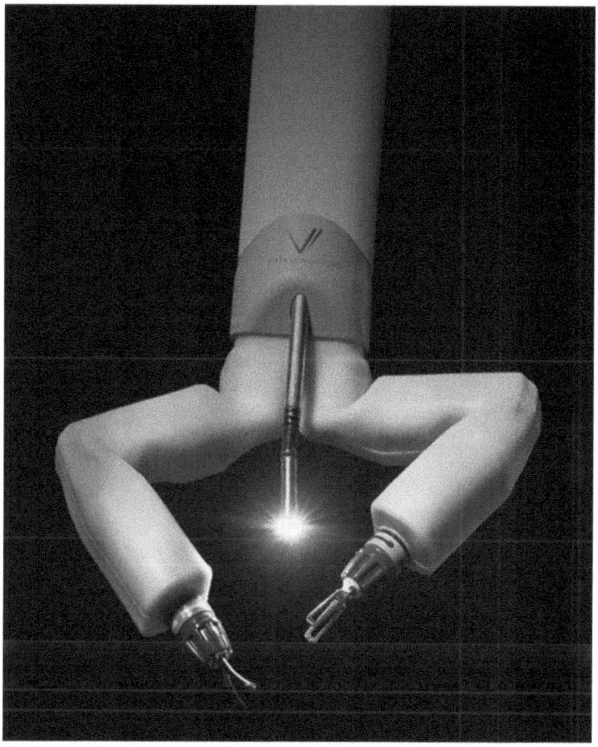

FIG. 1.14 University of Nebraska laparoscopic single-incision robot (courtesy of Virtual Incision Corp.).

Single-Incision Robot (Fig.1.14). Since its beginning, this robot is under continuous growth and different models of evolution have been created. It was created with the aim to be completely inside in the abdomen during surgery, allowing for improved surgeon's dexterity and resulting in fewer incisions and less pain for the patient. Additionally, it can be quickly repositioned to enable multiquadrant access in the abdomen. It is composed of a base attached to two dexterous arms with varying DOFs, allowing for grasping and cautery within the body. Each arm operates using a master–slave configuration. The particular technological advancement is the miniaturization of robotic arms and the motors that drive them, which allows for a reduced footprint and greater access to the patient relative to conventionally sized robotic-assisted platforms. It represents a step forward in laparoscopic single-incision surgery.

1.2.5 SPORT® Surgical System

Single Port Orifice Robotic Technology (SPORT®) Surgical System is developed by Titan Medical Inc. (Toronto, ON, Canada) (https://titanmedical.com) (Fig.1.15). It is a single-port surgical robot with a deployable 3D camera

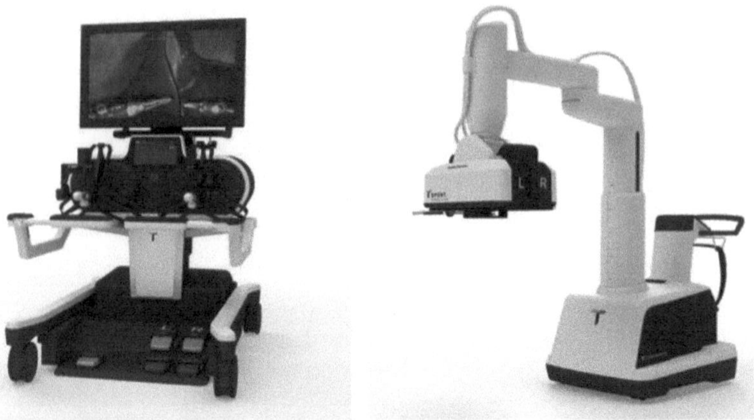

FIG. 1.15 Single port orifice robotic technology surgical system (courtesy of Titan Medical Inc.).

and two replaceable wristed motion instruments. It is mounted in a single stalk on a mobile tower with a boom. The workstation allows the surgeon to interact with the robotic platform, including a 3D HD endoscopic view via fiber-optic illuminated images displayed on a high-definition flat screen monitor. The design utilizes a collapsible system that can be inserted into the body through a 25-mm incision.

1.2.5.1 MiroSurge

MiroSurge is a robot under development at the Robotics and Mechatronics Center (RMC) (Oberpfaffenhofen-Webling, Germany) (https://dlr.de) (Fig.1.16), a multiinstitutional cluster entity of the Federal Republic of Germany national aeronautic, space research, and project management center. The system

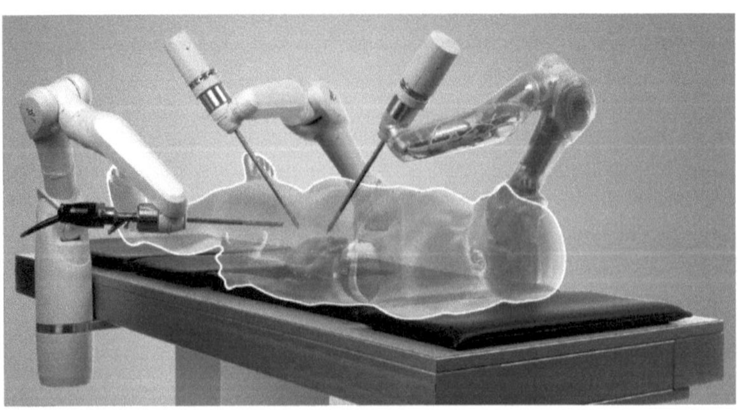

FIG. 1.16 MiroSurge (courtesy of DLR, German Aerospace Center).

incorporates 3–5 individual instruments carrying minimally invasive robot-assisted (MIRO) arms, which permits multiple modes of control, 7 DOFs, capacity to utilize various devices, and ability to be mounted to various locations on table rails. The instruments are equipped with haptic devices, and the surgeon is able to view all information via the master console HD 3D display.

1.2.5.2 Versius

The Versius robot is developed by Cambridge Medical Robotics (CMR) Surgical Ltd. (Cambridge, UK) (https://cmrsurgical.com) (Fig.1.17). It has independent modular arms easy to set up. Each of the robot arms has flexible joints like a human arm, which are controlled by a surgeon sitting at a console with a 3D HD screen. Haptic feedback from instrument to controller provides force feedback to the operator. Several variants of robotic arms using 5-mm instruments have been developed in order to reduce incision size.

1.2.6 Mazor robotics

Medtronic (Parsippany, NJ, United States) invested substantial resources in Mazor Robotics (Caesarea, Israel) (https://medtronic.com), to distribute its surgical robotic guidance system, the Mazor X, and its image-based guidance system, the Renaissance (Fig.1.18). Mazor X has a background in image-based, preplanned robotic guidance for spine and brain surgery.

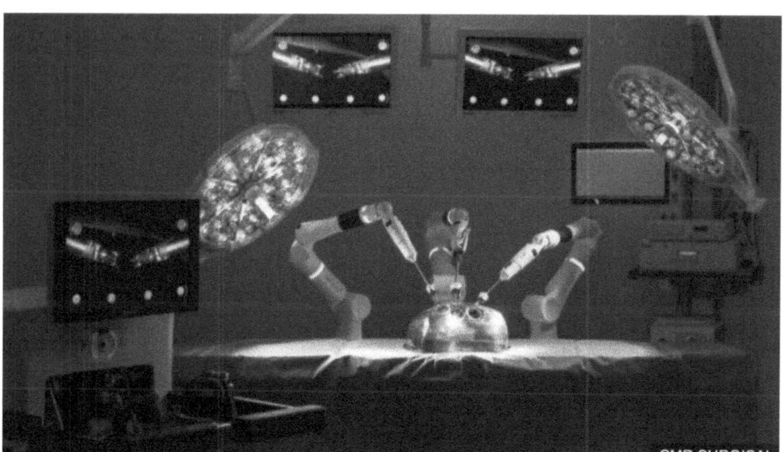

FIG. 1.17 Versius robot (courtesy of Cambridge Medical Robotics Ltd.).

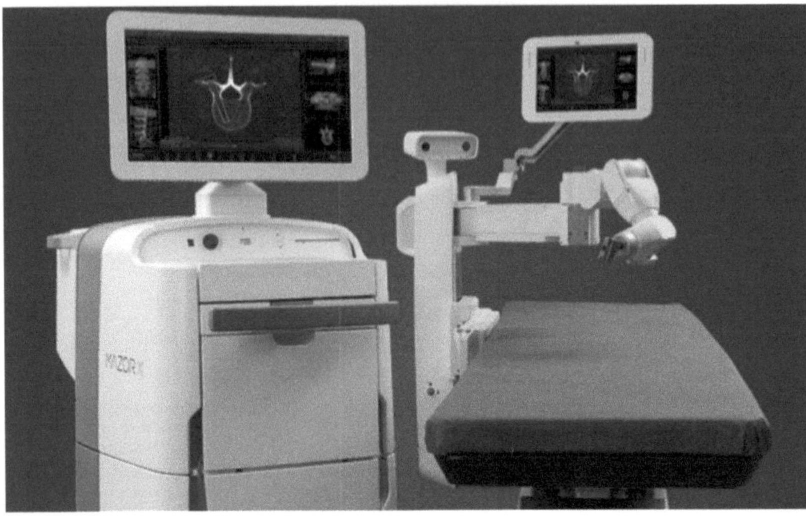

FIG. 1.18 Mazor X robot (courtesy of Medtronic).

1.2.7 Einstein

The Einstein surgical robot is under development by Medtronic (Minneapolis, MN, United States) (https://medtronic.com) in conjunction with a variety of partnerships including Mazor, and German Aerospace Center (DLR).

1.2.8 Verb surgical

Ethicon, subsidiary of Johnson and Johnson (New Brunswick, NJ, United States), is invested in a joint venture with Verily Life Sciences, a part of Alphabet Inc. (Mountain View, CA, United States) parent company of Google, called Verb Surgical (Verb) (http://verbsurgical.com). Given Alphabet's background in data analytics and machine learning and Ethicon's background in medical device manufacturing, this is an intelligent partnership. The goal of Verb Surgical has been reported not to be the introduction of a new robotic surgical platform so much as the introduction of a new category of digital surgery. Termed Surgery 4.0, Verb is attempting to create the successor to open, minimally invasive, and first-generation surgical robotic platforms, Surgeries 1.0, 2.0, and 3.0, respectively. The prototype has been reported to combine robotics and data-driven machine learning in a way to reduce surgical costs and to expand the use of such devices to a larger array of surgeons.

During the invention and growth of the robots described above, some other robotic tools have been developed as well. The objective remains always the improvement of the different advantages of MIS. Some of them are already known and here described.

1.2.9 FreeHand

FreeHand v1.2 is produced by FreeHand 2010 Ltd. (Guildford, UK) (http://freehandsurgeon.com) (Fig.1.19). It is a laparoscopic scope and camera controller which robotically manipulates a laparoscope through three axes, after it has been mounted on railings around the operative table. It is composed of a lockable articulating arm, an electronic control box, and a robotic motion assembly unit. It allows the surgeon to directly control the pan, tilt, and zoom movement of the scope without requiring to take the hands off the tools the surgeon is working with [4]. The benefits are as follows: a tremor-free image, allows solo surgery, frees up camera assistant, reengages laparoscope in precisely the same position should the scope need cleaning.

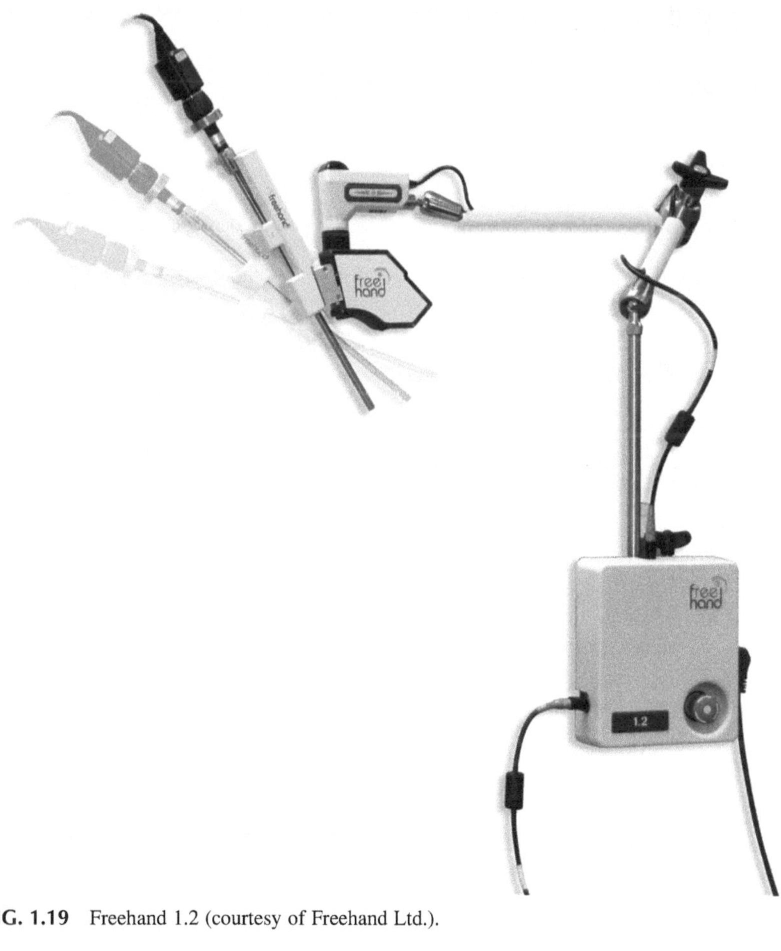

FIG. 1.19 Freehand 1.2 (courtesy of Freehand Ltd.).

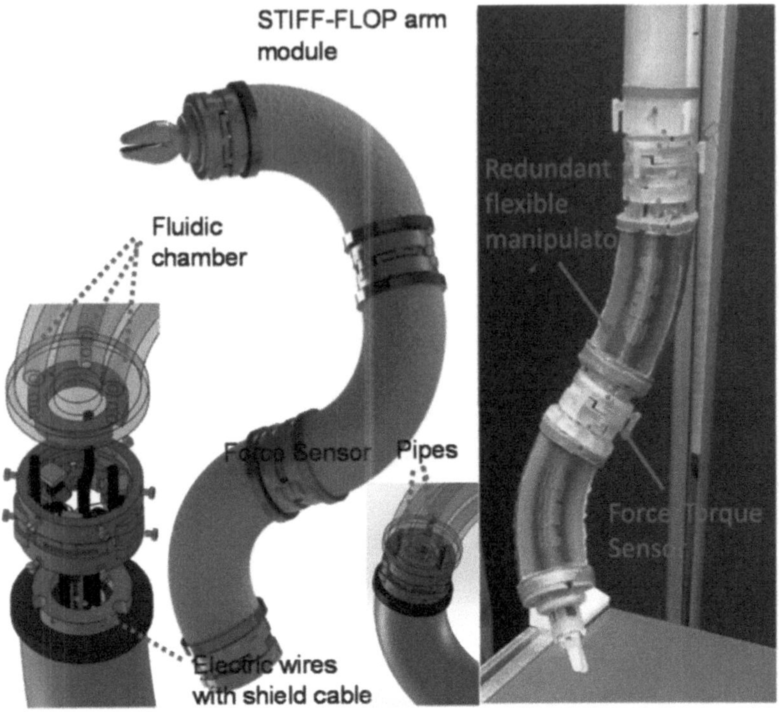

FIG. 1.20 Stiffness controllable flexible and learn-able manipulator for surgical operations (STIFF-FLOP) (courtesy of School of Advanced Studies—Pisa, Italy).

1.2.10 Stiffness controllable flexible and learnable manipulator for surgical operations (STIFF-FLOP)

The design of this robotic tool (Fig.1.20) was inspired by the fluid but string motion of an octopus arm (http://sssa.bioroboticsinstitute.it/projects/STIFF-FLOP). It is built to eliminate the division between force and dexterity and make the two together, creating a robot capsule of being used as both a retractor and a grasper in MIS. This robotic tool is cylindrical, consisting of three fluid-actuated inner chambers housed in a larger elastomeric cylinder, complete with braiding to prevent outward expansion and promote elongation. With improvements in the jamming mechanisms, it could allow for improved agility in surgery with enough strength to still perform surgical tasks, allowing probably surgery in hard-to-reach body cavities.

1.2.11 A miniature robot for retraction tasks under vision assistance in MIS

The robotic tool for retraction tasks under vision assistance (Fig.1.21) integrates brushless motors to enable additional DOFs to that provided by magnetic

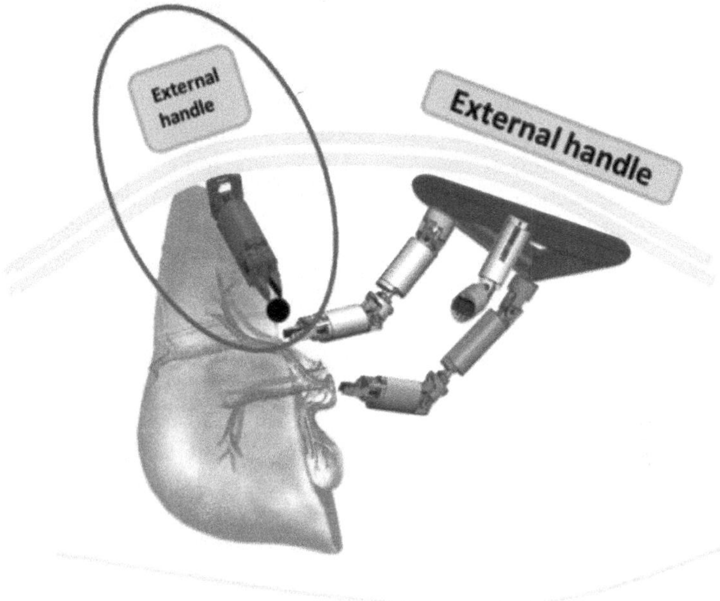

FIG.1.21 A miniature robot for retraction tasks under vision assistance (courtesy of Dr. Tortora, from Tortora et al. [5]).

anchoring, improving the dexterity of the overall platform [5]. The retraction robot can be easily controlled to reach the target organ and to apply a retraction force of up to 1.53 N. Additional freedom degrees can be used for smooth manipulation and grasping of the organ.

1.3 Clinical applications

RAMIS has been growing in different surgical fields and moreover has become the preferred approach for certain surgical procedures. Out of some surgical fields like urology, gynecology, and pelvic colorectal surgery, the superiority of RAMIS to conventional laparoscopy is still limited. This slow adoption can probably be addressed to some drawbacks like the cost of the robot and instruments, the requirement to utilize at a high rate to make up for operational costs, the inability to quickly switch instruments during a procedure, the size of the system, the increased setup time, and the lack of haptic feedback (which gives the user a sense of touch) [6,7].

Urological procedures that are now being routinely performed robotically are as follows: radical prostatectomy, radical cystectomy, renal procedures (mainly partial nephrectomy), and pyeloplasty, as well as ureteric reimplantation. For robot-assisted radical prostatectomy, there seems to be an advantage in terms of continence and potency over laparoscopic surgery. Robot-assisted radical cystectomy seems equal in terms of oncological outcome, but with lower complication rates. However, the effect of intracorporeal urinary diversion has hardly been examined. Robotic partial nephrectomy has proven safe and is most likely superior to conventional laparoscopic surgery, whereas there does not seem to be a real advantage for the robot in radical nephrectomy. For reconstructive procedures, e.g., pyeloplasty and ureteric reimplantation, there seems to be advantages in terms of operating time. Finally, in light of the significant costs and because high-quality data from prospective randomized trials are still missing, conventional urological laparoscopy is still alive [8].

Relative to open approaches, the robot has been shown to enhance intraoperative visualization and precision, mitigate surgeon tremor, hasten postoperative recovery, and shorten length of hospital stay for certain indications. However, it has also been associated with higher costs, longer operative times, and limited outcomes data [9].

In gynecology, RAMIS is more expensive than MIS, and the learning curve may facilitate more efficient MIS learning and adoption in novice surgeons compared with laparoscopic or vaginal techniques [10]. RAMIS is applicable and capable of offering an adequate treatment to selected patients [11]. Currently, robotic surgery is used for a variety of indications in the treatment of benign gynecological diseases as well as malignant ones. The most common procedures include benign hysterectomy, myomectomy, radical hysterectomy, lymph node dissections, or sacrocolpopexy.

In pelvic colorectal surgery, most robotic rectal surgeries are recommended for middle and lower rectal cancer. The learning curve has been shown to be shortened if compared to that of laparoscopic approach [12,13]. The most common obstacles appeared to be the longer operation time and the higher cost [14]. A meta-analysis suggested that complex rectal cancer such as obesity, male patients, preoperative chemoradiation, and distal rectal cancer may be good indications for robotic approach [15]. Other outcomes like recovery of bowel function, length of stay, morbidity, mortality and number of lymph nodes, surgical resection margins have been reported to be similar to laparoscopic surgery [16–19].

1.4 Conclusions

In this chapter, the most common available robots, the known companies involved in the development of surgical robots, and also the knowledge of some robotic tools under development have been described. Currently, there are other models and ideas under the stage of research and development, which cannot be described because the available information is scarce or not yet known. Hence, the world of robots remains under great expansion and continues to grow as the researchers create robots capable of more advanced procedures and interventions. Especially, the future of RAMIS is promising in regard to surgical access, vision, emerging competition, and intelligent systems.

References

[1] So DM, Hanuschik M, Kreaden U. Chap 7: The da Vinci surgical system. Rosen J, Hannaford B, Satava SR, editors. Surgical robotics-systems, applications, and visions. Springer: In; 2011.

[2] Stark M, Pomati S, D'Ambrosio A, Giraudi F, Gidaro S. A new telesurgical platform–preliminary clinical results. Minim Invasive Ther Allied Technol 2015;24(1):31–6.

[3] Remacle M. M N Prasad V, Lawson G, Plisson L, Bachy V, Van der Vorst S. Transoral robotic surgery (TORS) with the Medrobotics flex system: first surgical application on humans. Eur Arch Otorhinolaryngol 2015;272(6):1451–5.

[4] Stolzenburg JU, Franz T, Kallidonis P, Minh D, Dietel A, Hicks J, Nicolaus M, Al-Aown A, Liatsikos E. Comparison of the FreeHand(R) robotic camera holder with human assistants during endoscopic extraperitoneal radical prostatectomy. BJU Int 2011;107(6):970–4.

[5] Tortora G, Ranzani T, De Falco I, Dario P, Menciassi A. A miniature robot for retraction tasks under vision assistance in minimally invasive surgery. Robotics 2014;3(1):70–82.

[6] Oleynikov D. Robotic surgery. Surg Clin North Am 2008;88(5):1121–30.

[7] Hanly EJ, Talamini MA. Robotic abdominal surgery. Am J Surg 2004;188(4ASuppl):19S–26S.

[8] Schwaibold H, Wiesend F, Bach C. The age of robotic surgery - is laparoscopy dead? Arab J Urol 2018;16(3):262–9.

[9] Singla N, Singla AK. Evolution of robotic technology in urologic surgery. Urol Listy 2014;12 (3):13–6.

[10] Varghese A, Doglioli M, Fader AN. Updates and controversies of robotic-assisted surgery in gynecologic surgery. Clin Obstet Gynecol 2019;62(4):733–48.

[11] Rivas-López R, Sandoval-García-Travesí FA. Robotic surgery in gynecology: review of literature. Cir Cir 2020;88(1):107–16.

[12] Kim YW, Lee HM, Kim NK, Min BS, Lee KY. The learning curve for robot-assisted total mesorectal excision for rectal cancer. Surg Laparosc Endosc Percutan Tech 2012;22(5):400–5.

[13] Sng KK, Hara M, Shin JW, Yoo BE, Yang KS, Kim SH. The multiphasic learning curve for robot-assisted rectal surgery. Surg Endosc 2013;27(9):3297–307.

[14] Keller DS, Senagore AJ, Lawrence JK, Champagne BJ, Delaney CP. Comparative effectiveness of laparoscopic versus robot-assisted colorectal resection. Surg Endosc 2014;28 (1):212–21.

[15] Scarpinata R, Aly EH. Does robotic rectal cancer surgery offer improved early postoperative outcomes? Dis Colon Rectum 2013;56(2):253–62.

[16] Baik SH, Kwon HY, Kim JS, Hur H, Sohn SK, Cho CH, Kim H. Robotic versus laparoscopic low anterior resection of rectal cancer: short-term outcome of a prospective comparative study. Ann Surg Oncol 2009;16(6):1480–7.

[17] Kwak JM, Kim SH, Kim J, Son DN, Baek SJ, Cho JS. Robotic vs laparoscopic resection of rectal cancer: short-term outcomes of a case-control study. Dis Colon Rectum 2011;54 (2):151–6.

[18] Kang J, Yoon KJ, Min BS, Hur H, Baik SH, Kim NK, Lee KY. The impact of robotic surgery for mid and low rectal cancer: a case-matched analysis of a 3-arm comparison—open, laparoscopic, and robotic surgery. Ann Surg 2013;257(1):95–101.

[19] Kim JC, Yang SS, Jang TY, Kwak JY, Yun MJ, Lim SB. Open versus robot-assisted sphincter-saving operations in rectal cancer patients: techniques and comparison of outcomes between groups of 100 matched patients. Int J Med Robot 2012;8(4):468–75.

Chapter 2

Artificial intelligence for medical robotics

Erwin Loh[a] **and Tam Nguyen**[b]
[a]*St Vincent's Health Australia, Melbourne, Australia,* [b]*St Vincent's Hospital Melbourne, Melbourne, Australia*

2.1 Background

Artificial intelligence (AI) applications in medicine are becoming more common in terms of automation and diagnostics. AI aims to minimize and mitigate human error while keeping a high level of accuracy and precision. However, AI in medical robotics is an area still underdeveloped due to a number of technical limitations as well as concerns from the regulators regarding automated or semi-automated surgery.

The first use of robot in surgery was in 1978 by the system called programmable universal machine for assembly (PUMA) [1]. A decade later, it was used for neurosurgical biopsies and then by urologists who called it a surgeon-assistant robot for prostatectomy (SARP) [2]. Five years later, Wang and colleagues developed the automated endoscopic system for optimal positioning (AESOP) which was the first Food and Drug Administration (FDA)-approved surgical robot [3].

Two decades after the first robot was used in real surgery, Defense Advanced Research Project Agency (DARPA) in 1998 developed ZEUS, a new robot capable of reproducing the movements of the arms of the surgeon. This was a system that was later bought by the company which developed the widely used da Vinci master–slave robotics platform that went on to receive CE certification in 1999 and full FDA approval in 2001.

Currently, robotics surgery devices are used in many areas of medicine including robotic-assisted (RA) surgery in many subspecialties including ophthalmology, neurosurgery, orthopedic surgery, cardiac surgery, and minimally invasive surgery (MIS) in gastrointestinal, hepatobiliary, pancreatic, urological, and gynecological surgery, as described later.

Surgery is one of the most complex medical domains, with decision making required on the spot coupled with intense time pressures and a high potential for

Endorobotics. https://doi.org/10.1016/B978-0-12-821750-4.00002-5

detrimental consequences. This makes it a great challenge for robotics; hence, AI featured in surgery today is predominantly involved with passively assisting human surgeons to better execute their tasks.

Most medical robotics applications are used for precision operations during noninvasive surgery, which limits the ability to fully automate procedures with AI. In cases where machine learning (ML) models are employed, the robot will require to be extensively trained with labeled videos of surgeries. Now, changes in regulations on Software as a Medical Device (SaMD) are on the right path to reduce liability; however, further development is required before the software can be used to fully automated surgeries.

2.1.1 Eye surgery

The incorporation of graphical processing units into ML tasks has helped to improve diagnostics and prediction of outcomes for many prevalent ophthalmological conditions, notably glaucoma and retinal conditions [4].

2.1.2 Neurosurgery

ML was used to predict mortality outcomes after spinal tumor surgery [5], and there have been positive medical team perspectives, based on a cross-sectional survey, on the use of AI used in neurosurgery [6]. There has also been successful pituitary gland surgery, also called transsphenoidal surgery, using AI, realistic motion tracking, and telesurgery [7].

2.1.3 Cardiac Surgery

Six AI models have been used to assess the preoperative risk-of-death rate in patients with congenital heart defects undergoing cardiac surgery [8]. Some of the variables calculated were height, BMI (body mass index), and arterial oxygen saturation index.

An AI model using coronary CT angiography has been successfully implemented to profile perivascular fat attenuation index and predict cardiac risk in patients from three studies including patients who had undergone cardiac surgery, those who experienced major adverse cardiac events from the previous angiography, and patients with acute myocardial infarction. In particular, their assessment revealed a biomarker, called fat radiomic profile, which predicted improved cardiac risk [9].

2.1.4 Orthopedic surgery

An AI-assisted system has been used for postoperative follow-ups in the orthopedic patient cohort [10]. The study compared the AI system to the manual follow-up method and assessed factors of cost-effectiveness and feedback from patients. The study found that the AI method was not inferior to the manual method of follow-up. The AI method also saved on human resource costs.

However, a systematic review (study protocol published) on AI and ML trends and applications of orthopedic interventions has found a large number of small uncontrolled studies and articles with small subsets that describe actual applications of AI or ML. Therefore, the authors suggested that cohort studies and large randomized control trials are further needed [11].

2.1.5 Minimally invasive surgery (MIS)

A Mazor Spine Assist robot was approved by the US FDA to assist with placement of pedicle screws [12] and shown to increase the accuracy of pedicle screw placement and decrease radiation exposure to surgeons.

The first force-feedback-enabled virtual reality simulator, NOViSE, for natural orifice transluminal endoscopic surgery (NOTES) training has also been reported [13]. NOViSE could potentially facilitate NOTES development and contribute to its wider use by keeping practitioners up-to-date with this novel surgical technique and improve the educational experience without putting patients at risk.

2.2 AI in robotic surgery

Physical, mental, and technical variables dictate the performance of the surgeon and these factors affect the outcome. Surgical robots have the advantage of tremor cancellation, scalable motion, insusceptibility to fatigue, and greater range of axial movement that should, in turn, positively impact the quality of surgical care rendered.

Combining AI (ML, natural language processing, artificial neural networks, and computer vision) with surgical robots may reduce technical and human errors, operative time, and rates of complications and improve the outcome as an ultimate end point.

The robots can be taught specific procedures either by directly programming it (explicit learning) or by having the robot observe a surgeon or video directly (implicit learning), or be trained in simulation or virtual reality [1].

2.3 AI in diagnosis (pathology/radiology systems)

Artificial intelligence is quite effective in the diagnosis of different clinical conditions. In ophthalmology, an AI-based grading algorithm was able to screen fundus photographs from diabetic patients and identify to determine cases that should be referred to an ophthalmologist for further evaluation and treatment [14].

In dermatology, scientists trained a neural network and tested it against 21 board-certified dermatologists and found that the AI was able to classify skin cancer with the same competence as trained dermatologists [15].

In radiology, a deep learning algorithm used magnetic resonance imaging of the brain neonates to predict the diagnosis of autism in high-risk children at 24-month of age, with a positive predictive value of 81% [16]. In another study, a ML based on a single amyloid positron emission tomography scan obtained an accuracy of 84% of diagnosing dementia within 24-month [17].

In pathology, AI and computational pathology spearheading advancement of diagnosis through the use of clinical informatics, genomics, bioinformatics, and deep learning that lead to faster and higher accuracy of diagnosis, prognosis, and treatment of disease [18].

2.4 AI in virtual reality (VR) and simulation

Simulation training has becoming an important role in optimizing the technical and nontechnical skills for surgical trainees. Simulation training can move the learning curve of the trainee surgeons away from the patient in operating theater to a more controlled environment in the simulation room. Training addresses dimensional awareness, eye-hand coordination, and depth perception addressing the time-efficient and patient safety [19].

A recent systemic review demonstrated that for surgical tasks and activities, simulation with haptic feedback leads to an improved sense of realism and led to a reduced learning curve for trainees when compared to simulators without haptic feedback [20].

A number of VR trainers such as MIMIC or LAP Mentor have been used in training surgeons to acquire skills before entering the operating room. The MIMIC virtual reality trainer provides a platform for VR robotic training in a three-dimensional format. It currently offers training through a series of tasks, which can be evaluated by metric analysis to discriminate the expert from the novice. Future developments of this simulator will include improved simulation of suturing and procedure-based exercises [19].

The LAP Mentor™ simulator for laparoscopic cholecystectomy training is able to give a representative recreation of the intra-abdominal cavity and organs. User proficiency can then be determined by comparing against the metric feedback of the simulator, validated used curriculum, and modified versions of the General Surgical Skills Global Rating and Operation Specific Surgical Skills Global Rating [19].

2.5 AI in teaching and training

The virtual reality simulations mentioned already also include technical skill assessments. Technical skills of robotic surgery, as described in examples from thyroidectomy, are assessed by operation time, the number of performed operations, complication rate, intraoperative blood loss, and length of hospital stay [21].

For the da Vinci Residency and Fellowship Training Program, technical skills are rehearsed using the actual da Vinci Xi System (dVXS) in either an inanimate or an animate laboratory setting and under the guidance of a trained proctor. While the program provides suggested "scripts" of what should be done in the laboratory and what measures should be recorded, there are no specific goals or objectives outlined [21].

Simulation-based training is increasingly being used for training and assessment in surgery and medicine. The application of AI/ML technologies allows large data set to be implemented for training and education purposes.

The virtual operative assistant [22] is an explainable AI tool for simulation-based training in surgery and medicine. The tool establishes a basis for the potential role of integrating AI and virtual reality simulation into surgical educational teaching. The potential of linking expertise classification, objective feedback based on proficiency benchmarks, and instructor input creates a novel educational tool by integrating these three components into a formative educational paradigm.

2.6 AI in surgical planning and robotic-assisted surgery

A review on the potential of ML in improving cost and efficiency in 3D printing of anatomical models [23] indicated that through using more refined algorithms, cost reduction may occur by decreasing the need to pay for segmentation specialists and proprietary software. A successful example of this was recently shown where an open source and free to the public software was used to generate a 3D mandibular models used in the field of maxillofacial surgery.

Furthermore, ML could be employed to make recommendations for each patient, with these recommendations continually improving as the algorithm gathers more data. It could also potentially be used to decrease the amount of radiation needed to generate a CT image. Additionally, ML has the potential to vastly improve productivity by creating automatic image segmentation and consequently reducing the need for a full-time segmentation specialist.

2.7 Limitations

The AI and ML have the potential dramatically changed the way surgery is taught and practiced, but they do not solve all the challenges in surgery—in some cases, they do not always improve results and therefore should be closely assessed using cost–benefit analysis.

The AI and ML can introduce systematic biases in clinical data collection, thereby making the patterns they recognize or predictions they make prone to potential error and inaccuracy.

Technical limitations described for the da Vinci system include the lack of tactic (haptic) feedback (which is beneficial for certain surgery tasks such as knot tying). The surgeons are forced to rely on visual cues for position, velocity, and acceleration of the instruments [21]. Additionally, this also impacts "judgment of force on internal structures and characteristics of manipulated tissues, such as compliance, texture, or elasticity" [21].

Other limitations include cost, size of the currently available medical robots, reusability, and specialized training required often with steep learning curve experienced by users [21,24,25].

Reusability is another limitation, where the da Vinci system has up to 10 uses for most instruments [24].

In robot-assisted eye surgery, in particular to the vitrectomy procedure, there is a reported limitation in the robot's awareness throughout surgery, whereas its end-effector velocity and tilting angle remain task-specific with the macula as a preferable target site [4]. The authors suggested that these limitations may be addressed in the future with enhanced sensor input and ML incorporation.

Tremor removal is a function to remove the surgeon's undirected hand movement transferring to the instruments during surgery and causing harm to the patient. This function is not widely found in robot-assisted systems, therefore suggesting limited literature on the topic [24].

The VR simulation training is currently limited in graphic processing, haptic technology, and AI. Robotic VR simulators, employing a console similar to robot-assisted surgery console, are not yet made available for complex surgical training that requires high-fidelity simulation. Instead, they are limited to basic surgical tasks such as camera driving, targeting, pattern cutting, suturing, knot tying, basic use of surgical energy, and tissue manipulation [26].

Another limitation is in terms of the legality and logistics of installing a robotic surgical assistant, where the concern surrounding fully automated surgical procedures and the resulting healthcare regulations that may prohibit it. This challenge will likely be overcome as the technology becomes more reliable and the public is more comfortable with allowing a robot to operate on them without human assistance. Furthermore, data scientists and ML experts may still be developing the method for training a ML model to learn surgical procedures with the right surgical footage labeled according to all present body structures and accurate movement or pulsation of those structures. This would also include visible mechanical structures such as the robot's arms or a surgical implant.

2.8 Conclusions

Medical robotics currently has limited use cases in terms of fully automated surgery or other medical procedures. The main reason is due to regulations that dictate that a recognized professional must be administering these procedures.

Key challenges with the use of AI in medicine include the issue of legal liability and attribution of negligence when errors occur.

Now, changes in regulations on SaMD are on the right path to reduce liability; however, further flexibility is required before the software can be used to fully automated surgeries.

References

[1] Ozmen MM, Ozmen A, Koç ÇK. Artificial intelligence for next-generation medical robotics. In: Atallah S, editor. Digital surgery. Cham: Springer International Publishing; 2021. p. 25–36.

[2] Kwoh YS, Hou J, Jonckheere EA, Hayati S. A robot with improved absolute positioning accuracy for CT guided stereotactic brain surgery. IEEE Trans Biomed Eng 1988;35 (2):153–60. https://doi.org/10.1109/10.1354.

[3] Satava RM. Robotic surgery: from past to future—a personal journey. Surg Clin North Am 2003;83(6):1491 500. https://doi.org/10.1016/S0039-6109(03)00168-3.

[4] Urias MG, Patel N, He C, Ebrahimi A, Kim JW, Iordachita I, et al. Artificial intelligence, robotics and eye surgery: are we overfitted? Int J Retina Vitreous 2019;5:52. https://doi.org/10.1186/s40942-019-0202-y.

[5] DiSilvestro KJ, Veeramani A- ->, McDonald CL, Zhang AS, Kuris EO, Durand WM, Daniels AH. Predicting postoperative mortality after metastatic intraspinal neoplasm excision: development of a machine-learning approach. World Neurosurg 2021;146:e917–24. https://doi.org/10.1016/j.wneu.2020.11.037.

[6] Layard Horsfall H, Palmisciano P, Khan DZ, Muirhead W, Koh CH, Stoyanov D, Marcus HJ. Attitudes of the surgical team toward artificial intelligence in neurosurgery: international 2-stage cross-sectional survey. World Neurosurg 2021;146:e724–30. https://doi.org/10.1016/j.wneu.2020.10.171.

[7] Soldozy S, Young S, Yagmurlu K, Norat P, Sokolowski J, Park MS, Syed HR. Transsphenoidal surgery using robotics to approach the Sella turcica: integrative use of artificial intelligence, realistic motion tracking and telesurgery. Clin Neurol Neurosurg 2020;197:106152. https://doi.org/10.1016/j.clineuro.2020.106152.

[8] Chang M, Canseco JA, Nicholson KJ, Patel N, Vaccaro AR. The role of machine learning in spine surgery: the future is now. Front Surg 2020;7. https://doi.org/10.3389/fsurg.2020.00054.

[9] Oikonomou EK, Williams MC, Kotanidis CP, Desai MY, Marwan M, Antonopoulos AS, Antoniades C. A novel machine learning-derived radiotranscriptomic signature of perivascular fat improves cardiac risk prediction using coronary CT angiography. Eur Heart J 2019;40 (43):3529–43. https://doi.org/10.1093/eurheartj/ehz592.

[10] Khanna V, Pförringer D, Bian Y, Xiang Y, Tong B, Feng B, Weng X. Artificial intelligence-assisted system in postoperative follow-up of orthopedic patients: exploratory quantitative and qualitative study. J Med Internet Res 2020;22(5). https://doi.org/10.2196/16896.

[11] Maffulli N, Rodriguez HC, Stone IW, Nam A, Song A, Gupta M, Gupta A. Artificial intelligence and machine learning in orthopedic surgery: a systematic review protocol. J Orthop Surg Res 2020;15. https://doi.org/10.1186/s13018-020-02002-z.

[12] Staub BN, Sadrameli SS. The use of robotics in minimally invasive spine surgery. J Spine Surg 2019;5(Suppl 1):S31–40. https://doi.org/10.21037/jss.2019.04.16.

[13] Korzeniowski P, Barrow A, Sodergren MH, Hald N, Bello F. NOViSE: a virtual natural orifice transluminal endoscopic surgery simulator. Int J Comput Assist Radiol Surg 2016;11 (12):2303–15. https://doi.org/10.1007/s11548-016-1401-8.

[14] Gargeya R, Leng T. Automated identification of diabetic retinopathy using deep learning. Ophthalmology 2017;124(7):962–9. https://doi.org/10.1016/j.ophtha.2017.02.008.

[15] Esteva A, Kuprel B, Novoa R, Ko J, Swetter S, Blau H, et al. Dermatologist-level classification of skin cancer with deep neural networks. Nature 2017;(542):115–8. https://doi.org/10.1038/nature21056.

[16] Hazlett HC, Gu H, Munsell B, Kim SH, Styner M, Wolff J, Fonov V. Early brain development in infants at high risk for autism spectrum disorder. Nature 2017;542:348–51. https://doi.org/10.1038/nature21369.

[17] Mathotaarachchi S, Pascoal T, Shin M, Benedet A, Kang MS, Beaudry T, Rosa-Neto P. Identifying incipient dementia individuals using machine learning and amyloid imaging. Neurobiol Aging 2017;59:80–90. https://doi.org/10.1016/j.neurobiolaging.2017.06.027.

[18] Cui M, Zhang DY. Artificial intelligence and computational pathology. Lab Invest 2021;101 (4):412–22. https://doi.org/10.1038/s41374-020-00514-0.

[19] Patel HRH, Joseph JV. Simulation training in laparoscopy and robotic surgery. 1 Aufl. London: Springer Verlag London Limited; 2012.

[20] Atallah S. Digital surgery. Springer; 2020.

[21] Kim KC. Robotics in general surgery. 2014 ed. New York, NY: Springer New York; 2013.

[22] Mirchi N, Bissonnette V, Yilmaz R, Ledwos N, Winkler-Schwartz A, Del Maestro RF. The virtual operative assistant: an explainable artificial intelligence tool for simulation-based training in surgery and medicine. PLoS One 2020;15(2). https://doi.org/10.1371/journal.pone.0229596.

[23] Huff TJ, Ludwig PE, Zuniga JM. The potential for machine learning algorithms to improve and reduce the cost of 3-dimensional printing for surgical planning. Expert Rev Med Devices 2018;15(5):349–56. https://doi.org/10.1080/17434440.2018.1473033.

[24] Longmore SK, Naik G, Gargiulo GD. Laparoscopic robotic surgery: current perspective and future directions. Robotics 2020;9(2):42. https://doi.org/10.3390/robotics9020042.

[25] Spinoglio G. Robotic surgery current applications and new trends. 1st ed. Milano: Springer Milan; 2015.

[26] Tsuda ST, Kudsi OY. Robotic-assisted minimally invasive surgery: a comprehensive textbook. Springer; 2019.

Chapter 3

Colonoscopy robots

Yu Huan and Gastone Ciuti
The BioRobotics Institute, Scuola Superiore Sant'Anna, Pisa, Italy, Department of Excellence in Robotics & AI, Scuola Superiore Sant'Anna, Pisa, Italy

3.1 Introduction

Colorectal cancer (CRC), which is the development of cancer in the colon or rectum, is the third most commonly diagnosed cancer and the second cause of cancer death across the world. Based on GLOBOCAN 2020, a report produced by the International Agency for Research on Cancer (IARC), CRC occupies about 10% of the 19 million new cancer cases and 9.4% of the 10 million cancer deaths worldwide [1]. CRC incidence has been steadily rising globally every year, especially in developing countries, that are gradually following the westernized diet and lifestyle [2]. The potential risk factors for CRC growth include obesity, a diet with a low intake of fruits and vegetables, less physical exercise, alcohol drinking, and smoking [3,4]. To prevent and reduce the incidence of CRC, aside from having a healthy lifestyle, regular colonoscopy screening also plays an important role in the detection of early-stage CRC. As a matter of fact, colonoscopy screening together with colonoscopic polypectomy lowers the incidence of CRC and mortality [5], and colonoscopy is popularized across the world as one of the most commonly performed and valuable procedures in clinical interventions [6]. Colonoscopy is considered the gold standard for the detection of precancerous lesions, in particular for serrated polyps [7].

Conventional colonoscopy (CC) system is composed of a flexible colonoscope featuring a bendable tip with 2 degrees of freedom (DOFs), and the distal portion of the tip is equipped with a camera, light source, an instrument port, together with nozzles for air and water. During the colonoscopic procedure, the physician introduces the scope through the anus of the patient, and then by combined manipulation of rotation knobs at the handle (2 DOFs for bending at the distal tip) and the manual control of the probe (2 DOFs for insertion and rotation), the colonoscope can navigate along the colonic tract to perform the screening. If needed, the colonoscope also allows for therapeutic actions, e.g., a dedicated tool could be introduced through the instrument port to perform

Endorobotics. https://doi.org/10.1016/B978-0-12-821750-4.00003-7

a biopsy or removal of cancerous lesions [8]. An alternative to CC is a wireless capsule endoscopy (WCE), which is less invasive and thus causes less discomfort to the patients [9]. WCE utilizes a capsule equipped with a camera, white light LED-based illumination, an onboard battery, and wireless communication for image streaming. However, due to the size constraint, it is challenging to also implement active locomotion or lens cleaning capabilities to the capsule. As a result, the motion (stop or move) and accurate localization (position and orientation) of the capsule cannot be achieved within the colonic tract [10], which may cause a low detection rate of colonic pathologies in specific areas [11]. Another limitation is that WCE is not featured with therapeutic means, so if treatments are needed, CC would have to be performed. So far, CC is considered to be the most effective approach for the diagnosis and treatment of CRC, and it represents the gold standard for the evaluation of various colonic pathologies. Today, colonoscopy could be performed with an emerging and promising approach, i.e., colonoscopy robots (CRs), which have been developed and introduced recently to further enhance CC.

The development of robotics is growing rapidly in the last decades; it is an interdisciplinary field that integrates both computer science and engineering [12]. A robot is a machine programmed by a computer, which is capable of carrying out a complex series of actions with different levels of autonomy [13]. The ultimate goal in developing robots has always been to enhance human performance (e.g., precision and speed) and to partially or completely substitute human operators in applications where hazardous or physical constraints are posed in front of humans [14]. Nowadays, the adoption of robots has been spreading among various applications, and especially with the public's growing attention to health care, robotics techniques have been introduced also in the medical and surgical fields [15]. Surgical robots are generally endowed with actuation mechanisms, and they are able to perform surgical tasks either through surgeon teleoperation or potentially with autonomous control. CR, as one type of surgical robot dedicated to the colonoscopic procedure, have been investigated and developed to circumvent the main issues of CC [16]. During CC procedures, physicians push the colonoscope along the colonic tract through the anus of a patient, which might induce trauma to patients due to excessive pushing/stretching against the colonic wall or arise difficulty for further insertion of the colonoscope due to looping events [17].

To address the aforementioned clinical issues, CRs adopt robotic techniques to CC systems, which leverage computer-aided actuation mechanisms, sensing modalities, control algorithms, and intelligent vision to facilitate colonoscopy so as to achieve safe (no trauma) and effective (no looping) surgical interventions [18]. Compared to conventional devices, CRs are able to provide locomotion with less pain and higher precision than manual insertion, have intelligent vision systems for lesion identification and classification, and improved ergonomics to ease the physical and cognitive workload of physicians. As highlighted by Ciuti et al., the difference between CRs and conventional ones is nestled in their intrinsic capability to enable and perform controlled assisted actions or autonomous procedures in an unstructured deformable environment, such as in the colonic tract [19]. In order to realize such systems, efforts from both hardware and software aspects are necessary. From the hardware point of view, vital packages such as actuation mechanisms, sensors, and high-definition vision modules are needed to endow "physical" capabilities for the CRs to navigate, perceive, and see inside the colonic tract. From the software point of view, the corresponding motion controllers and artificial intelligence (AI)-based algorithms then serve as "minds" to process the sensed information and transmit commands to enable semi-/fully autonomous navigation and perform surgical tasks.

In this context, many potential benefits have been highlighted; thanks to the adoption of CRs. In particular, with the advancement and maturity of fabrication, actuation and control technologies, alongside AI algorithms, enhanced teleoperation or even autonomous colonoscopy will not be a fantasy anymore, but an envisioned next-generation clinical scenario.

The chapter is structured as follows: Section 3.2 describes CRs with marketing qualifications; Section 3.3 describes CRs with innovative features under research development; Section 3.4 describes the implementation of AI to enhance the performance of CRs; and lastly, Section 3.5 raises discussions and gives conclusions of this chapter.

3.2 Commercially certified robotic colonoscopes

To improve CC, emerging robotic techniques have been utilized to endow active locomotion to flexible colonoscopes. Currently, there are five CR systems that have obtained marketing qualifications (e.g., CE mark, FDA or CFDA approval), even if some of them are not available on the market anymore.

The NeoGuide Endoscopy System (NeoGuide Systems, Inc., Los Gatos, CA, USA) features a computer-aided scope composed of 16 individual electromechanical-actuated segments, each of which allows for 2 DOFs bending (Fig. 3.1A). Moreover, the system is sensorized: a position sensor is integrated at the tip of the scope to monitor the steering commands of the endoscopist and an external position sensor is placed at the anus to monitor the insertion depth [20,21]. During the operation, the endoscopist manually inserts the scope and maneuvers steering dials on the handle, under the real-time visualization via a conventional charge-coupled device (CCD). The controller extracts the depth and tip steering commands datasets thanks to the aforementioned sensors,and then actuates each bending segment to automatically articulate in a "follow-the-leader" manner during the scope insertion (active mode), thus reducing looping rates and forces applied to the colonic wall. Apart from this active mode, in which the shape of the scope is actively controlled, the system also features a passive mode where its function and stiffness of scope resemble conventional colonoscopes [20]. A clinical trial of the system on

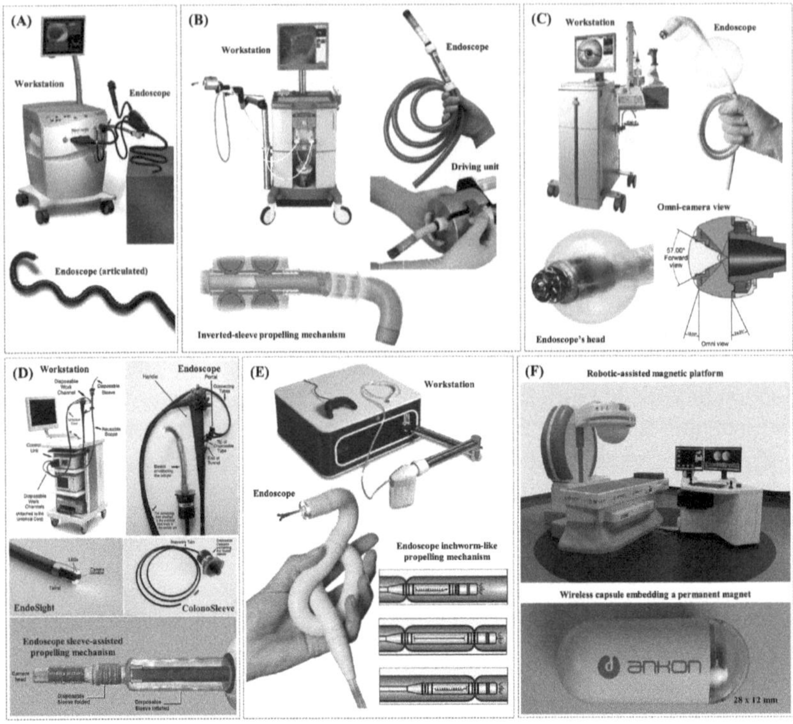

FIG. 3.1 Commercially certified endoscopy robots [19]: (A) NeoGuide Endoscopy System (NeoGuide Endoscopy System Inc., Los Gatos, CA USA); (B) Invendoscope™ SC40 (Invendo Medical GmbH, Weinheim, Germany); (C) Aer-O-Scope System (GI View Ltd., Ramat Gan, Israel); (D) ColonoSight (Stryker GI Ltd., Haifa, Israel); (E) Endotics System (ERA Endoscopy Srl, Pisa, Italy); and (F) NaviCam® (Ankon Technologies Co, Ltd. Wuhan, Shanghai, China).

10 patients was performed in [21], whereas the cecum was reached in all cases and the average procedure time, including therapeutic interventions, was 34 min (range: 24–60 min). No adverse events (e.g., hematoma, perforation, and submucosal petechial lesions) occurred during the intervention and the withdrawal of the scope, and the post-intervention assessment after 48 h and 30 days revealed no complications nor adverse effects. All the participated patients were willing to partake in the intervention with the system again, and the overall reported satisfaction of the physician with the system was superior (i.e., 100%). The system was initially developed in 2001; then it obtained FDA approval in 2006 and was acquired by Intuitive Surgical, Inc. (Sunnyvale, CA, USA) in 2009 [18], enabling the development of Ion, a robotic-assisted endoluminal platform for a minimally invasive peripheral lung biopsy.

The Invendoscope™ SC40 (Invendo Medical GmbH, Weinheim, Germany) features a deflectable (up to 180 degrees) endoscopic tip which is electrohydraulic actuated and an inverted sleeve mechanism electromechanically driven by eight wheels, allowing for the colonoscope to propel forward or backward (Fig. 3.1B). The colonoscope has an inner sheath, which is covered by the sleeve. Below the deflectable tip, the sleeve is inverted and firmly fixed by a locking mechanism, and the other end of the inverted sleeve can move forward/backward to unfold/fold itself. Thanks to this mechanism, the colonoscope can "grow/shrink" just below the distal tip. During the operation, the physician uses a handheld control unit to perform all the endoscopic and software functions. To be more specific, the handheld control unit includes keys with functions to move the sleeve forward or backward and a joystick to deflect the tip and steer the colonoscope in any direction. The vision module consists of three white light LEDs and is equipped with an advanced complementary metal-oxide-semiconductor (CMOS) imaging chip for image visualization. Standard functions such as suction, irrigation, and insufflation are also provided, together with a working channel for the insertion of a biopsy instrument [18]. A clinical trial was performed among 34 healthy volunteers, whereas the cecum was reached in 82% of the cases, and the acceptance mean score was 1.96 (range 1–6; 1 = no discomfort). No immediate or delayed side effects or complications occurred in any of the procedures [22]. The Invendoscope™ SC40 was CE marked in 2008 and then substituted by its successor version Invendoscope™ SC200, which features a handheld control interface and is manually inserted. Invendoscope™ SC200 (as the colonoscope of Invendoscopy™ E200 system) was CE marked in 2016. In 2017, Invendo Medical GmbH was acquired by Ambu A/S (Ballerup, Denmark), who announced the FDA clearance for the successor version Invendoscope™ SC210 (as the colonoscope of Invendoscopy™ E210 system) in 2018 [19]. Moreover, Ambu A/S announced on its website the launch of a single-use and sterile gastroscope and colonoscope in 2021 [23].

The Aer-O-Scope System (GI View Ltd., Ramat Gan, Israel) features a disposable, self-propelling, self-steering, and pneumatically driven robotic colonoscope (Fig. 3.1C). During the insertion of the scope, a proximal balloon is inflated to seal the anus, and the physician gently maneuvers the scope into the colon. After inflation of a distal balloon located proximal to a deflectable tip, CO_2 is inflated to fill the space between the two balloons and generates pneumatic pressure (less than 54 mbar [24]) to push the distal balloon forward, which reduces the needed exertion force from the physician and allows the flexible scope to navigate. Once the distal balloon reaches the cecum and the scope needs to be retracted, the CO_2 between the proximal and distal balloons is vented, and CO_2 is inflated between the cecum and the distal balloon to push the distal balloon backward. A 360 degrees omni-directional CMOS camera, a dedicated optical module, and LEDs are integrated into the distal deflectable tip, whose bending direction is remotely controlled via a handheld interface by the physician. A clinical trial was performed with 56 completed subjects, whereas the cecum was reached in 55 cases with a 98.2% cecal intubation rate

(CIR), and no mucosal damage or adverse events were reported [25]. The system was initially developed in 2007 and received FDA clearance in 2014, and then received FDA clearance for a successor version with therapeutic functionalityin 2016 [26].

The ColonoSight (Sightline Technologies Ltd., Haifa, Israel) features a reusable colonoscope with LEDs and a camera at the tip, enfolded by a disposable plastic sleeve with integrated operating channels and distal air-pressure-powered engine (IntraPull technology) components (Fig. 3.1D). During the procedure, compressed air is pumped into the sleeve by pushing a foot pedal, which generates a push force to the tip of the colonoscope and extends the folded sleeve, enabling the self-propelling of the colonoscope inside the colon. The sleeve isolates the colonoscope from direct contact with any colonic contents and, therefore, disinfection is not needed. A comprehensive trial was performed with 178 participants, whereas the cecum was reached for 161 participants (90%) with an average (±SD) time of 11.2±6.5 min. No complications, bleeding, nor perforation occurred during all the procedures [27]. The system obtained FDA approval in 2004 and then was acquired by Stryker GI Ltd. (Haifa, Israel) in 2006.

The Endotics System (ERA Endoscopy Srl, Pisa, Italy) features a disposable, self-propelling, and pneumatically driven robotic colonoscope (Fig. 3.1E). At the active segment of the robotic colonoscope, a distal and proximal clamping mechanism and a central extensive module form the locomotion components. With programmed actuation cycles, the whole colonoscope could navigate along the colon tract with inchworm-inspired locomotion. To be more specific, the actuation phases are 1) firstly, the proximal clamp anchors itself to the mucosal surface via pneumatic suction; 2) then, the central extensive module elongates via pneumatic inflation; 3) the distal clamp is thus brought forward along the colon tract, and it anchors itself to the mucosa surface via pneumatic suction, as well; 4) the proximal clamp is then deactivated and detaches itself from the mucosa; 5) lastly, the central module contracts and the proximal clamp is thus brought forward along the colon tract, resulting in a forward travel of the whole colonoscope; and 6) finally, the actuation cycle is repeated [18]. The steerable tip contains a CMOS camera with a 140 degrees field of view, a water and air channel for cleaning/drying the lens and for insufflation, and a 3 mm operative channel. The bending and propelling motions of the steerable tip are remotely controlled by a handheld interface through an electropneumatic control box, and the tip can achieve high-precision 2 DOFs bending up to 180 degrees. Moreover, the system is able to guarantee advanced endoscopic imaging techniques by means of electronic chromoendoscopy [19]. An experimental study with 40 enrolled patients evaluated the stress pattern (i.e., force vs time acted on the bowel) differences between the CR, i.e., Endotics System, and the CC; the results showed that the stress pattern related to the CR was 90% lower than the one of the CCs. Additionally, CR demonstrated a higher diagnostic accuracy since it was able to visualize small polyps and angiodysplasias not seen during CC, thanks to the lower insufflation rate. All patients rated CR as virtually painless compared to CC, ranking pain and discomfort as 0.9 and 1.1, respectively, on a scale of 0 to 10, versus 6.9 and 6.8, respectively, for the CC [28]. Another study was conducted on 71 patients with clinical or familial risk of colonic polyps/carcinomas. The cecum was reached in 81.6% of examinations with CR and 94.3% with CC; the average time (± SD) was 45.1 ± 18.5 and 23.7 ± 7.2 min for CR and CC, respectively. No patient required sedation during the CR examination, compared with 19.7% of patients undergoing CC. Finally, the sensitivity and specificity of the Endotics System for detecting polyps were 93.3% and 100%, respectively, with a positive predictive value of 100% and a negative predictive value of 97.7% [29]. In another retrospective study, 276 CT examinations were performed on patients, who had undergone CC and failed cecum intubation. The cecum intubation rate was then addressed in 102 of 276 patients, applying the exclusion criterion. Overall, Endotics System was successful in 93.1% of the incomplete CC cases (95% performance) [30]. Recently, a pilot study was performed to assess an experienced endoscopist training progress on CR. The results of two consecutive blocks of 27 (Group A) and 28 (Group B) were compared: CIR increased from 92.7% to 100%, and cecal intubation time (CIT) significantly decreased from 55 to 22 min. Moreover, most patients judged the procedure as mild or no distress, with a high willingness (92.7%) to repeat the CR [31]. The system was CE marked in 2007 and then obtained an extension for the European mark in 2017, which would be valid until 2022.

Apart from the five aforementioned commercially certified CR systems, an embodiment of robot-aided endoscopic platform, i.e., NaviCam® (Ankon Technologies Co, Ltd., Wuhan, Shanghai, China), is worth mentioning even though it is only currently applicable only for a gastroscopy (Fig. 3.1F). The platform is composed of a robot-aided external magnetic field generator, a teleoperation console, and an ingestible wireless capsule [32]. During the procedure, the capsule could navigate and inspect within the stomach under the control of an external magnetic field. To be more specific, the capsule is featured with an internal permanent magnet, an on-board battery; it also has a compact size (OD: 12 mm, L: 28 mm), 5 DOFs (three translational, two rotational), accurate positioning (5 mm), together with a vision module composed of LEDs and a camera (140 degrees field of view and 0–60 mm depth of field) for high-quality and real-time imaging. A second-generation capsule includes an enhanced vision module to further optimize the diagnosis [33]. The platform received approval by the CFDA mark with a Class III medical device registration certificate titled "Magnetically Controlled Robotic Capsule Endoscope," and was granted a De Novo classification in 2020 by the FDA.

Table 3.1 summarizes the five commercially certified CRs based on their marketing qualifications, actuation principles, technical features, and corresponding clinical references.

TABLE 3.1 Summary of commercially qualified CRs in terms of marketing qualifications, availability on the market, actuation means for locomotion, technical features, and clinical references.

<table>
<tr><th>System</th><th>Marketing qualifications | availability on the market</th><th>Actuation</th><th>Technical features</th><th>Clinical trial</th></tr>
<tr><td>NeoGuide</td><td>2006 FDA approval | not available</td><td>Electromechanical and manual</td><td>– 16 individual actuated segments, each with 2 DOFs
– Computer-aided “follow-the-leader” locomotion</td><td>[20,21]</td></tr>
<tr><td>Invendoscope</td><td>2008 CE mark (SC40) 2016 CE mark (SC200) 2018 FDA clearance (SC210) | not available</td><td>Electromechanical</td><td>– Electrohydraulic actuated deflectable tip
– “Grow/shrink” mechanism</td><td>[22]</td></tr>
<tr><td>Aer-O-Scope</td><td>2014 FDA clearance 2016 FDA clearance | not available</td><td>Pneumatic and manual</td><td>– Self-propelling and self-steering
– Inflated CO_2 to generate pneumatic pressure
– Distal and proximal balloons</td><td>[25,34]</td></tr>
<tr><td>ColonoSight</td><td>2004 FDA approval | not available</td><td>Pneumatic</td><td>– Self-propelling
– Distal air-pressure-powered engine
– No need for disinfection, thanks to isolation sleeve</td><td>[27]</td></tr>
<tr><td>Endotics</td><td>2007 CE mark 2017 CE mark extension | available</td><td>Electropneumatic</td><td>– Self-propelling
– Inchworm-like locomotion
– Distal and proximal mucosal clampers</td><td>[28–31]</td></tr>
</table>

3.3 Research-oriented colonoscopy robots

Apart from the commercially certified CR systems, several research groups are endeavoring to integrate novel actuation mechanisms for CRs to realize painless yet effective navigation. Based on different actuation principles, CRs can be classified into (1) electric-, (2) hydraulic- or pneumatic-, and (3) magnetic-actuated.

Electric actuation leverages electric actuators, e.g., motors together with motion transmission mechanisms, to generate sufficient locomotion power for the robot to navigate along the anatomical path. This approach might face issues related to the constrained design space available for the tip of a CR, due to the size of the motors (if embedded within the robot) or motion transmission components (if the motors are embedded outside).

The research group from the School of Aerospace and Mechanical Engineering, Korea Aerospace University led by Prof. Byungkyu Kim contributed to the development of three types of CR systems: (1) an inchworm-like pneumatic-actuated system, (2) a reel mechanism-based electric-actuated system, and (3) a self-propelled caterpillar-based electric-actuated system. The inchworm-like mechanism consists of a central extensive/compressive bellow module, together with a front and a rear clamper. Each clamper features a six-leg mechanism that can be folded (detach from the colon) or unfolded (anchor to the colon) based on a synchronized pneumatic spring trigger. The forward movement is achieved, thanks to the following actuation cycle: (1) the front clamper unfolds its six-leg mechanism and anchors itself to the colon; (2) the central bellow is then vacuumed to shrink its length and bring the folded rear clamper forward; (3) the rear clamper then unfolds and anchors to the colon, and the central bellow starts to be inflated and brings the folded front clamper forward; (4) then negative pressure is applied so that the front clamper unfolds to anchor and the rear clamper folds to detach from the colon, meanwhile the central bellow shrinks its length, bringing the whole device a stroke forward; and (5) finally, the actuation cycle is repeated. In that way, the device can navigate along the colon tract [35]. The reel mechanism-based system mainly features a sliding distal six-leg clamper that could be folded (detach from the colon) or unfolded (anchor to the colon) based on a push/pull trigger, a counter spring, an antagonistic pair of pulling wire with a pushing spring, and a reel actuated by a stepper motor. The locomotion of the device is achieved with the following actuation cycle: (1) the reel rotates when the motor is turned on; (2) the pulling wire then overcomes the antagonistic spring force and pulls the unfolded sliding clamper, overcoming the force components of counter spring and colon-clamper resistance; (3) the cylindrical body of the device is thus brought forward; (4) the motor then rotates in the reverse direction, resulting in the clamper to fold itself and slide back to the distal position of the cylindrical body; and (5) finally, the actuation cycle is repeated. In that way, the device can move forward. Locomotion tests were performed and the robot achieved high mobility (66.04 ± 5.97 mm/s with a motor speed of 200 r/min), and the error between the theoretical and experimental results was less than 1% [36]. Based on the same system, the authors further integrated soft silicone on the six legs of the clamper to test maneuverability and safety of the robot with an *in vitro* porcine colon. As a result, the static and dynamic force acting on the colon is less than the bursting force of a human colon. In addition, the maneuverability of the robotic colonoscope is not hindered even with the soft cover [37] (Fig. 3.2A). Recently, the group further integrated a cable-driven steering module (2 DOFs bending) distally to the locomotion robot and a feeding module to assist the insertion and navigation within the colon [38] (Fig. 3.2B). The self-propelled caterpillar-based system features a robot body

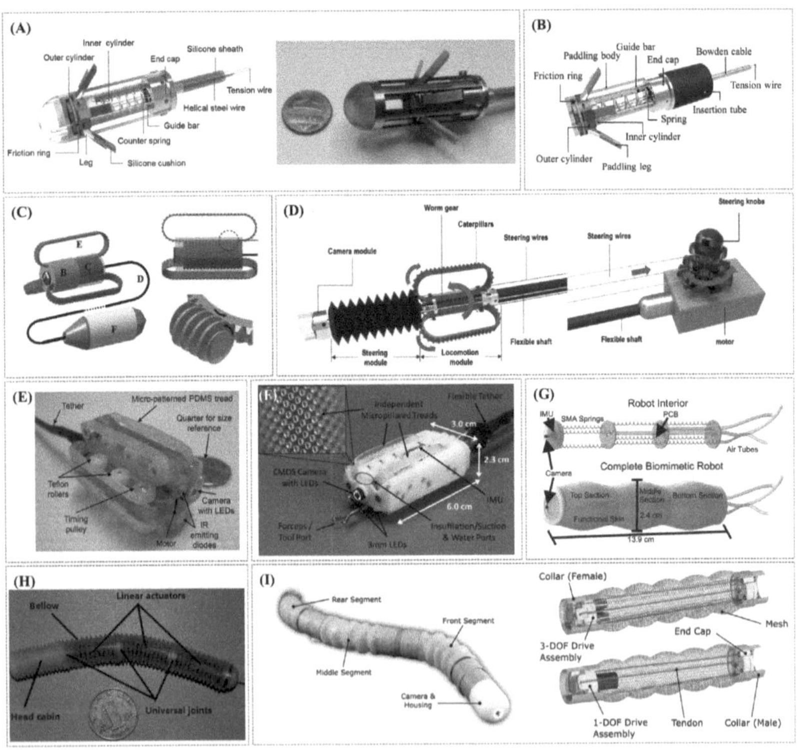

FIG. 3.2 Research-oriented CRs based on electric actuation: (A) a reel mechanism-based robotic colonoscope with enhanced features for safety and maneuverability [37]; (B) a CR system integrating feeding and steering mechanisms with self-propelled paddling locomotion [38]; (C) a flexible caterpillar-based robotic colonoscope driven by an external motor via a flexible shaft [39]; (D) an elastic caterpillar-based self-propelled robotic colonoscope with enhanced safety and mobility [40]; (E) a tethered robotic capsule endoscope using micro-patterned treads [41]; (F) Endoculus: an improved robotic capsule based on micro-patterned treads [42]; (G) a SMA-actuated biomimetic robot with soft skin [43]; (H) an improved version of the squirm CR tested with *in vivo* porcine colon [44]; and (I) a CR meshworm with segment-bending anchoring [45].

composed of three elastic caterpillars simultaneously rotated via a worm gear mechanism, which is actuated by an integrated miniaturized motor [46]. The group then substituted the previous integrated motor with a higher power external motor, which enables faster locomotion speed and reduces the risk of short circuits. The mechanical transmission from the motor rotation to the worm gear rotation is achieved through a flexurally flexible but torsionally stiff shaft [39] (Fig. 3.2C). Centered on caterpillar-based locomotion, the group further improved the system by integrating a cable-driven steering mechanism at the distal portion of the locomotion module. A mobility test was performed in an excised porcine colon, and the system demonstrated reliable locomotion performance with forward and backward velocities of 5.0 ± 0.4 mm/s and 9.5 ± 0.9 mm/s, respectively (forward velocities of 6.1 ± 1.1 mm/s and 4.7 ± 0.7 mm/s in case of 30 and 60 degrees inclined angles, respectively). An *in vivo* test was performed in a live mini pig under sedation; the robot managed to arrive at the distal transverse colon (600 mm from the anus) but it was not able to move further, due to the high extensibility/elasticity of the mucosa and the interference with the fecal materials [40] (Fig. 3.2D).

The research group from the University of Colorado led by Prof. Mark E. Rentschler contributed to the development of a tethered robotic capsule endoscope, utilizing micro-patterned treads. The design of this system is integrated at the distal capsule housing, whereas the housing is designed to host the drivetrain (including a DC motor, gear mechanism, and timing pulleys), four pairs of micro-patterned polydimethylsiloxane (PDMS) treads, a vision module, and an illumination module [41] (Fig. 3.2E). The external surface of each PDMS tread features a micro-pattern consisting of 140-μm diameter, 70-μm high circular pillars. The pillars were equally spaced with a center-to-center distance of 245 μm, which provided sufficient traction on the bowel tissue while minimizing tissue damage [47]. A complementary research was then carried out to investigate the normal adhesive response between a synthetic tissue substrate and a smooth PDMS surface. Such adhesion response serves a critical mobility design parameter of the robotic capsule relying on PDMS treads to travel along the colon tract [48]. Recently, the advanced version of this robotics capsule, namely Endoculus, has been developed. Endoculus is composed of multi-DOFs sensors (i.e., inertial measurement unit, magnetometer, motor encoders, and motor current sensors), two independent micropillered treads actuated via two separate motors (allowing for 2 DOFs skid-steering), a CMOS camera with adjustable-intensity LEDs, working channels for insufflation, irrigation, and a tool port for endoscopy instruments (e.g., forceps). An *in vivo* test was performed with a live pig, which successfully demonstrated endoscopic functionalities and promising results in terms of locomotion. Endoculus was not able to achieve consistent traction in the sigmoid area, whereas the main reason is due to the significantly tight essence of the tested subject. In addition, an *ex vivo* test was performed with a 40-cm excised porcine colon, and the results demonstrated a forward/reverse locomotion speed up to 40 mm/s on the colon mucosa, for both insufflated and collapsed cases, which is faster than the previous prototype (maximum 3 mm/s) [42] (Fig. 3.2F). Apart from the micro-patterned tread-based capsule, this group also developed a CR featuring three soft modular sections, whereas each section features 3 DOFs (one axial translation and two rotations). Each section adopts three independently actuated shape memory alloy (SMA) springs, which shrinks their length under joule heating. A silicone rubber skin wraps around the whole robot to provide a passive elastic recovery force to stretch the SMA springs to their original relatively long length when joule heating is not applied to the SMA springs. To reduce the cool time of SMA springs and increase actuation bandwidth, three air tubes are embedded for each section to provide forced convection. A parametric finite element analysis (skin curvature and thickness) was performed to maximize the recovery force without hindering the SMA spring contraction performance. A controller unit was designed and implemented for each of the sections, enabling the robot to achieve ±90 degrees 2 DOFs bending in less than 4 s with almost zero steady-state error. The robot adopts a peristaltic propel motion, and tests demonstrated a maximum and average speed of 4 mm/s and 0.37 mm/s in a transparent 40 mm outer diameter (OD) polyvinyl chloride (PVC) tube [43] (Fig. 3.2G).

The research group from the Shanghai Jiaotong University led by Prof. Kundong Wang proposed a three-segment squirm robotic endoscope with a compact size of 13 mm in diameter and 105 mm in length and a lightweight of 22.3 g. Each segment of the robot body is composed of a linear actuator with a micromotor, a turbine-worm, and a wire wrapping-sliding mechanism. Then, based on a contact biomechanical model targeting safe and effective locomotion in soft tissues, a soft monolithic bellow was designed and wrapped around the robot's body. The bellow increased the static friction and decreased the kinetic friction when shifting its contact state with the colon, and thus during each locomotion cycle: (1) the distal segment firstly slides forward with low kinetic friction, while the central and proximal ones serve as anchors due to high static friction; (2) then the distal segment contracts and central segment extends; (3) then the central segment contracts and proximal segment extends; (4) eventually, when the proximal contracts, the other two segments serve as an anchor and thus a stroke is generated; and (5) finally, the actuation cycle is repeated. The robot was tested with an *in vitro* porcine bowel [49] and then in an *in vivo* porcine colon [44] (Fig. 3.2H).

Bernth et al. from King's College London proposed a worm-inspired, three-segment CR with multiple DOFs movements. Each segment of the robot is composed of an elastic mesh structure (polyester material) that can elastically deform upon actuation of antagonistic tendon pairs, which are individually/universally pulled/released by the pulley's mechanism and DC motors. The distal and proximal segments are actuated by three motors, thus enabling 1 DOF extension/contraction and 2 DOFs bending, while the central segment is actuated by one motor, thus only allowing 1 DOF extension/contraction.

The robot takes advantage of its steering capability of distal and proximal segments so as to selectively anchor themselves with the colonic wall to increase friction (bent tip press against the colonic wall to jam) when needed. This design is highly compatible with various colon diameters, thanks to adjustable bending angles. Moreover, the device was tested in a simulated colon made of a flexible plastic tube, achieving an average locomotion speed of 1.21 mm/s [45] (Fig. 3.2I).

An SMA-based CR was developed under the European FP7 project, namely "Colonic Disease Investigation by Robotic Hydro-colonoscopy (CODIR project)" [50], and under the European H2020 project, namely "Compliant Actuation Robotic Platform for Flexible Endoscopy (CARPE project)" [51]. Under the frameworks of these two projects, Manfredi et al. developed a CR based on modular miniaturized compliant joints [52]. Each joint has 2 DOFs bending capability enabled by two pairs of antagonistically arranged SMA wires. By applying joule heating to each SMA wire, it contracts and generates rotational motion. The implementation of an SMA-based closed-loop controller could generate sinusoidal motorless locomotion of the CR, which would travel by its intrinsic snake-like locomotion from the anus to the cecum.

Hydraulic or pneumatic actuations are extensively explored and adopted in soft robots, and thanks to the intrinsic safety, they have also been studied for medical applications. With this actuation approach, the CR feature soft chambers that could be inflated or deflated hydraulically or pneumatically. Although many benefits of this actuation method have been highlighted, it also needs complementary tubing and a pneumatic/hydraulic source, with requirements and challenges related to the sealing.

Dehghani et al. developed a semi-autonomous CR with minimally invasive locomotion [53]. The robot is composed of a distal tip, a middle latex tube, and a proximal rectal fixture. In particular, the latex tube can be inflated with compressed air and unwind the portion of the latex tube coiled within the distal tip, acting as a "growing" CR (Fig. 3.3A). *Ex vivo* tests were performed, and the robot was able to reach 1.5 m of a porcine colon with an average speed of 28 mm/s. The locomotion essence of this robot effectively prevents the formation of loops during the procedure.

Chen et al. developed an inchworm-inspired and pneumatic-driven CR [54]. The robot consists of a central propulsion segment featuring two linear balloons with reinforced springs and two gripper segments, each of which features two diagonally twisted balloons (Fig. 3.3B). All of the three segments can be inflated/deflated via pneumatic actuation, and by properly programming the actuation sequences, the robot can realize inchworm-like locomotion. Locomotion tests were successfully performed through pipes with diameters ranging from 22 to 32 mm, and 12 mm/s locomotion speed was reached in a 24-mm diameter acrylic pipe. Moreover, locomotion test and colon imaging inspection were qualitatively evaluated with an *ex vivo* swine colon, proving the feasibility of the robot for colonoscopy.

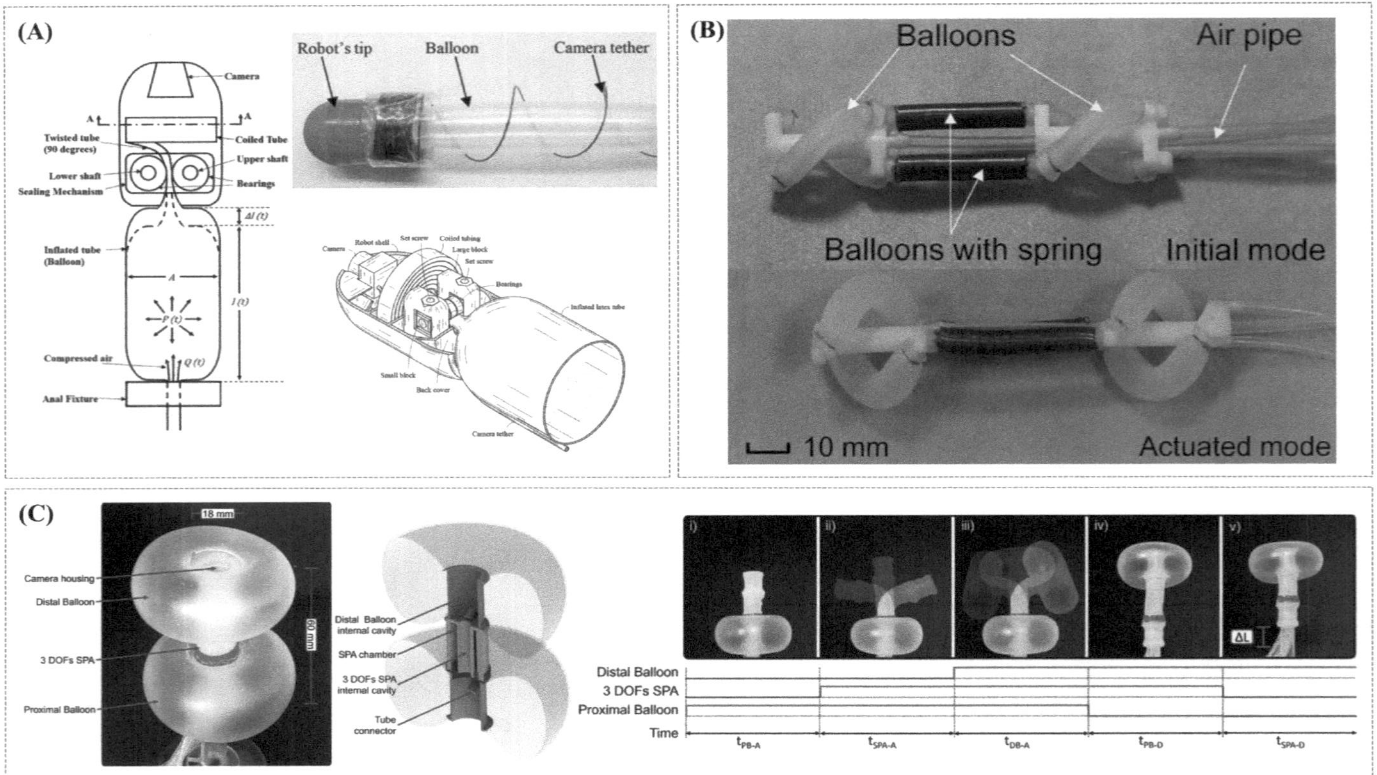

FIG. 3.3 Research-oriented CRs based on pneumatic actuation: (A) a self-steering pneumatically driven CR [53]; (B) a bio-inspired self-propelling CR [54]; and (C) a soft pneumatic inchworm double balloon (SPID) colonoscope [55].

Another novel pneumatically actuated soft robot has been developed by Manfredi et al. [55]. The CR, named SPID (i.e., soft pneumatic inchworm double balloon), consists of two soft balloons distally and proximally located in the robot serving as colonic wall anchors, together with a central pneumatic actuator with 3 DOFs (2 DOFs bending and 1 DOF axial extension) (Fig. 3.3C). The robot is compact in size (18 mm in diameter and 60 mm in length) and lightweight (10 g). SPID has been tested in a deformable synthetic colon simulator, resembling the shape and sizes of a human colon. It efficiently reached the cecum after traveling 1.4 m along the simulator with an average velocity of 2.8 mm/s. After cecal intubation, it could then be retracted by manually pulling the air tubes.

Apart from embedding robotic locomotion mechanisms directly within the colonoscope, magnetic actuation is also extensively studied as an external and active robotic locomotion approach. Magnetic locomotion utilizes an external magnetic (either permanent or electromagnetic) field, mounted as a robotic manipulator's end effector, which applies magnetic force and remotely controls the position/orientation of a permanent magnet, embedded at the tip of the colonoscope. Table. 3.2 provides a complementary pros & cons comparison between permanent magnet and electromagnet.

TABLE 3.2 Pros and cons comparison table between permanent magnets and electromagnets.

Magnetic field sources	Pros	Cons
Permanent magnet	– Generates persistent magnetic field standalone – Generates higher force considering a bulk with similar size – Customizable sizes, shapes, and magnetization directions	– Less flexible, cannot be "turned off" – Strength of magnetic field cannot be modulated
Electromagnet	– Can be "turned on/off" as needed – Strength of magnetic field can be modulated – Can generate various types of magnetic field under cooperation with several electromagnets	– Higher cost, needs additional power supply and cooling equipment – Generates lower force considering a bulk with similar size of a permanent magnet – Needs electric current control strategies to modulate the generated magnetic fields – Physical limit due to the presence of large number of coil winding and cooling equipment

Magnetic actuation-based CR systems were initially explored under the framework of a European FP6 project, namely "Versatile Endoscopic Capsule for gastrointestinal TumOr Recognition and therapy (VECTOR project)," which was coordinated by Novineon Healthcare Technology Partners GmbH (Tuebingen, Germany) [56]. Under the framework of the VECTOR project, Ciuti et al. proposed a wireless robotic capsule endoscopy system and carried out a series of studies. The system utilizes an external permanent magnet mounted as the end-effector of a 6 DOFs anthropomorphic robotic manipulator and an internal permanent magnet endowed in the capsule. The capsule also integrated a CMOS camera with four white light LEDs and a triaxial accelerometer enabling feedback control based on inertial sensing [57]. Then, a comparison study was performed to evaluate the magnetic capsule steering performance of robot-assisted and manual approaches. The results demonstrated that in *ex vivo* conditions, robot-assisted was superior to manual approach in terms of targets reached ($87\% \pm 13\%$ vs $37\% \pm 14\%$), while inferior in terms of trial completion time (423 ± 48 s vs 201 ± 24 s). Under *in vivo* conditions, the robot-assisted approach exhibited higher precision of movement and better reliability compared to its manual counterpart [58]. Another comparison study was performed, whereas 11 experts and 11 trainees performed complete colonoscopy by robotic magnetic capsule and by conventional colonoscopy in an *ex vivo* phantom simulator with installed artificial targets mimicking polyps. The results showed that complete colonoscopy was feasible in all cases with both approaches. The robotic magnetic approach was feasible and showed adequate accuracy, i.e., the successful target detection rates were 80.9% and 85.8% for robotic and conventional approaches, respectively [59]. Apart from the aforementioned wireless approach, tethered magnetic-driven robotic capsule colonoscopy has also been explored. Valdastri et al. proposed a solution as a compromise between conventional and wireless capsule colonoscopy, by endowing magnetic-driven capsule to a compliant tether with multilumen channels enabling insufflation, suction, irrigation, tool insertion, and lens cleaning [60] (Fig. 3.4A). A model was then proposed to predict the resistance force applied on the capsule and experimentally validated by performing drag force experiments, considering different capsule geometrical parameters and tissue properties [63]. In another study, a comprehensive analytical model (applicable for axial and diametrical magnetization) was derived to understand the magnetic forces and torques applied to a cylindrical (solid or annular) permanent magnet under an arbitrary external magnetic field [64]. Recently, Martin et al. developed a magnetic-driven colonoscope that utilizes intelligent machine vision to realize autonomous control, allowing for effective colonic tract navigation under *in vivo* conditions even by nonexpert users [61] (Fig. 3.4B). This solution took advantage of both "front-wheel" magnetic actuation and multifunctionalities of traditional tethered colonoscopes, resulting in a CR with a painless locomotion approach and versatile therapeutic equipment. Compared with the wireless capsules [58] (Fig. 3.4C), the tethered approach

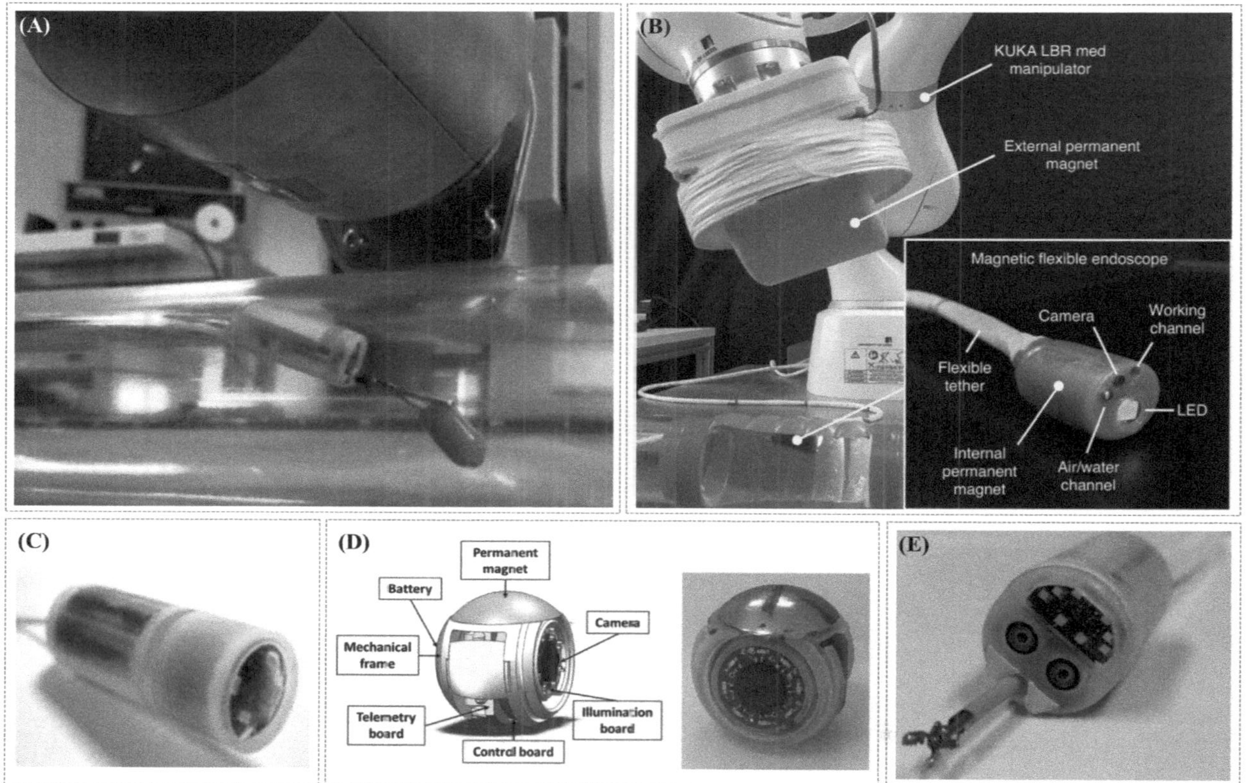

FIG. 3.4 Research-oriented capsule CRs based on magnetic actuation: (A) a tethered magnetic capsule robotic system [60]; (B) autonomous control of a magnetic-driven capsule based on computer vision strategies [61]; (C) a wireless magnetic capsule for colonoscopy inspection [58]; (D) SUPCAM EU project spherical magnetic capsule for colonoscopy [62]; and (E) Endoo EU project tethered magnetic capsule. *(Courtesy of Ciuti et al., scientific and technical manager of the Endoo EU project.)*

allows for routing of electric cables and functional tubes, which opens the possibility to embed multiple functional features to the capsule. Furthermore, complementary studies are performed to advance tethered magnetic-driven capsule colonoscopy in terms of localization [65,66] and control [67–69].

A spherical colonoscopy capsule to be navigated via an external electromagnetic field was designed under the framework of a European FP7 project, namely "new cost-effective and minimally invasive endoscopic device able to investigate the colonic mucosa, ensuring a high level of navigation accuracy and enhanced diagnostic capabilities (SUPCAM project)" [70], coordinated by S.E.D. Srl (Certaldo, Italy) and supervised by Dr. Alessandro Tozzi, who originally came up with the omni-vision spherical capsule concept. The research outcomes include: (1) the design and optimization of the magnetic interactions (forces and torques), provided in between an adjustable external electromagnetic field and an internal permanent magnet embedded in the 26-mm diameter spherical capsule [62] (Fig. 3.4D), (2) test-bench experimental evaluation in *in vitro* and *ex vivo* conditions [71], together with (3) a dedicated wireless charging system for the capsule [72].

A novel derivative soft-tethered magnetic-actuated colonoscope was developed under the framework of a European H2020 project, namely "endoscopic versatile robotic guidance, diagnosis, and therapy of magnetic-driven soft-tethered endoluminal robots (Endoo project)," coordinated by the Scuola Superiore Sant'Anna (Pisa, Italy) [73] (Fig. 3.4E). The robotic system is composed of a soft-tethered stereoscopic capsule, a medical workstation, and an industrial collaborative robot (COMAU SpA, Turin, Italy) [74]. The capsule features a permanent magnet, white light illumination, a high-resolution imaging module, custom-made binocular lens, and a working channel for the insertion of therapeutic instruments. In particular, under the framework of the Endoo project, fully convolutional neural networks were adapted to identify and segment polyps in colonoscopy images, and the obtained segmentation accuracy and detection precision are 73.61% and 86.31%, respectively [75]. Another related contribution was on contactless measurements (location and size of the lesion) using a computer vision approach based on neural networks [76].

Table 3.3 reports a comparative table highlighting the pros and cons of the three actuation methods.

3.4 AI in colonoscopy

During CC screening, the successful detection and resection of neoplastic polyps directly reduce colorectal cancer mortality, yet the polyp miss rate could reach as high as 20% [77], and a prospective trial of optical biopsy of small colon polyps using narrowband imaging showed that the accuracy of physicians was only 80% in diagnosing detected polyps as adenomas [78]. To cover such gap, computer-aided algorithms could be implemented with CR platforms, e.g., intelligent vision module, so as to facilitate the surgical procedure in terms of

TABLE 3.3 Pros and cons comparison table among electric, hydraulic or pneumatic, and magnetic actuations.

Actuation principles	Pros	Cons
Electric	– High operation bandwidth – Precise motion control – High force/toque output	– Risk of electrical short circuit – Potentially high-power consumption – Limited design space – Increased size and weight
Hydraulic or pneumatic	– Normally embedded with soft robots, assure safe and gentle interaction with tissue – Compact size and lightweight – Disposable and low cost	– Reduced structural robustness due to soft materials – Reduced precision in motion and shape control
Magnetic	– Cost-effective – Front-wheel locomotion, which induces less pain	– Prone to magnetic interference – Complex and unintuitive manipulation without a capsule localization strategy

pathological prediction and classification and to further assist decision-making for endoscopists or even play a role in training nonexpert endoscopists. With the development of technologies in areas of endoscopic vision (e.g., higher definition, larger field of view, etc.), AI starts to play an important role and shows potential to significantly enhance colonoscopy screening procedures. Different from the traditional modeling approach, the AI approach utilizes a large amount of annotated polyp image data for training and optimizing the algorithm to recognize or characterize polyps, allowing it to become more robust and reliable for surgical application if extensive and continuous data training would be provided.

One of the main applications of AI in colonoscopy is pathological prediction (i.e., lesion detection and classification), e.g., colorectal polyp detection and characterization [79,80], identification of colonic neoplasms [81] or flat neoplasia [82]. Bernal et al. addressed the automatic polyp detection sub-challenge as a part of the endoscopic vision challenge at the international conference on Medical Image Computing and Computer-Assisted Intervention (MICCAI) in 2015. They reported the results of comparative evaluation of various polyp detection methods and the results showed that convolutional neural networks were the state of the art [83]. Indeed, computer-aided diagnostic (CAD) systems using advanced AI techniques represent an emerging technology with high potential to promote a paradigm shift in colonoscopy, and real-time decision support software has been developed to serve the purpose of optical biopsy

[84–86]. Some of the current CAD tools could even compete with the performance of human experts, thanks to the development in AI [87].

Although AI exhibited the promising potential to enhance colonoscopy screening performance, challenges still remain. One of the main challenges is associated with the annotation of the polyps, which requires manual labeling of the polyp position, shape, and classification (Fig. 3.5A) [88,89]. As anticipated, since the training of CAD algorithms needs a large amount of training data to gain accuracy and robustness, the low bandwidth of human annotation hinders the acquisition of input training data. As a matter of fact, the amount of training images used in colonoscopy is in the order of several thousands, which is not comparable to similar algorithms in other vision applications. This is mainly due to this dedicated medical application field, where only eligible human observers (e.g., endoscopists or pathologists) are qualified to perform trustable and accurate annotation of polyps. To overcome this issue, possible solutions arise such as open datasets [90,91], involvement of professional data enrichment and annotation companies [92], the establishment of image database dedicated for colonoscopy [93], as well as more invested efforts in unsupervised

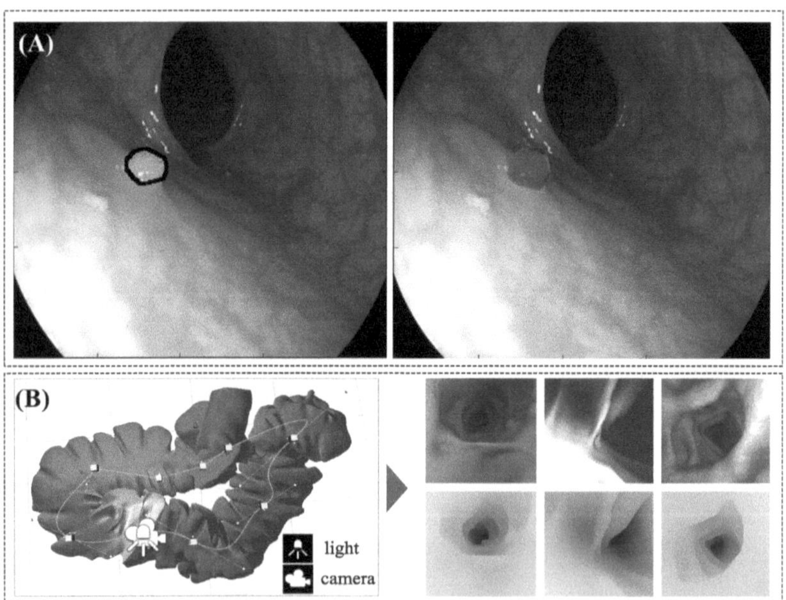

FIG. 3.5 The utilization of AI on endoscopic images [19]. (A) Demonstrates an AI algorithm being able to automatically detect and mark potential polyp regions in the endoscopic images, whereas original endoscopic images are in the upper row and the corresponding ones with AI markers are in the lower row. (B) Illustrates a 3D geometry reconstruction of a virtual colon extracted from CT. The image on the left shows a segment of colon surface mesh within Unity environment. The image on the right is examples of rendered RGB images with corresponding depth maps generated along camera path.

learning [94]. Although challenges still remain, several startup companies are gradually bringing AI-based polyp detection systems into mature medical products, e.g., ai4gi [95], Odin Vision [96], and Wision A.I. [97]. Moreover, major medical instrument companies such as Medtronic Inc. [98] and PENTAX Medical [99], together with photographic film manufacturer Fujifilm [100] are also endeavoring to commercialize AI-based systems.

Another potential application of AI in colonoscopy could be to facilitate the navigation procedure and the anatomical reconstruction of the colon, fostered by the fusion of endoscopic image, position sensor, together with diagnostic imaging modalities. The implementation of AI algorithms may enable quantitative measurement of observed tissue/structure (e.g., surface shape and morphology), shape reconstruction of the colonic tract, and camera pose estimation therein. Indeed, this need is not only addressed specifically in colonoscopy, but also among the research or application of minimally invasive surgery in general [101]. Mountney et al. reviewed various optical methods from laparoscopic/endoscopic cameras to extract tissue morphology for robotic guidance [102]. However, those classical methods have been gradually substituted by deep learning-based anatomical geometry (i.e., depth) estimation, which exhibited accurate and real-time performance [103,104]. Supervision can be realized via an anatomical simulator or *ex vivo* scenes where ground truth can be generated using medical imaging modalities, e.g., CT scan or structured light [105]. In addition, some derived tasks, e.g., 3D depth prediction from endoscopic images have been tackled by training conditional generative adversarial networks (Fig. 3.5B) [106]; a binary polyp size estimation has been achieved based on deep neural network architecture, which determines whether the dimension of a polyp reaches to the criterion for resecting [107].

3.5 Discussion and conclusions: What is now and what's next?

Although CC has shown its effectiveness in terms of both diagnostic and therapeutic needs during colonoscopy procedures, the manual insertion of the flexible colonoscope still faces occasional issues, such as self-buckling or stretching of the colonic tissues, which induce trauma or discomfort to patients. Thanks to the development of technologies in the different aspects of fabrication, actuation, sensors, control algorithms, and AI. CC systems can evolve to CR systems, which have already been or have the potential to be endowed with painless and effective front-wheel locomotion (or self-propelling), accurate shape control/estimation within the tortuous colonic tract, together with AI-based intelligent vision. These technical features, if properly developed and integrated, would definitely lead to colonoscopic procedures with better clinical outcomes, less discomfort for the patients, and thus increased acceptance. Currently, there is still a lack of many needed efforts to bring robust and safe CRs into the market, let alone wide adoption of such systems around the world. However, we have witnessed the potential and have foreseen the massive contributions that surgical robotic systems could bring to the healthcare system in general, which significantly revolute the quality, approach, and experience of clinical interventions. The da Vinci robotic system (Intuitive Surgical Inc., CA, USA), as one of the most successful instances, has demonstrated its capability and applicability in urology, cardiothoracic, gynecology, and abdominal surgery. On one hand, the system benefits the patients in terms of higher precision, shorter hospitalization, faster recovery, and less trauma. On the other hand, it benefits the surgeons in terms of tremor canceling, 3D immersive and vivid view, enhanced dexterity, and improved ergonomics. Furthermore, nowadays, with the maturity in telecommunication techniques and possible popularity of 5G network, remote robotic surgery would be enabled to allocate medical resources in a more humanized and reasonable manner.

As a matter of fact, scientists, researchers, and engineers are endeavoring to invent intelligent robotic systems not only in the medical field, but also in many other industrial fields, e.g., welding, precision farming, and construction. Humans cannot compete physically with robots in terms of precision, vision, computation, dexterity, load capacity, robustness, fatigue, and operation bandwidth. Nowadays, with the exploration and adoption of AI, even human experiences might become less substantial, e.g., AlphaGo developed by DeepMind Technologies (then acquired by Google) can beat the best professional Go game players, in a strong manner. The fact that AI can beat individuals in the Go game is significant since Go is considered a strategic game that has been extensively researched and accumulated by humans for more than 2500 years. Moreover, although the basic rules of Go are simple, yet its complexity is unimaginable (with 10^{170} possibilities). The success of AI in Go game shows great potential and hopes to tackle global challenges from disease treatment to climate change. Implementing AI in robots would even endow more capabilities for robots to "think," or as least efficiently support decisions, and make strategic assessments to solve challenges beyond the physical and mental constraints of humans.

The development of technology is a double-edged sword, and the current intelligent robotic techniques must be developed and utilized in compliance with human ethical standards so that technology could benefit mankind and serve society. Coming back to the core topic of this chapter, CRs, which represent an important and successful application of intelligent robotic systems, have been researched, developed, and adopted to improve this specific clinical field. Even if promising features, such as painless navigation, 3D immersive and vivid vision, and miniaturized robotic biopsy tools, could enhance the colonoscopy procedure, the maturity and robustness of CRs still need to be improved. These require concerted efforts from multiple forces, such as biomedical researchers, engineers, software and algorithm developers, instrument manufacturers, and healthcare institutions together with medical staff. Eventually, these combined inputs will pave the way for the new era of colonoscopies and more importantly for robotic surgery in general, leading to a better outcome and experience for patients, thus contributing to improving the quality of health care toward the whole society.

Acknowledgments

The authors would like to thank Alexander Breschi for the English proofreading.

Conflicts of interest

The authors declare no conflict of interest.

References

[1] Sung H, et al. Global cancer statistics 2020: GLOBOCAN estimates of incidence and mortality worldwide for 36 cancers in 185 countries. CA Cancer J Clin Feb. 2021. https://doi.org/10.3322/caac.21660, caac21660.

[2] Rawla P, Sunkara T, Barsouk A. Epidemiology of colorectal cancer: incidence, mortality, survival, and risk factors. Przeglad Gastroenterologiczny 2019;14(2):89–103. https://doi.org/10.5114/pg.2018.81072. Termedia Publishing House Ltd.

[3] Center MM, Jemal A, Smith RA, Ward E. Worldwide variations in colorectal cancer. CA Cancer J Clin Nov. 2009;59(6):366–78. https://doi.org/10.3322/caac.20038.

[4] Keum NN, Giovannucci E. Global burden of colorectal cancer: emerging trends, risk factors and prevention strategies. Nat Rev Gastroenterol Hepatol 01 Dec 2019;16(12):713–32. https://doi.org/10.1038/s41575-019-0189-8. Nature Research.

[5] Kaminski MF, Robertson DJ, Senore C, Rex DK. Optimizing the quality of colorectal cancer screening worldwide. Gastroenterology Jan. 2020;158(2):404–17. https://doi.org/10.1053/j.gastro.2019.11.026.

[6] Dekker E, Rex DK. Advances in CRC prevention: screening and surveillance. Gastroenterology May 2018;154(7):1970–84. https://doi.org/10.1053/j.gastro.2018.01.069.

[7] IJspeert JEG, et al. CT-colonography vs. colonoscopy for detection of high-risk sessile serrated polyps. Am J Gastroenterol Apr. 2016;111(4):516–22. https://doi.org/10.1038/ajg.2016.58.

[8] Rex DK, et al. Quality indicators for colonoscopy. Gastrointest Endosc Jan. 2015;81(1):31–53. https://doi.org/10.1016/j.gie.2014.07.058.

[9] Rex DK, et al. Accuracy of capsule colonoscopy in detecting colorectal polyps in a screening population. Gastroenterology May 2015;148(5):948–957.e2. https://doi.org/10.1053/j.gastro.2015.01.025.

[10] Koulaouzidis A, Iakovidis DK, Karargyris A, Rondonotti E. Wireless endoscopy in 2020: will it still be a capsule? World J Gastroenterol May 2015;21(17):5119–30. https://doi.org/10.3748/wjg.v21.i17.5119.

[11] Eliakim R, et al. Evaluation of the PillCam Colon capsule in the detection of colonic pathology: results of the first multicenter, prospective, comparative study. Endoscopy Oct. 2006;38(10):963–70. https://doi.org/10.1055/s-2006-944832.

[12] Brooks RA. New approaches to robotics. Science (80–) Sep. 1991;253(5025):1227–32. https://doi.org/10.1126/science.253.5025.1227.

[13] Robot—Wikipedia. https://en.wikipedia.org/wiki/Robot#cite_note-2 (accessed 17.03.21).

[14] Polygerinos P, et al. Soft robotics: review of fluid-driven intrinsically soft devices; manufacturing, sensing, control, and applications in human-robot interaction. Adv Eng Mater Dec. 2017;19(12). https://doi.org/10.1002/adem.201700016, 1700016.

[15] Cleary K, Nguyen C. State of the art in surgical robotics: clinical applications and technology challenges. Comput Aided Surg Jan. 2001;6(6):312–28. https://doi.org/10.3109/10929080109146301.

[16] Redelmeier DA, Katz J, Kahneman D. Memories of colonoscopy: a randomized trial. Pain Jul. 2003;104(1–2):187–94. https://doi.org/10.1016/S0304-3959(03)00003-4.

[17] Loeve A, Breedveld P, Dankelman J. Scopes too flexible and too stiff. IEEE Pulse Nov. 2010;1(3):26–41. https://doi.org/10.1109/MPUL.2010.939176.

[18] Yeung CK, Cheung JLK, Sreedhar B. Emerging next-generation robotic colonoscopy systems towards painless colonoscopy. J Dig Dis 2019;20(4):196–205. https://doi.org/10.1111/1751-2980.12718.

[19] Ciuti G, et al. Frontiers of robotic colonoscopy: a comprehensive review of robotic colonoscopes and technologies. J Clin Med May 2020;9(6):1648. https://doi.org/10.3390/jcm9061648.

[20] Eickhoff A, Jakobs R, Kamal A, Mermash S, Riemann JF, van Dam J. In vitro evaluation of forces exerted by a new computer-assisted colonoscope (the NeoGuide Endoscopy System). Endoscopy Dec. 2006;38(12):1224–9. https://doi.org/10.1055/s-2006-945014.

[21] Eickhoff A, et al. Computer-assisted colonoscopy (the NeoGuide Endoscopy System): results of the first human clinical trial ('PACE Study'). Am J Gastroenterol Feb. 2007;102(2):261–6. https://doi.org/10.1111/j.1572-0241.2006.01002.x.

[22] Rösch T, et al. A motor-driven single-use colonoscope controlled with a hand-held device: a feasibility study in volunteers. Gastrointest Endosc Jun. 2008;67(7):1139–46. https://doi.org/10.1016/j.gie.2007.10.065.

[23] "Single-Use Gastroenterology Endoscopes." https://www.ambu.com/endoscopy/gastroenterology (accessed 19.04.21).

[24] Li Z, Chiu PWY. Robotic endoscopy. Visc Med 01 Feb 2018;34(1):45–51. S. Karger AG, https://doi.org/10.1159/000486121.

[25] Gluck N, Melhem A, Halpern Z, Mergener K, Santo E. A novel self-propelled disposable colonoscope is effective for colonoscopy in humans (with video). Gastrointest Endosc May 2016;83(5):998–1004.e1. https://doi.org/10.1016/j.gie.2015.08.083.

[26] Boškoski I, Costamagna G. Endoscopy robotics: current and future applications. Dig Endosc Mar. 2019;31(2):119–24. https://doi.org/10.1111/den.13270.

[27] Shike M, et al. Sightline ColonoSight system for a disposable, power-assisted, non-fiber-optic colonoscopy (with video). Gastrointest Endosc Oct. 2008;68(4):701–10. https://doi.org/10.1016/j.gie.2007.12.062.

[28] Cosentino F, Tumino E, Passoni GR, Morandi E, Capria A. Functional evaluation of the endotics system, a new disposable self-propelled robotic colonoscope: in vitro tests and clinical trial. Int J Artif Organs 2009;32(8):517–27. https://doi.org/10.1177/039139880903200806.

[29] Tumino E, Sacco R, Bertini M, Bertoni M, Parisi G, Capria A. Endotics system vs colonoscopy for the detection of polyps. World J Gastroenterol Nov. 2010;16(43):5452–6. https://doi.org/10.3748/wjg.v16.i43.5452.

[30] Tumino E, et al. Use of robotic colonoscopy in patients with previous incomplete colonoscopy. Eur Rev Med Pharmacol Sci 2017. https://pubmed.ncbi.nlm.nih.gov/28272700/. (Accessed 10 January 2021).

[31] Trecca A, Catalano F, Bella A, Borghini R. Robotic colonoscopy: efficacy, tolerability and safety. Preliminary clinical results from a pilot study. Surg Endosc Mar. 2020;34(3):1442–50. https://doi.org/10.1007/s00464-019-07332-6.

[32] Zou W-B, et al. Magnetic-controlled capsule endoscopy vs. gastroscopy for gastric diseases: a two-center self-controlled comparative trial. Endoscopy Jan. 2015;47(06):525–8. https://doi.org/10.1055/s-0034-1391123.

[33] Jiang B, et al. Second-generation magnetically controlled capsule gastroscopy with improved image resolution and frame rate: a randomized controlled clinical trial (with video). Gastrointest Endosc Jun. 2020;91(6):1379–87. https://doi.org/10.1016/j.gie.2020.01.027.

[34] Gluck N, Fishman S, Melhem A, Goldfarb S, Halpern Z, Santo E. A novel colonoscope with panoramic visualization detected more simulated polyps than conventional colonoscopy in a live swine model. Endosc Int Open Oct. 2015;03(06):E642–5. https://doi.org/10.1055/s-0034-1393080.

[35] Park HJ, Kim D, Kim B. A robotic colonoscope with long stroke and reliable leg clamping. Int J Precis Eng Manuf Aug. 2012;13(8):1461–6. https://doi.org/10.1007/s12541-012-0192-2.

[36] Lee D, Joe S, Jung JH, Kim JU, Kim B. A simple and reliable reel mechanism-based robotic colonoscope for high mobility. Proc Inst Mech Eng Part C J Mech Eng Sci Aug. 2018;232 (16):2753–63. https://doi.org/10.1177/0954406217723941.

[37] Lee D, Joe S, Kang H, An T, Kim B. A reel mechanism-based robotic colonoscope with high safety and maneuverability. Surg Endosc Jan. 2019;33(1):322–32. https://doi.org/10.1007/s00464-018-6362-2.

[38] Kang M, Joe S, An T, Jang H, Kim B. A novel robotic colonoscopy system integrating feeding and steering mechanisms with self-propelled paddling locomotion: a pilot study. Mechatronics Feb. 2021;73:102478. https://doi.org/10.1016/j.mechatronics.2020.102478.

[39] Kim D, Lee D, Joe S, Lee BI, Kim B. The flexible caterpillar based robotic colonoscope actuated by an external motor through a flexible shaft. J Mech Sci Technol Nov. 2014;28 (11).4415–20. https://doi.org/10.1007/s12206-014-1009-2.

[40] Lee D, Joe S, Choi J, Lee BI, Kim B. An elastic caterpillar-based self-propelled robotic colonoscope with high safety and mobility. Mechatronics Nov. 2016;39:54–62. https://doi.org/10.1016/j.mechatronics.2016.08.002.

[41] Sliker LJ, Kern MD, Schoen JA, Rentschler ME. Surgical evaluation of a novel tethered robotic capsule endoscope using micro-patterned treads. Surg Endosc 2012;26(10): 2862–9. https://doi.org/10.1007/s00464-012-2271-y.

[42] Formosa GA, Prendergast JM, Edmundowicz SA, Rentschler ME. Novel optimization-based design and surgical evaluation of a treaded robotic capsule colonoscope. IEEE Trans Robot Apr. 2020;36(2):545–52. https://doi.org/10.1109/TRO.2019.2949466.

[43] Alcaide JO, Pearson L, Rentschler ME. Design, modeling and control of a SMA-actuated biomimetic robot with novel functional skin. In: Proceedings—IEEE international conference on robotics and automation; Jul. 2017. p. 4338–45. https://doi.org/10.1109/ICRA.2017.7989500.

[44] Wang K, Ma J, Wang F, Wang Z, Yan G, Zhou Y. Full-driving soft robotic colonoscope in compliant colon tissue. J Med Eng Technol Nov. 2017;41(8):662–9. https://doi.org/10.1080/03091902.2017.1394387.

[45] Bernth JE, Arezzo A, Liu H. A novel robotic meshworm with segment-bending anchoring for colonoscopy. IEEE Robot Autom Lett Jul. 2017;2(3):1718–24. https://doi.org/10.1109/LRA.2017.2678540.

[46] Kim D, Lee D, Kim B, Lee BI. A self-propelled robotic colonoscope using elastic caterpillars. In: IEEE; 2013. https://doi.org/10.1109/ISR.2013.6695672.

[47] Sliker LJ, Wang X, Schoen JA, Rentschler ME. Micropatterned treads for in vivo robotic mobility. J Med Devices Trans ASME Dec. 2010;4(4). https://doi.org/10.1115/1.4002761.

[48] Kern MD, Ortega Alcaide J, Rentschler ME. Soft material adhesion characterization for in vivo locomotion of robotic capsule endoscopes: experimental and modeling

results. J Mech Behav Biomed Mater Nov. 2014;39:257–69. https://doi.org/10.1016/j.jmbbm.2014.07.032.

[49] Wang K, Wang Z, Zhou Y, Yan G. Squirm robot with full bellow skin for colonoscopy. In: 2010 IEEE international conference on robotics and biomimetics, ROBIO 2010; 2010. p. 53–7. https://doi.org/10.1109/ROBIO.2010.5723302.

[50] "Colonic Disease Investigation by Robotic Hydro-colonoscopy (CODIR Project)." https://cordis.europa.eu/project/id/268519 (accessed 26.02.21).

[51] "Compliant Actuation Robotic Platform for Flexible Endoscopy (CARPE Project)." https://cordis.europa.eu/project/id/665696 (accessed 26.02.21).

[52] Manfredi L, Cuschieri A. Design of a 2 DOFs mini hollow joint actuated with SMA wires. Materials (Basel) Oct. 2018;11(10):2014. https://doi.org/10.3390/ma11102014.

[53] Dehghani H, et al. Design and preliminary evaluation of a self-steering, pneumatically driven colonoscopy robot. J Med Eng Technol Apr. 2017;41(3):223–36. https://doi.org/10.1080/03091902.2016.1275853.

[54] Chen Z, Liu J, Wang S, Zuo S. A bio-inspired self-propelling endoscopic device for inspecting the large intestine. Bioins Biom Oct. 2019;14(6). https://doi.org/10.1088/1748-3190/ab45c9, 066013.

[55] Manfredi L, Capoccia E, Ciuti G, Cuschieri A. A soft pneumatic inchworm double balloon (SPID) for colonoscopy. Sci Rep Dec. 2019;9(1):1–9. https://doi.org/10.1038/s41598-019-47320-3.

[56] "Vector." (accessed 26.02.21). [Online]. Available: http://www.vector-project.com/.

[57] Ciuti G, Valdastri P, Menciassi A, Dario P. Robotic magnetic steering and locomotion of capsule endoscope for diagnostic and surgical endoluminal procedures. Robotica Mar. 2010; 28(2):199–207. https://doi.org/10.1017/S0263574709990361.

[58] Ciuti G, et al. Robotic versus manual control in magnetic steering of an endoscopic capsule. Endoscopy 2010;42(2):148–52. https://doi.org/10.1055/s-0029-1243808.

[59] Arezzo A, et al. Experimental assessment of a novel robotically-driven endoscopic capsule compared to traditional colonoscopy. Dig Liver Dis 2013;45(8):657–62. https://doi.org/10.1016/j.dld.2013.01.025.

[60] Valdastri P, et al. Magnetic air capsule robotic system: proof of concept of a novel approach for painless colonoscopy. Surg Endosc Dec. 2012;26(5):1238–46. https://doi.org/10.1007/s00464-011-2054-x.

[61] Martin JW, et al. Enabling the future of colonoscopy with intelligent and autonomous magnetic manipulation. Nat Mach Intell Oct. 2020;2(10):595–606. https://doi.org/10.1038/s42256-020-00231-9.

[62] Lucarini G, Ciuti G, Mura M, Rizzo R, Menciassi A. A new concept for magnetic capsule colonoscopy based on an electromagnetic system. Int J Adv Robot Syst Mar. 2015;12(3):25. https://doi.org/10.5772/60134.

[63] Sliker LJ, Ciuti G, Rentschler ME, Menciassi A. Frictional resistance model for tissue-capsule endoscope sliding contact in the gastrointestinal tract. Tribol Int Oct. 2016; 102:472–84. https://doi.org/10.1016/j.triboint.2016.06.003.

[64] Li J, et al. Magnetically-driven medical robots: an analytical magnetic model for endoscopic capsules design. J Magn Magn Mater Apr. 2018;452:278–87. https://doi.org/10.1016/j.jmmm.2017.12.085.

[65] Salerno M, et al. A discrete-time localization method for capsule endoscopy based on on-board magnetic sensing. Meas Sci Technol Nov. 2012;23(1). https://doi.org/10.1088/0957-0233/23/1/015701, 015701.

[66] Taddese AZ, Slawinski PR, Pirotta M, De Momi E, Obstein KL, Valdastri P. Enhanced real-time pose estimation for closed-loop robotic manipulation of magnetically actuated

capsule endoscopes. Int J Rob Res Jul. 2018;37(8):890–911. https://doi.org/10.1177/0278364918779132.

[67] Ciuti G, et al. A comparative evaluation of control interfaces for a robotic-aided endoscopic capsule platform. IEEE Trans Robot Apr. 2012;28(2):534–8. https://doi.org/10.1109/TRO.2011.2177173.

[68] Slawinski PR, Simaan N, Taddese AZ, Obstein KL, Valdastri P. Sensitivity ellipsoids for force control of magnetic robots with localization uncertainty. IEEE Trans Robot Oct. 2019;35(5):1123–35. https://doi.org/10.1109/TRO.2019.2917817.

[69] Barducci L, Pittiglio G, Norton JC, Obstein KL, Valdastri P. Adaptive dynamic control for magnetically actuated medical robots. IEEE Robot Autom Lett Oct. 2019;4(4):3633–40. https://doi.org/10.1109/LRA.2019.2928761.

[70] "The SUPCAM." http://supcam.eu/ (accessed 20.03.21).

[71] Lucarini G, Mura M, Ciuti G, Rizzo R, Menciassi A. Electromagnetic control system for capsule navigation: novel concept for magnetic capsule maneuvering and preliminary study. J Med Biol Eng Aug. 2015;35(4):428–36. https://doi.org/10.1007/s40846-015-0055-2.

[72] Tortora G, Mulana F, Ciuti G, Dario P, Menciassi A. Inductive-based wireless power recharging system for an innovative endoscopic capsule. Energies Sep. 2015;8(9):10315–34. https://doi.org/10.3390/en80910315.

[73] "Endoo Project Home | Endoo Project." http://www.endoo-project.eu/ (accessed 26.02.21).

[74] Bianchi F, et al. An innovative robotic platform for magnetically-driven painless colonoscopy. Ann Trans Med Nov. 2017;5(21). https://doi.org/10.21037/atm.2017.09.15. AME Publishing Company.

[75] Brandao P, et al. Fully convolutional neural networks for polyp segmentation in colonoscopy. In: Medical imaging 2017: computer-aided diagnosis, 10134; Mar 2017. p. 101340F. https://doi.org/10.1117/12.2254361.

[76] Di Iakovidis K, Dimas G, Karargyris A, Bianchi F, Ciuti G, Koulaouzidis A. Deep endoscopic visual measurements. IEEE J Biomed Heal Inform Nov. 2019;23(6):2211–9. https://doi.org/10.1109/JBHI.2018.2853987.

[77] Kumar S, et al. Adenoma miss rates associated with a 3-minute versus 6-minute colonoscopy withdrawal time: a prospective, randomized trial. Gastrointest Endosc Jun. 2017;85(6):1273–80. https://doi.org/10.1016/j.gie.2016.11.030.

[78] Rees CJ, et al. Narrow band imaging optical diagnosis of small colorectal polyps in routine clinical practice: the detect inspect characterise resect and DISCARD 2 (DISCARD 2) study. Gut Apr. 2016;66(5):887–95. https://doi.org/10.1136/gutjnl-2015-310584.

[79] Mori Y, et al. Artificial intelligence for colorectal polyp detection and characterization. Curr Treat Options Gastroenterol 2020;18(2):200–11. https://doi.org/10.1007/s11938-020-00287-x.

[80] J. K. Min, M. S. Kwak, and J. M. Cha, "Overview of deep learning in gastrointestinal endoscopy," Gut and Liver, vol. 13, no. 4. Editorial Office of Gut and Liver, pp. 388–393, 2019, doi: https://doi.org/10.5009/gnl18384.

[81] Kudo SE, et al. Artificial intelligence-assisted system improves endoscopic identification of colorectal neoplasms. Clin Gastroenterol Hepatol Jul. 2020;18(8):1874–1881.e2. https://doi.org/10.1016/j.cgh.2019.09.009.

[82] Hassan C, Bhandari P, Antonelli G, Repici A. Artificial intelligence for non-polypoid colorectal neoplasms. Dig Endosc Jan. 2021;33(2):285–9. https://doi.org/10.1111/den.13807.

[83] Bernal J, et al. Comparative validation of polyp detection methods in video colonoscopy: results from the MICCAI 2015 endoscopic vision challenge. IEEE Trans Med Imaging 2017;36(6):1231–49. https://doi.org/10.1109/TMI.2017.2664042.

[84] Ahmad OF, Stoyanov D, Lovat LB. Human-machine collaboration: bringing artificial intelligence into colonoscopy. Frontl Gastroenterol 2019;10(2):198–9. https://doi.org/10.1136/flgastro-2018-101047. BMJ Publishing Group.

[85] Mori Y, et al. Computer-aided diagnosis for colonoscopy what can we expect of CAD in clinical automated detection of colorectal polyps. Endoscopy 2017;813–9.

[86] Ahmad OF, et al. Artificial intelligence and computer-aided diagnosis in colonoscopy: current evidence and future directions. Lancet Gastroenterol Hepatol 2019;4(1):71–80. https://doi.org/10.1016/S2468-1253(18)30282-6.

[87] de Groof AJ, et al. Deep-learning system detects neoplasia in patients with Barrett's esophagus with higher accuracy than endoscopists in a multistep training and validation study with benchmarking. Gastroenterology 2020;158(4):915–929.e4. https://doi.org/10.1053/j.gastro.2019.11.030.

[88] Y. Mori and S. E. Kudo, "Detecting colorectal polyps via machine learning," Nat Biomed Eng, vol. 2, no. 10. NLM (Medline), pp. 713–714, Oct. 2018, https://doi.org/10.1038/s41551-018-0308-9.

[89] Mori Y, et al. Real-time use of artificial intelligence in identification of diminutive polyps during colonoscopy. Ann Intern Med Sep. 2018;169(6):357. https://doi.org/10.7326/M18-0249.

[90] "Gastrointestinal Image ANAlysis Challenge—Grand Challenge." https://giana.grand-challenge.org/ (accessed 25.02.21).

[91] "EAD2019—Grand Challenge." https://ead2019.grand-challenge.org/ (accessed 25.02.21).

[92] "iMerit." (accessed 25.02.21). [Online]. Available: https://imerit.net/.

[93] van der Sommen F. Gastroenterology needs its own ImageNet. J Med Artif Intell Dec. 2019;2:23. https://doi.org/10.21037/jmai.2019.11.03.

[94] Fu Y, et al. More unlabelled data or label more data? a study on semi-supervised laparoscopic image segmentation. In: Lecture notes in computer science (including subseries lecture notes in artificial intelligence and lecture notes in bioinformatics), 11795 LNCS; Oct. 2019. p. 173–80. https://doi.org/10.1007/978-3-030-33391-1_20.

[95] "ai4gi | AI Solutions for Colon Polyp Detection and Differentiation." https://ai4gi.com/ (accessed 25.02.21).

[96] "Odin Vision | Artificial Intelligence Enhanced Colonoscopy Procedures." https://odin-vision.com/ (accessed 25.02.21).

[97] "Wision A.I" (accessed 25.02.21). [Online]. Available: https://www.wision.com/.

[98] "GI Genius™ Intelligent Endoscopy Module | Medtronic." https://www.medtronic.com/covidien/en-gb/products/gastrointestinal-artificial-intelligence/gi-genius-intelligent-endoscopy.html# (accessed 25.02.21).

[99] "Gastroenterology | PENTAX Medical (EMEA)." https://www.pentaxmedical.com/pentax/en/95/2/DISCOVERY-new (accessed 25.02.21).

[100] "CAD EYE | Fujifilm Europe." https://www.fujifilm.eu/eu/cadeye (accessed 25.02.21).

[101] Stoyanov D. Surgical vision. Ann Biomed Eng 2012;40(2):332–45. https://doi.org/10.1007/s10439-011-0441-z.

[102] Mountney P, Stoyanov D, Yang GZ. Three-dimensional tissue deformation recovery and tracking. IEEE Signal Process Magazine 2010;27(4):14–24. https://doi.org/10.1109/MSP.2010.936728.

[103] Mahmood F, Durr NJ. Deep learning and conditional random fields-based depth estimation and topographical reconstruction from conventional endoscopy. Med Image Anal Aug. 2018;48:230–43. https://doi.org/10.1016/j.media.2018.06.005.

[104] Visentini-Scarzanella M, Sugiura T, Kaneko T, Koto S. Deep monocular 3D reconstruction for assisted navigation in bronchoscopy. Int J Comput Assist Radiol Surg Jul. 2017; 12(7):1089–99. https://doi.org/10.1007/s11548-017-1609-2.

[105] Lin J, Clancy NT, Stoyanov D, Elson DS. Tissue surface reconstruction aided by local normal information using a self-calibrated endoscopic structured light system. In: Lecture notes in computer science (including subseries lecture notes in artificial intelligence and lecture notes in bioinformatics), vol. 9349. Springer Verlag; 2015. p. 405–12.

[106] Rau A, et al. Implicit domain adaptation with conditional generative adversarial networks for depth prediction in endoscopy. Int J Comput Assist Radiol Surg Jul. 2019;14(7): 1167–76. https://doi.org/10.1007/s11548-019-01962-w.

[107] Itoh H, et al. Towards automated colonoscopy diagnosis: binary polyp size estimation via unsupervised depth learning. In: Lecture notes in computer science (including subseries lecture notes in artificial intelligence and lecture notes in bioinformatics), vol. 11071 LNCS; Sep. 2018. p. 611–9. https://doi.org/10.1007/978-3-030-00934-2_68.

Chapter 4

Soft robotic systems for endoscopic interventions

Sara-Adela Abad[a,b], Alberto Arezzo[c], Shervanthi Homer-Vanniasinkam[a], and Helge A Wurdemann[a]

[a]*Department of Mechanical Engineering, University College London, London, United Kingdom,* [b]*Institute for Applied Sustainability Research, Quito, Ecuador,* [c]*Department of Surgical Sciences, University of Torino, Turin, Italy*

4.1 Introduction

Open surgery has been the traditional way to obtain direct access to internal organs via a large incision for many years. Intraoperative palpation can be performed by the clinician's finger(*tip*)s to differentiate tumorous and unhealthy tissue from healthy tissue. This facilitates the surgeon to completely resect identified tissue, i.e., to perform a surgical excision, using surgical instruments such as scalpels, forceps, or curettes [1], or remove small tissue samples (biopsies) for histology examination. However, open surgery is associated with slow patients' recovery time, postoperative pain, increased rate and severity of postoperative complications, blood loss, and immunological stress response of the tissue, resulting in less invasive surgery techniques becoming more preferred [2,3].

With the invention of the endoscope in the early 19th century for visualization of the esophagus and stomach [4], medical procedures have been transformed into minimally invasive interventions for diagnosis (cystoscope, 1878) and therapy (polypectomy, 1969). This concept was adapted to surgery when laparoscopic instruments were introduced into the organism: In minimally invasive surgery (MIS), also called laparoscopic or keyhole surgery, miniature video cameras and tools mounted at the end of long rigid rods are inserted through small incisions ranging from 3 to 12mm [5,6]. These miniature cameras provide visual feedback to the clinician of the internal organs of the patient. At the same time, surgical instruments/tools can be used to probe tissue surface (through rigid tool-soft tissue interaction) or to provide treatment such as cutting and cauterization. From the mid-1980s, MIS has become increasingly

Endorobotics. https://doi.org/10.1016/B978-0-12-821750-4.00004-9

popular worldwide. Surgery became even less invasive when the use of natural orifice was demonstrated in so called natural orifice transluminal endoscopic surgery (NOTES).

Compared to open surgery, MIS offers several benefits including shortened postoperative recovery; better therapeutic outcome; reduced tissue trauma, postoperative pain, and scarring; and less immunological stress response of the tissue [2]. However, compared to open surgery, MIS is more demanding because the clinicians' skills may be affected by the limited vision of the operative site, and motion constraints, resulting in the reduction of intuitiveness and unnatural repetitive movements [4].

In the last decade, surgical robots have been introduced into the operating theater to provide the surgeon with advanced dexterity by using teleoperated laparoscopic tools with increased degrees of freedom (DOFs) compared to manually operated laparoscopic instruments. This implies controlling the robotic surgical system through a master console that can be located nearby the patient or in a remote location. Surgeons can perform remote precise surgical procedures due to 3D high-definition endoscopic vision systems and control systems that can scale surgical instruments' motion and reduce tremor of the hand.

With the introduction of wireless capsule endoscopes, flexible endoscopes, laparoscopic tools, and robotic systems, examinations, biopsies, and treatments can be delivered in a minimally invasive way. Some of these systems have associated risks such as tissue/organ perforation, bleeding, or incomplete resection of adenomas in endoscopic procedures. These are mainly caused by the instruments' rigidity and the limited touch feedback of surgeons during MIS [7].

Recently, soft robots for minimally invasive surgery have been proposed. Properties such as flexibility, squeezability, and biocompatibility of their materials make them suitable to avoid damaging tissues inside the body while navigating through a narrow opening with obstacles [7]. Moreover, their variable stiffness control allows the tool holder to become stiff to have better transmission of the force while a less stiff state allows reorienting the tools [8–11]. This makes soft robotic systems more ergonomic for the surgeon and easier for navigating inside the patient [12].

Soft robots are also less prone to fulcrum effect problems and absence of triangulation that are commonly observed in MIS [3]. The mechanical constraints of the laparoscope at the incision point cause fulcrum effect; this effect includes scaling and inversion of both the motion of the laparoscope and the environment-endoscope interaction forces [13]. On the other hand, triangulation refers to the triangle arrangement of the laparoscope and two surgical instruments that maximizes the view, and the operation of the instruments in the area of interest [14]. This convergence is achieved by having the laparoscope at the top and one surgical instrument at each of the lower points of the triangle minimizing the interference between instruments and the endoscope.

This chapter focuses on the application of soft robotic systems for endoscopic procedures. To understand the challenges associated with endoscopic procedures, the analysis starts by discussing common endoscopic interventions, their implications on the patient, and possible complications. Then, commercially available endoscopic tools are examined to determine their limitations and how their mechanical properties contribute to elevating potential risks or complications during endoscopic procedures. This analysis is used as the foundation to discuss how soft robotics research has proposed solutions to overcome these aforementioned challenges over the last decade. In the end, this leads to the identification of shortcoming and future research challenges in soft robotics for endoscopic applications.

4.2 Overview of endoscopic procedures

Medical procedures deploying endoscopic instruments involve three categories, i.e., examination, confirmation of a diagnosis, and treatment. This section provides an overview and context of the differences and similarities in the requirements, risks, and limitations of these endoscopic procedures.

4.2.1 Visual examination and diagnostic procedures

Visual information is of paramount importance to understand the preliminary health condition of soft tissue and organs. This is confirmed by the World Health Organization (WHO). In 2000, WHO proposed a classification of tumors of the digestive system that takes into account visual characteristics [15]. Table 4.1 summarizes the color, presence of vessels, and surface pattern of colorectal lesions' characteristics of the tissue for the invasion states from 1 to 3, while Fig. 4.1 contains narrow-band imaging examples that highlight the corresponding characteristics described in the table [16–18]. For instance, the figure in the third-row first column of Fig. 4.1 illustrates the highest invasion state (Type 3 Carcinoma) where the color of the tissue has changed to dark brown in certain areas.

An examination procedure is commonly carried out through natural orifices while the patient is awake to investigate the cause of symptoms. Local anesthesia may be given to numb a specific area such as the throat or the anus. Depending on the procedure, a sedative may also be used to relax the patient and reduce awareness.

On the other hand, a biopsy is a procedure where a small sample of organ or tissue is taken to examine, i.e., the sample cell's type, internal activity, and shape. A flexible endoscope is inserted through an incision or a natural orifice such as the throat or anus. Once the distal end of the endoscope has reached the

TABLE 4.1 Narrow-band imaging (NBI) colorectal endoscopic classification [16].

	Color	Vessels	Surface patterns
Type 1 Hyperplastic	Lighter or similar than surrounding tissue	None or isolated vessels vessels	Absence of patterns or homogenous dark or white spots
Type 2 Adenoma	Browner than surrounding tissue	White structures surrounded by brown vessels	Branched white, tubular or oval structures surrounded by brown vessels
Type 3 Carcinoma	Brown to dark brown	Missing vessels	Amorphous or absent pattern

desired location, the surgeon will take the sample using small cutting tools. The anesthesia depends on the endoscope insertion point and the investigated area inside the body. Biopsies allow assessing the severity of a condition that has already been diagnosed. These imply defining the degree of inflammation or the invasion state of cancer [16]; this information is profoundly important for the decision-making of future treatment [19].

It is worth noting that in recent years surgeons have used artificial intelligence and a visual examination to provide a diagnosis. This approach has been proved to be as useful and reliable as biopsy [20]. Last year, the Food and Drug Administration (FDA) designated machine learning algorithms of AI Medical Service Inc. (Tokyo, Japan) as *breakthrough devices* analyzing thoroughly and rapidly the images from an endoscopic examination of the entire digestive tract [21]. It provides support to clinicians in the diagnosis of conditions such as gastric cancer. Another commercially available artificial intelligence (AI) algorithm is Lunit INSIGHT MMG (Lunit, Seoul, Korea) [22]. This CE certified algorithm can improve breast cancer detection using mammography images. For lung cancer detection, this algorithm has shown a higher sensitivity than radiologists [23]. In future years, AI applied to medical imaging is likely to replace biopsy when lesions are visible.

Common endoscopic examination procedures include the following:

- **Esophagogastroduodenoscopy (EGD)** or upper endoscopy is the examination of the esophagus, stomach, and duodenum using an endoscope that is inserted through the mouth. An endoscope comprises lenses and light to obtain visual feedback via an external monitor. It requires patient sedation. By using other tools, it can also be employed for biopsies, performing cell

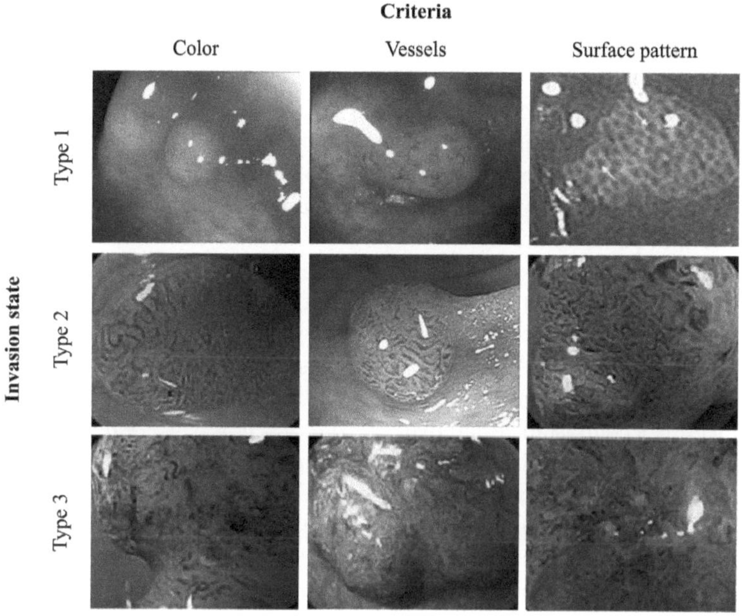

FIG. 4.1 Narrow-band imaging (NBI) colorectal endoscopic classification (presented in Table 4.1) considering visual information that can be provided by an endoscope. The criteria are based on the color, amount of vessels, and surface patterns that can be observed in the examined tissue. *These figure samples are taken from Hewett DG, Kaltenbach T, Sano Y, Tanaka S, Saunders BP, Ponchon T, Soetikno R, Rex DK. Validation of a simple classification system for endoscopic diagnosis of small colorectal polyps using narrow-band imaging. Gastroenterology 2012;143 (3):599–607.e1; Iwatate M, Ikumoto T, Hattori S, Sano W, Sano Y, Fujimori T. NBI and NBI combined with magnifying colonoscopy. Diagnostic and therapeutic endoscopy. 2012;2012.*

tests (which requires a small brush), or for treating abnormalities such as polyps, strictures, and bleeding ulcers [24]. Complications may include bleeding or gastrointestinal tract tearing.

- **Colonoscopy** comprises examining the inside of the colon from the rectum until the cecum (where the small bowel attaches to the large bowel) using a specific endoscope called colonoscope. It may require patient sedation. Similarly to upper endoscopy, a colonoscope provides visual feedback and can be used to obtain a biopsy. Sometimes, this procedure can be used for treating abnormalities such as polyps, strictures, or bleeding. This method is commonly used to evaluate the colon of patients with conditions such as iron deficiency anemia, colorectal cancer (CRC), postpolypectomy, and postcancer resection surveillance; complications may imply colon perforation, postpolypectomy bleeding, and missed or incomplete resected adenomas [25]. Colonoscopy is more sensitive but has less patient acceptance than sigmoidoscopy [26].

- **Enteroscopy** allows visualizing the small bowel. Enteroscopy can be performed using (a) a longer conventional endoscope to examine the upper part of the small bowel or (b) a swallowable capsule endoscopy that comprises a wireless camera sending information to a recording device worn by the patient; after finishing, the data are downloaded for a clinician's analysis; patient sedation is not required; this approach is only for exploration. Depending on the approach, the complications may imply bleeding, perforation, or pancreatitis.
- **Endoscopic retrograde cholangiopancreatography (ERCP)** uses a technique in the treatment and examination of the ducts involving the liver, pancreas, and gallbladder. A specialized endoscope together with a side-mounted camera that facilitates passing a catheter into the bile and pancreatic ducts is deployed. The catheter inserted through the endoscope allows injecting a contrast material (dye) that outlines those ducts when X-rays are taken. This endoscopic examination's complexity and length require patient sedation and significant training and experience of the clinician. ERCP is used to detect blockages that may be caused by cancerous tissue, narrowing of the ducts, or other alterations. This endoscopic procedure helps to treat conditions such as pancreatic cancer, pancreatitis, and biliary strictures. Adverse conditions related to this procedure are cholangitis, bleeding, pancreatitis, or perforation [27].
- During **bronchoscopy**, a thin endoscope, called bronchoscope, is inserted through the mouth or nose allowing the examination of breathing passages in the lungs. This procedure requires fasting prior to the procedure, local anesthesia, patient sedation, and may include taking a biopsy. Cause of symptoms such as noisy breathing, persistent cough, or coughing up blood can be investigated. It has therapeutic applications such as tracheal intubation, retrieving an inhaled foreign body, or suction of sputum [28]. Complications such as bleeding can occur depending on the level of comorbid diseases [29]. Between 6 and 12h after the procedure, one in three patients may develop fever or sweat. Approximately one in ten patients, who required a transbronchial biopsy, suffers from lung wall perforation caused by the bronchoscope [30].
- **Laryngoscopy** is an examination of the larynx and vocal cord movement using an endoscope, which might require the patient to receive local anesthesia. The clinician can choose between a flexible endoscope, called fiber optic laryngoscope, and a rigid optical instrument, called direct laryngoscope. Laryngoscopy allows for performing biopsies, microsurgical procedures such as removing polyps in the vocal cord or urgent endotracheal intubation. Laryngoscopy is used for investigating hoarseness [31], trouble breathing or swallowing, persistent earache or sore throat, or a symptom related to cancer. Rare complications including pain, infection, bleeding, and hoarseness can occur. Furthermore, comorbidities and operators' expertise might increase the number of attempts required for urgent endotracheal intubation by direct laryngoscopy [32].

- During **cystoscopy**, clinicians reach the bladder through the urethra to examine its internal volume and lining using a thin endoscope, called the cystoscope. Similarly to laryngoscopy, cystoscopy can be performed using a flexible or rigid endoscope. Depending on the necessity of treatment and, hence, type of the optical instrument, the patient may need anesthesia, e.g., a spinal or general anesthesia. Cystoscopy is used to investigate the cause of problems such as frequent tract infections, blood in the urine, or persistent pelvic pain; to take a biopsy; or to provide treatments, i.e., injecting a medicine, removing or inserting a stent, or removing bladder stones. Cystoscopy is considered a safe procedure; however, risks can include urinary tract infections, not being able to urinate after the procedure, and potential damage to the bladder caused by the cystoscope [33].
- **Hysteroscopy** is the gold standard for examining the inside of the uterus [34] inserting a narrow rigid small endoscope, called hysteroscope, into the vagina and through the cervix to reach the uterus. This procedure is used to investigate the cause of health issues related to miscarriages, heavy periods, or pelvic pain; to diagnose conditions, i.e., polyps and fibroids; and to provide treatment such as removal of polyps, fibroids, or scar tissue. The patient may need local or general anesthesia (in particular, when treatment is provided). Although uncommon, the main complications might be uterus or cervix damage, excessive bleeding, or uterus infection [35].
- **Endoscopic ultrasound** (**EUS**) examination requires a special endoscope fitted with a small ultrasound device on the distal end. This endoscope is inserted through the mouth or anus. Patient sedation is required. This examination allows visualizing layers of the gastrointestinal tract wall and surrounding organs such as the spleen, liver, pancreas, gallbladder, and adrenal glands. The clinician can gain more understanding about cancer stages or diseases of the bile duct, pancreas, and gallbladder [36]. Complications may comprise bleeding, throat or intestinal wall perforation, or pancreatitis.

4.2.2 Endoscopic treatments

After a diagnosis is confirmed, the clinician will understand and suggest the treatment for the patient. This may imply invasive, open interventions such as *operative procedures* that would comprise manual tissue manipulation with one hand or *surgical procedures* that would require tissue handling with both hands. Endoscopes can offer the opportunity to deliver treatments in a minimally invasive way. During minimally invasive surgery (MIS), direct treatments such as removing polyps, cauterizing a bleeding vessel, or removing an inflamed organ can be delivered.

MIS has become the preferred method over open surgery because of smaller incisions needed to introduce surgical tools, less postoperative pain for the patient, lower risk of infection, shorter hospital stay and recovery time, and reduced scarring and blood loss [2]. Here, endoscopes can be inserted through small incisions or natural orifices using either rigid laparoscopic optical instruments or flexible endoscopes.

MIS is performed using long thin instruments inserted through small 3–15 mm incisions (i.e., trocar ports) in the abdominal area [1,6]. Prior to the procedure, the patient will receive anesthesia. On the one hand, single-port MIS comprises using an endoscope with the surgical instruments being passed through its working channel. On the other hand, several small incisions are required for multiport MIS to introduce a single laparoscopic instrument through each trocar port. Whereas single-port MIS is considered to be less invasive, challenges in triangulation [37] in combination with the fulcrum effect might occur. In comparison, multiport surgery offers enhanced maneuverability and less physical interference between the laparoscopic instruments.

Another minimally invasive procedure is natural orifice transluminal endoscopic surgery (NOTES). Here, a flexible endoscope can be navigated to reach other organs or tissues with reduced postoperative pain, minimal or little skin scarring, and faster recovery time. Furthermore, among the procedures through the various natural orifices, transvaginal NOTES is the preferred [38] access point operating on female patients because of its safe entry, simple closure, and easy to decontaminate [39].

There are two types of NOTES: On the one hand, pure NOTES uses a flexible endoscope passed through a natural orifice (i.e., urethra, rectum, vagina, or mouth) [40] to perform a surgical intervention. Therefore, the surgery could be performed without any incision in the abdomen [38]. On the other hand, surgeons use flexible or rigid endoscopes that are passed through trocars and natural orifices simultaneously during hybrid NOTES [41]. A common procedure is transvaginal gallbladder removal, in which an endoscope introduced through an umbilical port provides visual feedback.

Common minimally invasive endoscopic (laparoscopic, NOTES, or hybrid) procedures for treatment include the following:

- **Peroral endoscopic myotomy (POEM)** is performed to treat achalasia which is a condition where the lower esophageal sphincter (LES, muscular ring that closes the esophagus from the stomach) becomes bulky, eventually fails to open during swallowing, and food does not reach the stomach. During POEM, an endoscope incises the muscle fibers of the LES to decrease LES' resting pressure unblocking the pass of food from the esophagus to the stomach. POEM is considered pure NOTES, in which the patient receives general anesthesia [42]. This procedure is generally safe. However, some complications might include perforation of the esophagus, bleeding during the procedure, and infection.
- In **Percutaneous endoscopic gastrostomy (PEG)**, an endoscope assists in placing a flexible feeding tube into the stomach through a small incision in the abdominal wall. This requires patient sedation. Patients with insufficient

oral intake or swallowing problems might benefit from PEG. Complications such as bowel perforation or hemorrhage may occur [43].

- **Endoscopic laser ablation (ELA)** requires an endoscope with a therapeutic laser. It is used in cases such as twin-to-twin transfusion syndrome (TTTS) [44], prostate problems [45], or lung cancer [46]. According to [47], TTTS risks such as tissue damage and premature rupture of the membrane are mainly caused by the poor adaptability of the semirigid or rigid instruments employed.
- **Transanal endoscopic microsurgery (TEM)** requires a specific microscope and instruments. Incisions are not needed because the rectum is reached through the anus. It is widely applied in procedures such as removing benign polyps, repairing complex rectovisceral fistulas, or excision of rectal neoplasms [48]. TEM risks are lower than using open surgery, but some of the complications include bleeding, infection, or pelvis inflammation.
- **Transanal minimally invasive surgery (TAMIS)** is an alternative for removing benign polyps and early-stage rectal cancers [49]. The difference between TAMIS and TEM is the former uses a disposable port while the latter employs a resterilized reusable port. TAMIS is more accessible and cost-effective than TEM [50].

4.3 Review of commercially available solutions for endoscopic procedures

Endoscopes deployed for minimally invasive procedures that allow examining and/or provide visual feedback during treatments of, e.g., the brain, joints, lungs, or bladder must meet strict requirements with regard to: material biocompatibility for avoiding adverse reactions of the body, stability of the distal tip for avoiding any overshooting or problem related to inertia, maneuverability for navigating through the soft tissue and organs, reliability and durability, depth perception, and localization of the distal tip.

As stated in Section 4.2, endoscopes are used for examination, diagnostic confirmation (by taking biopsies), and treatment. An examination can be performed using optical tools, while biopsies and treatments use minimally invasive tools (in particular, laparoscopic tools, flexible endoscopes, or even robotic platforms). Therefore, this review of commercially available endoscopic instruments is divided into optical tools and instruments for therapeutic endoscopy.

4.3.1 Instruments for optical examination

A wireless capsule endoscope can provide visual feedback of the internal tissues or organs for investigating the cause of a problem such as stomach ulcers, or abdominal pain. This untethered single-use approach is a less invasive alternative to standard endoscopy. Before swallowing the capsule, the patient needs to fast and take purgatives and liquids for bowel preparation. Then, the capsule moves passively through the digestive system using peristalsis. The sensors placed on the patient's body allow the wireless recorder to collect video data transmitted from the capsule. The information can be downloaded onto a computer and analyzed by medical personnel. In the end, the capsule is excreted naturally from the body.

Capsule endoscopes do not require patient sedation (though the patient might experience minimal discomfort). This is one reason why capsule endoscope has become standard for diagnosing Crohn's disease [51].

The commercially available PillCam capsule (Medtronic, Dublin, Ireland) is used for monitoring lesions that may be related to Crohn's disease, iron deficiency anemia, or obscure bleeding. It can also be used for visualizing the small bowel. EndoCapsule 10 (Olympus, Tokyo, Japan) can provide a 160 degree view angle and has a battery life of up to 12h [52]. This allows monitoring the entire gastrointestinal tract with minimal disruption to the patient and endoscopist.

Nevertheless, the absence of control for, e.g., orientating, navigating, and maneuvering discards this method for investigating specific areas [53]. To overcome this limitation, the C-Scan endocapsule (Check-Cap Ltd., Isfiya, Israel) employs X-ray to construct 2D and 3D maps of the inside of the colon [54]. On the other hand, MiroCam (IntroMedic, South Korea) and NaviCam (Ankon Medical Technologies, Wuhan, China) improved the maneuverability and controllability of the capsule endoscope system using external magnetic fields. NaviCam can advance and change the viewing angle in steps of 2mm and 3mm, respectively. Additionally, the ANKON ESNavi software provides dimensions of visualized lesions [55].

Flexible endoscopes can also provide visual feedback of the internal organs or tissues for investigating the causes of problems. They commonly comprise a working channel where surgical instruments are introduced to provide a treatment or confirm a diagnosis while performing an examination. Therefore, flexible endoscopes can also be employed for therapeutic endoscopy, so they are presented in Section 4.3.2.1.

4.3.2 Instruments for therapeutic endoscopy

In therapeutic endoscopy, an instrument is used to confirm a diagnosis or provide treatments such as removing polyps, a failing organ, or cauterizing bleeding vessels. After the patient has been sedated, small operative instruments are inserted through the working channel of an endoscope (single-port surgery) or through several incisions (multiport surgery). Simultaneously, visual information is provided to the clinician on a monitor.

Characteristics and requirements for an endoscope depend on the medical procedure and the number of ports and the tools employed. For instance,

TABLE 4.2 Requirements of a flexible endoscope used for an examination (camera) and operative (tools) procedures for single-port and multiport surgery: here, total mesorectal excision (TME).

	Single-port surgery		Multiport surgery	
Requirements	*Camera*	*Tools*	*Camera*	*Tools*
Tube stiffness	Variable for organ or tissue adaptability			
Degrees of freedom (DOFs)	At least 2 DOFs for bending +1 DOF for elongation (tip navigation)			
Tip diameter	Acceptable: 10 mm Ideally: 8 mm			
Max. trocar diameter	40 mm		Acceptable: 15 mm Ideally: ≤ 10 mm	
Overall length	350 mm		300 mm	Acceptable: 350 mm Ideally: ≥400 mm
Force at the tip	Low holding force	Acceptable: 3 N Ideally: 4–5 N	Low holding force	Acceptable: 3 N Ideally: 4–5 N

Table 4.2 summarizes the requirements of a therapeutic endoscope for single-port and multiport surgeries such as total mesorectal excision (TME). The maximum trocar diameter increases from 15 mm (multiport surgery) to 40 mm (single-port surgery). Since the camera has a lower interaction with the environment than a tool, the camera requires a low holding force while the tool needs 3–5 N.

4.3.2.1 Flexible endoscopes

Fundamental control interfaces and components of flexible endoscopes are illustrated in Fig. 4.2. Endoscopists use one hand to operate the controls such as suction valves or angulation of the distal end. The other hand manages torquing, pushing, or removal of the insertion tube.

The insertion tube is part of the endoscope that mainly changes across procedures. The insertion tube comprises a bending section at the distal end to facilitate the orientation of the camera. As Fig. 4.2C shows, this tube contains the following:

1. A visualization system that refers to the camera (image sensor) and cables for transmitting the video signal.

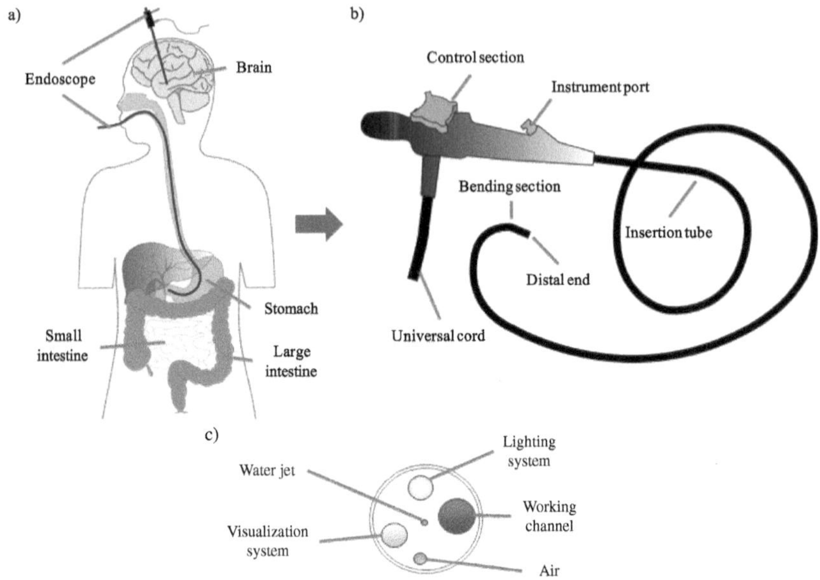

FIG. 4.2 Endoscope. (A) Endoscopes can be used for performing medical procedures in different parts of the body, such as the brain, stomach, intestines, and joints. Endoscopes can be inserted through incisions or natural orifices. (B) Fundamental components of a standard flexible endoscope, and (C) the distal end of the insertion tube. Endoscopes contain a working channel to increment its usability. Surgical instruments inserted through the instrument port allow using endoscopes to examine tissues or organs, confirm a diagnosis (by taking a biopsy), or provide treatment.

2. A lighting system for the distal end of the endoscope.
3. A working channel which is fundamental for many reasons including flushing, gas distention, suction, injection, bleeding control (energy or clipping), and eventually taking biopsies which can be done by forceps or by needle cutting or aspiration.
4. Channels for water and air feeding and water jet.
5. A bending mechanism including four wires to control the angulation of the distal tip of the endoscope.
6. Spiral metallic bands to provide flexibility to the tube, transmit the torques and subtle movements from the wrist of the endoscopist, and protects the internal components from external forces.
7. A cover made of polymer is biocompatible. The cover provides watertight capability, and its smoothness provides negligible damage to tissues when the endoscope is inserted.
8. Mechanical features that enhance the comfort of the patient and endoscopist and the ease and speed of its insertion in the patient.

The dimensions of an endoscope, such as length and external diameter, depend on the medical procedure. For instance, the dimensions of an endoscope for examining the stomach and the lungs (bronchoscope) are different due to the characteristics of the examined tissues.

Advantages of flexible endoscopes, compared to their stiff, rigid counterpart, are that flexible endoscopes do not present the fulcrum effect and there may be a lower risk of loss of triangulation. Furthermore, flexibility can improve the adaptability of the insertion tube to the organs and tissues. The risk of damaging the interior of the body during navigation is lower than that of the rigid endoscope. This adaptability facilitates reaching organs that are far from the incision or entry point. The tools needed for the interventional procedure are inserted through the working channel, which minimizes any damage caused to surrounding tissue or organs.

Flexible endoscopes can be employed for NOTES. While laparoscopic surgery may require up to five incisions [56], pure NOTES may not require any incision. This shortens the recovery time, trauma, and pain to the patient. Moreover, these instruments can be used for hybrid NOTES, where they work together with laparoscopic instruments.

Nevertheless, the flexibility affects the precision, the stability of the tip, and the force that can be applied. For instance, when pulling soft tissue, flexible endoscopes may bend; this dissipates the pulling force applied to the tissue [57]. In addition, there might be a risk of looping, and the reduced haptic feedback would not provide an accurate judgment of the force applied. These drawbacks increase the risk of tissue or organ damage.

4.3.2.2 Laparoscopic instruments

As illustrated in Fig. 4.3, a laparoscope is a long rigid variation of an endoscope presented in Section 4.3.2.1. It is made of a small stiff tube, which is inserted through a small incision, with a light source and camera at its tip to retrieve visual feedback. Clinicians insert a laparoscope to access the internal abdominal area and pelvis. Small surgical tools and a tube to inflate the abdomen with carbon dioxide are inserted through the endoscope's working channel or additional trocars. These so-called laparoscopic instruments allow the clinician to manipulate soft tissue and have various functionalities, including a needle driver for suturing, trocar, bowel grasper, and surgical mesh, for instance. After the procedure, the gas is evacuated, the small incision is sewed, and a dressing is applied. A similar procedure is also conducted for removing tissues samples (biopsy) or damaged organs.

Conventional laparoscopic surgery has several advantages over open surgeries (e.g., smaller incision, fewer tissues damage, and less blood loss). However, (a) stiff endoscopes make it challenging to navigate through the internal anatomy increasing the risk of injury to soft tissue or organs; (b) the motion

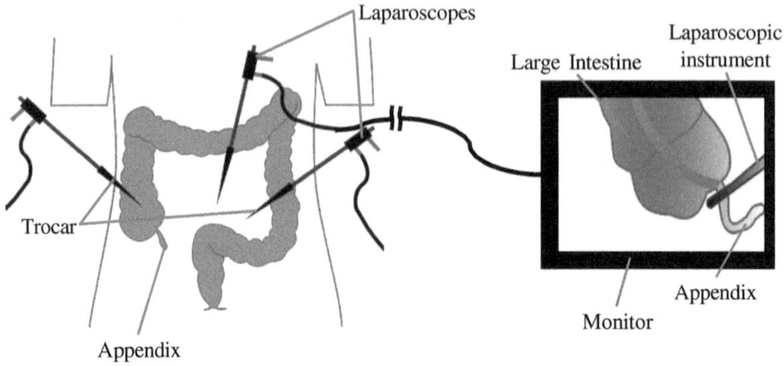

FIG. 4.3 Laparoscopic appendix removal surgery with three incisions (trocars). The laparoscope is inserted through the trocar in the umbilicus providing the surgical site's visual feedback to the clinician. The trocars on the left and right hand side are used to insert rigid laparoscopic instruments and, hence, maintain triangulation. Any clashing between instruments is minimal. The endoscope provides a clear view of the instruments and the surgical site.

range and dexterity are limited by the DOFs provided by the joints of the laparoscopic tools making it challenging for clinicians to work in small spaces; (c) 2D visual feedback provides a poor depth perception; (d) rigid tools can affect the judgment of the force applied to organs or tissues; (e) reduced tactile feedback makes delicate tasks such as tying sutures more challenging; (f) laparoscopic surgery is not ergonomic for clinicians because unnatural movements have to be performed repetitively; and (g) the fulcrum effect requires developing a nonintuitive motor skill as tool endpoints move in the opposite direction to the surgeon's hands. These limitations leave room for complications that may not be common. However, there are still risks, such as organ or major artery damage, gas bubbles entering into arteries or veins, or vein thrombosis that lead to pulmonary embolism, limiting the applicability of laparoscopic surgeries. Furthermore, to treat most of the complications, further surgical intervention might be required [58].

Robotic-assisted minimally invasive surgery (RMIS) comprises a robotic surgical system that has been designed to improve surgeons' predictability, accuracy, repeatability [56], and ergonomics. Before surgery, specialized nurses need to dedicate time to set up the slave side of the surgical robot inserting the instruments through trocars and connecting these to each robotic arm. During the procedure, the surgeon controls the robotic system on the slave side through a master console allowing the surgeon to navigate a 3D high-definition endoscopic camera as well as a number of robotic arms simultaneously. Robotic surgical systems are more ergonomic for the clinician as they maintain an upright position and can navigate and control the robotic surgical instruments with high dexterity intuitively [59]. Another advantage is that simulators can be employed for training and enhancing the surgeon's skills.

Current commercially available, robotic-assisted surgical platforms provide a 3D magnified high-definition image to the master console. These optical tools give a more detailed vision during interventions. The master–slave robotic system allows the clinicians to execute more precise movements filtering any vibrations/physiological tremor. Therefore, surgeons can perform procedures with higher precision.

A well-known surgical robot is the multiport da Vinci Surgical System and the single-port surgical robot da Vinci SP (Intuitive Surgical, Sunnyvale, CA, USA). The da Vinci is commonly used in procedures such as cardiac, colorectal, and general surgeries [60]. It is composed of the surgeon console (master side), the vision cart, and the patient cart (slave side). The surgeon console contains the controls of the surgical instruments and provides 3D visual feedback to the clinician. The vision cart provides image processing capability, information systems, power generation, and visual feedback to the operating team through a screen. The patient cart comprises four robotic arms that the surgeon can teleoperate via the console. The rigid straight laparoscopic instruments with articulated tips are cable-driven and/or actuated through traditional mechanical couplings enabling the clinician to deliver an intervention with high precision and stability [56]. On the other hand, the patient cart of the da Vinci SP contains a single robotic arm. This single-port surgical system includes three wristed elbowed instruments and a wristed elbowed da Vinci endoscope that are inserted through a 2.5 cm cannula. Da Vinci SP can be used for urological procedures and removing benign and T1 and T2 tumors by transoral otolaryngology surgical procedures in the oropharynx. This robotic system is not intended for general laparoscopic surgery procedures, yet [60]. Thus, due to the rigidity of the tools, these laparoscopic solutions are useful when the organ is close to the incision location.

The Versius Surgical Robotic System (CMR Surgical Ltd., CAM, UK) is a laparoscopic robot with a master–slave configuration similar to the da Vinci Surgical System. The major differences compared to the da Vinci are that the clinician operates the console in an upright position while standing or sitting (rather than looking slightly downward through the console binoculars) wearing 3D glasses and each robotic arm is installed on an individual cart [61]. Hence, repositioning of the robotic arms is simplified.

The Senhance Surgical System (TransEnterix Inc., NC, USA) is a laparoscopic platform as shown in Fig. 4.4A. The clinician teleoperates rigid tools mounted on individual carts in an upright position while sitting. 3D high-definition (HD) visualization is provided on the screen of the master console. Also, the Senhance Surgical System provides kinesthetic feedback to the clinician (pressure and tension between the instruments and the environment) [62].

The Monarch Platform (Auris Health, Inc., CA, USA) comprises a flexible robotic endoscope that allows entering the body through natural orifices. Once it reaches the desired location, a needle can take soft tissue samples. The console is fundamentally different from the laparoscopic surgical robotic systems as the

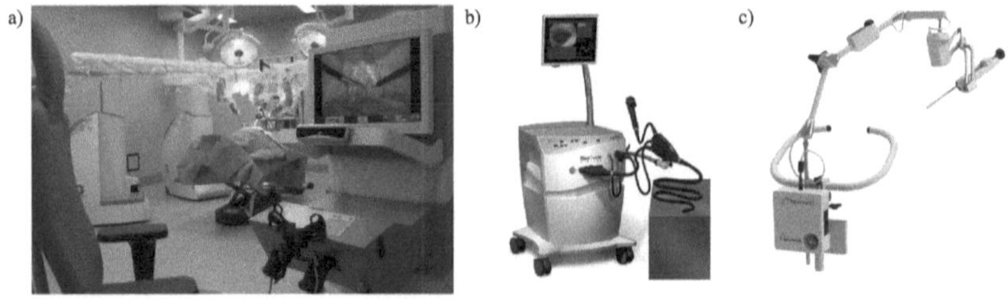

FIG. 4.4 Endoscopic robotic-assisted technologies that are commercially available and have FDA approval and/or CE certification. (A) the Senhance Surgical System uses rigid instruments, measures tool interaction with the environment, and provides kinesthetic feedback to the clinician; (B) the NeoGuide Endoscopy System is a computer-assisted colonoscope providing 3D real-time maps; and (C) the FreeHand Assisted Surgery supports standard laparoscopic surgeons by holding the laparoscopic camera steadily and, hence, providing comfort to the clinician. *(Panel (A) taken from Pappas T, Fernando A, Nathan M. Senhance surgical system: robotic-assisted digital laparoscopy for abdominal, pelvic, and thoracoscopic procedures. In Handbook of Robotic and Image-Guided Surgery. Elsevier; 2020. p. 1–14; Panel (B) taken from Eickhoff A, Jakobs R, Kamal A, Mermash S, Riemann JF and Van Dam J. In vitro evaluation of forces exerted by a new computer-assisted colonoscope (the NeoGuide Endoscopy System). Endoscopy; 2006; 38(12). pp. 1224–1229; Panel (C) Courtesy: Freehand, Copyright©2021 Freehand.)*

controller interface is held with both hands allowing clinicians to navigate the endoscope and reach peripheral nodules in the lung. Lung cancer diagnosis and treatment are the main medical applications [63].

Flex Robotic System (Medrobotics, MA, USA) is a compliant flexible endoscope surgical system with gravity compensation for single-port MIS [64]. It has an onboard HD visualization system. Compliance is achieved by its concentric mechanism together with its multiple linkages. Two working channels allow inserting flexible surgical instruments. The tip of the robot is steered via a parallel robotic joystick.

The NeoGuide Endoscopy System (NeoGuide Endoscopy System Inc., Los Gatos, CA, USA), illustrated in Fig. 4.4B, presents a modular compliant structure. It is made of 16 equally sized electromechanically actuated modules. The distal tip provides a wide range of movements in all directions. Its mapping approach to navigate through the natural curves of the colon and its looping reduction system [65] decrease any forces applied to the colon wall [66]. The system employs an external position sensor for localization measurements. In the system's passive mode, the endoscope is stiff to take a biopsy or deliver therapy. In the active mode, the system follows the commands of the surgeon given through the controller interface.

The InvendoScope (Invendo Medical GmbH, Germany—acquired by Ambu, Denmark, in 2017) is a single-use flexible colonoscope. The system is made of a colonoscope, a reusable handheld controller, and a processing unit. After the instrument is introduced into the rectum, the distal end advances using its propulsion mechanism based on an inverted sleeve mechanism with eight drive wheels gripping into the inner side of this sleeve; the rotation of this wheels allows growth or reduction at the tip [67]. Its DOFs provide a 180 degrees rotation in all directions; the maximum insertion length is 170 cm; its bending radius is 35 mm for retroflexion colon visualization; and the radius of its working channels is 3.1 cm [68].

A robotic approach occupying minimal footprint has been realized by the FreeHand Assisted Surgery, shown in Fig. 4.4C. This commercially available system gives the surgeon a "third" hand to stabilize the laparoscopic camera. Visual feedback is improved as the clinician can control the camera with head movements only and without the need to unhand any laparoscopic tools [69].

As mentioned earlier, robotic surgical systems can provide a number of benefits, including the surgeon's predictability, accuracy, repeatability [56], and ergonomics. Though a wide range of interventional applications may be suggested for these systems (leveraging the economic justification for healthcare providers), their deployment might not always be the ideal option when delivering treatment, and some clinicians would prefer open surgery [70].

4.3.3 Challenges in current endoscopic interventions

Technology has rapidly advanced over the last decades with many available and emerging surgical robotic platforms entering the market in the near future. Nevertheless, some challenges make open surgery the only option to complete or perform certain surgeries. Taking into account the various endoscopic procedures, the risks faced by clinicians and patients, and the current commercially available technologies discussed in previous sections, it can be stated that there are common challenges across endoscopic procedures.

Clinicians can understand the location and orientation of laparoscopic cameras once they have been inserted into the body through trocar ports as these are nonflexible, rigid rods. However, challenges occur when navigating these optical tools and additional laparoscopic instruments around organs to the point of interest. In some cases, new incisions have to be made to allow easier access.

When deploying flexible endoscopes, maneuvering and navigating the tip of these devices, which have multiple DOFs, in 3D space become challenging. In addition, endoscopes are usually advanced through natural orifices by pushing the tail. On the one hand, the insertion process can lead to undesired rubbing motions between the endoscope's body and surrounding soft tissue. On the other hand, poor navigation capability can result in incorrect pathways and trajectories. During colonoscopy for instance, lateral pressure of a bowed loop of the colonoscope against a stretched loop of the colon can cause substantial intestinal ruptures [4]. These cannot only lead to tissue damage (traumatic contact inside the lumen), but experienced endoscopists might also miss locations of polyps. Furthermore, the loss of dexterity might become critical in neurosurgical operations such as tumor extraction [71].

From the ergonomics point of view, colonoscopists commonly suffer hand, wrist, forearm, and shoulder injuries. These may be caused by the repetitive pinching, excessive use of upper limbs, torquing and gripping forces, and unnatural neck and body posturing [72] that are required when conducting endoscopic interventions (in particular, in operations when a rigid endoscope is deployed). It is of paramount importance to solve these ergonomic challenges because these procedures are a significant part of the endoscopist's workload.

It is worth highlighting that it is essential to maintain healthy ergonomics and comfort of endoscopists and patients for endoscopic examination and confirmation of diagnostic procedures. Improving ergonomics can increase efficiency in the examination (polyp detection), and an increasing number of patients may be keen to undergo an endoscopic procedure. Therefore, clinicians may be able to give an earlier and accurate diagnose together with their corresponding treatment.

Endoscopic systems with haptic (tactile and kinesthetic) feedback could also enhance the skills of clinicians. Compared to open surgeries, touch information is lost during robotic-assisted surgical procedures due to most master–slave configurations not having embedded haptic feedback solutions. In fact, the Senhance Surgical System provides force feedback, but training is required to accurately judge the received feedback. In procedures such as neurosurgical operations, an accurate judgment of the forces is essential to avoid damaging healthy tissue [2]. Therefore, haptic feedback systems should be intuitive to have a transparent interpretation of the information. Having feedback of the tissue stiffness is likely to improve precision, effectiveness, safety, and reliability of surgical procedures such as tumor excision [2]. In these medical interventions, there is a risk of requiring recurrent surgeries due to incomplete removal of the affected tissue.

Despite these challenges, lower tissue trauma, immunological stress response, postoperative pain, and patient recovery time make surgeons prefer minimally invasive approaches. During the last decade, research in soft robotics has resulted in promising solutions to overcome some of the aforementioned limitations. In fact, researchers have been successful in combining the accessibility of flexible endoscopes, the controllability of rigid robots, and safety and environmental adaptability of soft materials mechanisms [73].

4.4 Soft robotic systems for endoscopic procedures: State-of-the-art in research

With ongoing discussion about the definition of *soft robots*, the term *soft* is commonly referred to the inherent material and structural compliance [74]. In this chapter, we use this term to denote robots that are made of soft materials and might include mechanisms that change their body stiffness and shape [75].

Soft materials can include colloids, polymers, liquids, gels, foams, granular materials, and most soft biological materials [76]. According to [77], *soft robotics* comprises all the active and reactive compliant systems ranging from soft actuators and sensors, artificial muscles (i.e., electro-active polymers) up to soft electronics and soft energy harvesting. Soft robotics offer characteristics that make them suitable for healthcare application, including endoscopic interventions. A number of soft materials that soft robots are made of have different levels of biocompatibility. For instance, silicone elastomers (NuSil Technology, CA, USA) are safe to work inside the body for a limited period of time [78,79]. On-demand stiffness of soft robotic structures (e.g., to manipulate soft tissue) can be controlled using external stimulus, for instance, such as electric current [11,80], light, or injection or extraction of fluids [81,82]. This controllable stiffness capability, together with the inherent softness, squeezability, and flexibility of the materials, facilitates inherently safe navigation and maneuverability inside the body with a low risk of damaging tissue or organs. Moreover, soft robotic interaction with the environment and their elongation capabilities [82,83], which can be used for locomotion, can simplify the process for reaching distant organs. Commonly, the properties of the soft robot are defined by the soft actuator and sensors that the system is made of.

Soft actuators are usually driven by fluids [84,85], material jamming [86], tendons [87], or smart materials [80]. Flexible fluidic actuators are made of shrinkable and soft materials that are nonferromagnetic and, hence, magnetic resonance (MR)-compatible. Many flexible actuator structures have embedded hollow chambers that can produce active locomotion when inflated and/or deflated [83]. Inflation also increases the contact area with the environment and, therefore, can increase friction between the soft actuator and environment. Soft actuators are commonly lightweight because they are pneumatically actuated by an external source. Granular jamming actuators comprise an external membrane filled with granules. The stiffness of these types of actuators varies based on different levels of applied negative pressure, "jamming" the internal material affecting the relative speed of the particles of the filling material and its density inside the membrane. Therefore, the rigidity of these actuators will incrementally increase [88]. Other approaches include cable-driven actuators that can be lightweight and scalable if the electric components (motors) are externally located [8]. Smart materials such as shape memory alloys present a high energy density, corrosion resistance, MR-compatible (nonmagnetic behavior), and scalable [56].

Sensors provide feedback for controlling soft actuators and understanding physical interactions with the environment. Apart from application-driven requirements with regard to the sensing range, resolution, durability, and repeatability, sensing systems for soft actuators should be flexible and, in some cases, stretchable. Soft robotic applications can employ capacitive, resistive [89,90], magnetic, optoelectronic, or light intensity [91] sensing. Capacitive, resistive, miniature magnetic elements, or waveguides, respectively, are embedded inside the soft robotic body returning a variation of the sensed parameter depending on the applied strain. In optoelectronic sensing, electrical signals that vary with the applied strain as the distance the light has to travel changes are measured. Embedding these materials in elastomers of soft robotic bodies without affecting the overall stiffness is possible due to the use of nanomaterials or conductive elastomer composites, deterministic and ultrathin structures, liquid metals, or conductive gels.

Based on this overview of principles of soft actuators and sensors, the following section will present the state-of-the-art of current soft robotic systems for endoscopic applications that have been validated in *ex vivo* phantom and cadaver environments. *In vivo* animal or human clinical tests involving soft robotic systems for endoscopic applications have not been reported up to the time of writing this chapter to the best of the authors' knowledge.

It has to be highlighted that this review comprises entirely soft and hybrid (including soft and rigid parts) robots for endoscopic applications. Following the structure of previous sections, this review is divided into optical examination tools and instruments for therapeutic endoscopy.

4.4.1 Soft robotic instruments for optical examination

When performing visual examinations, endoscopic capsules are the least invasive, as mentioned in Section 4.3.1. Commercially available capsule endoscopes move through peristalsis inside the body. This approach is less invasive than conventional endoscopy. However, the lack of the capsule's orientation and navigation control hinders the diagnostic efficacy, especially in the stomach [92]. Additionally, frequent production of obstructed images and the inability to perform therapeutic procedures or biopsy decrease the usability of capsule endoscopes [93,94].

To address some of these limitations, recent research proposes using magnetically actuated soft capsule endoscopes [51]. The locomotion and rotation of the capsule and, hence, the camera view can be controlled externally via an interface. These capsules can then be steered toward soft tissue areas to collect biopsy samples acquiring stomach deep tissue biopsy samples through a fine-needle biopsy technique. After taking a sample, the capsule is retrieved by a tether to avoid any contamination. Nevertheless, reorientation of the patient is required to take biopsies in various locations (anterior, posterior, and lateral stomach surfaces). The advantage of this soft capsule endoscope over its commercially available counterparts is its ability to performing needle biopsy. Its soft Sarrus linkage allows naturally aligning the needle translation to the capsule's longitudinal axis, and the collapsing motion of its soft body exposes the needle for taking the sample. Therefore, examination and collection of samples for diagnosis confirmation can be performed simultaneously [51].

In recent years, there has been an increasing interest in inchworm like soft robots for colonoscopy [82,83,95]. They are characterized by advancing peristaltically and being navigated through tubular structures. In [95], an 18 mm external diameter Soft Pneumatic Inchworm Double balloon (SPID) mini-robot is presented. This robot is made of Vero-Clear, a transparent photopolymer, and Ecoflex 00–30, a soft, stretchable silicone material. It is worth noting that the

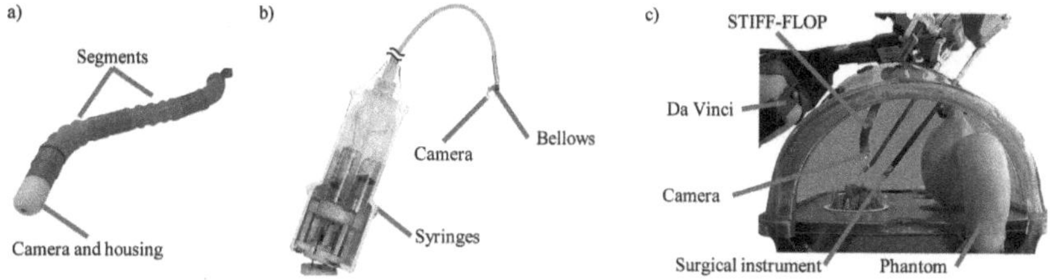

FIG. 4.5 Soft robots for endoscopic applications: (A) Worm-inspired robotic endoscope. (B) Low-cost disposable endoscope for visual examination. A high-definition camera and a number of LEDs mounted at the tip are connected to three rubber bellows. Retraction, elongation, and bending are controlled through the pressurization of these bellows using syringes. (C) The STIFF-FLOP robotic manipulator for MIS is inspired by the octopus arm—the first soft robotic instrument that performed procedural steps of colorectal surgery in a human cadaver. *(Panel (A) Courtesy: Julius Bernth, Bernth J, Arezzo A, Liu H. A novel robotic meshworm with segment-bending anchoring for colonoscopy. IEEE Robot Autom Lett 2017;2(3):1718–1724; Panel (B) This figure is a modification from Garbin N, Mamunes A, Sohn D, Hawkins R, Valdastri P, Obstein K. Evaluation of a novel low-cost disposable endoscope for visual assessment of the esophagus and stomach in an ex-vivo phantom model. Endosc In Open 2019;7(9):E1175.)*

actuation pressure is significantly lower than the internal pressure of the colon when it is inflated for a surgical procedure. Anchoring is provided by two inflatable balloons linked by a 3 DOFs soft pneumatic actuator. Pressurizing one or two of the three embedded chambers will result in bending behavior beyond 100 degrees, whereas simultaneous actuation of all three chambers will lead to overall elongation [95]. Bernth *et al.* proposed a three-module worm-inspired robotic endoscope illustrated in Fig. 4.5A [83]. Each module is made of an elastic mesh structure driven antagonistically by tendons. The minimum dimension of the tubular environment that the robot can move through is defined by the 26 mm outer diameter of the rigid collar of its modules. The endoscopic robot can move forwards and backwards as well as anchoring itself. Controlling the bending of its flexible modules adjusts the camera's orientation. It is worth noting that this robot is designed for tubular environments; hence, robot-environment interaction has been proposed to improve sensing capabilities and force generation [96].

To minimize the risk of cross-contamination or infections, that might occur in deploying reusable endoscopes for stomach and esophagus examinations, e.g., in low- and mid-income countries, a disposable low-cost endoscope has been presented in [97,98] as shown in Fig. 4.5B. The total cost of this 13.5 mm diameter endoscope is about $ 25. It comprises an HD camera and three LEDs for illumination at the tip. The tip bending, elongation, and retraction are produced by the pressurization of three rubber bellows arranged in parallel. Syringes control the internal pressure of these bellows. During the evaluation, novice, gastroenterology fellows, and gastroenterology attendings required less time to perform an examination task with a conventional endoscope than with this low-cost endoscope. However, it is worth noting that attendings and fellows stated that navigation of the endoscope bears a higher cognitive load compared to using conventional endoscopes, whereas novices found the operation of the device intuitive and easy to learn.

4.4.2 Soft robotic instruments for therapeutic endoscopy

Nature can provide valuable inspiration when creating soft robotic devices. For instance, the octopus arm has the ability to squeeze through narrow openings, bend and navigate around obstacles, and catch and manipulate prey through controlling the stiffness of its arm. The STIFF-FLOP project funded by the European Commission has created soft, stiffness-controllable endoscopic devices (as illustrated in Fig. 4.5C) inspired by capabilities of the octopus allowing these new medical instruments to be navigated and maneuvered inside the body with a low risk of damaging soft tissue or organs. The STIFF-FLOP manipulator has multiple embedded chambers that can be fluidically actuated. Elongation is achieved by pressurizing all chambers simultaneously, while bending can be performed by activating one or two chamber(s) (pairs) only. Furthermore, squeezability is a feature of the STIFF-FLOP robot as its body is made of soft silicone. Stiffness controllability has been achieved by integrated granular jamming chambers [99] or applying an antagonistic principle using air and tendon-driven actuation [87]. To improve the stability and transmission of forces of the robotic arm, Brancadoro *et al.* [86] proposed a fiber jamming approach to provide variable stiffness capabilities without increasing its length or external radius of 14.5 mm. The STIFF-FLOP device has been designed to have a free working channel allowing to deploy an endoscopic camera (e.g., with an integrated stiffness sensor [100,101]) or flexible instruments [3].

Russo *et al.* proposed a multiarticulated soft pop-up robotic arm. 3D structures were created by a hybrid soft pop-up manufacturing approach based on folding multiple layers of laminated rigid-flexible, biocompatible material. This robotic arm is integrated on the distal end of a flexible endoscope. The pitch joint of the 3 DOFs mechanism can actuate the end effector (a soft suction-based gripper) to manipulate soft tissue. The add-on device was evaluated inside an ex vivo porcine stomach, demonstrating that soft tissue can be handled without any visual obstruction in the camera image of the endoscope [78].

In Ref. [102], a manual platform for single-port surgery is proposed based on a variable stiffness mechanism using liquid metal to overcome challenges in limited payload and workspace, triangulation, and to address the fulcrum effect. The bidirectional solid-to-liquid conversion of the metal composite, activated by a variable temperature system, allows controlling the stiffness of the tubes holding the surgical instruments. In particular, effective force transmissions can be achieved in a solid state. On the other hand, reorientation of the tools can be accommodated when the metal is soft and liquid.

Using low-melting-point alloys (LMPAs) to stiffen soft continuum devices has been investigated for various applications in MIS [11,80,103]. They comprise a central flexible inner pipe covered by a heating wire in contact with the LMPA. The entire structure is embedded inside a silicone body. Changing the electric current in the heating wire will change the temperature, which again varies the stiffness. A high current leads to high heat dissipation, which stimulates the solid-to-liquid transition of the LMPA.

On the other hand, Runciman *et al.* proposed a soft approach to deploy flexible endoscopes and two flexible surgical instruments for performing endoscopic submucosal dissection with the aim to decrease any risk associated with the interaction between the endoscope and any soft tissue. While navigating, the soft robot, made of 120 μm thickness material, surrounds the flexible endoscope. Once the surgical site is reached, the soft robot inflates to increase its volume and radial stiffness. Force transmission cables are connected to a variable stiffness structure and hold any end effectors through which surgical instruments can pass. By controlling the length of these cables, the pose of the instruments can be controlled [104].

4.4.3 Soft haptic technologies for minimally invasive endoscopic procedures

Haptic refers to the sense of touch. Haptic feedback is generally defined as a combination of tactile and kinesthetic feedback. Tactile feedback refers to the mechanoreceptors' information (i.e., Pacinian, Merkel, Meissner, or Ruffini corpuscles) under the skin. They provide information about pressure distribution, stimulus location, deep static touch feature (shapes and edges), and stiffness [105]. Each mechanoreceptor is sensitive to vibrations in a particular range. Their concentration on human skin depends on the part of the human body. For instance, fingerpads are the places with the highest concentration of Meissner and Merkel corpuscles while that of the other two corpuscles is relatively low [106]. Kinesthetic feedback refers to the information given by sensors from muscles, tendons, and joints. They provide information regarding weight and joint angles of the arm, wrist, hand, and fingers. This information helps to determine an approximate weight and dimensions of objects with whom the hands are interacting. In general, haptic feedback supports clinicians on the decision-making process.

Haptic technologies for remote or robotic-assisted endoscopic applications aim to reproduce the physical interaction between the soft tissue and clinician's hands or fingertips that would have been experienced if the surgery had been performed in an open way. As [105] states, tactile feedback systems intend to provide mechanical properties (i.e., surface texture, viscosity, and compliance) of soft tissue. Surgeons use this feedback to differentiate healthy from cancerous tissue or to detect arteries [107]. On the other hand, kinesthetic feedback informs about the force applied to the tissue. The clinician uses this type of feedback in tasks such as suturing and pulling tissue, which is crucial for avoiding unnecessary trauma or damage to tissues or organs.

Several haptic modalities have been investigated: visual feedback [108], haptic tactile feedback, haptic force feedback, haptic stiffness feedback, and a combination of these. In this section, we highlight haptic stiffness feedback displays made of soft material structures. In [109], a 3D haptic display is proposed that can change its shape and stiffness. The display is made of an array of inflatable silicone pouches with embedded granular jamming material. Applying positive or negative pneumatic pressure will result in shape change or stiffening. Medical applications include palpation training.

In [110], a multifingered variable stiffness tactile feedback actuator is proposed. Similarly to [109], the variable stiffness mechanism is a combination of granular jamming that is located on top of a pneumatic chamber. The pneumatic chamber's role is to enhance stiffness discrimination while the granular jamming provides the variable stiffness stimulus. This solution can be applied to the index, middle, and ring fingers to improve the haptic feedback of surgeons while performing robotic-assisted minimally invasive procedures [110]. The INSTINCT project funded by the Engineering and Physical Sciences Research Council (grant agreement number EP/S014039/1) aims at integrating miniaturized variable stiffness tactile feedback actuators into the master console of surgical robots for MIS. The advantage of these haptic displays is that surgeons can use their fingertips to palpate soft tissue and assess their mechanical properties.

4.4.4 Technical challenges for emerging soft robotic, endoscopic tools

As soft robotic systems offer promising advantages over traditional rigid (flexible) endoscopic systems, their development has often been motivated by medical applications. However, only a small number of soft robotic medical devices have actually demonstrated their capabilities in anatomical phantom or cadaver environments. In general, technical requirements for these systems are set by specific endoscopic procedures. For instance, MR compatibility of medical devices is usually necessary for endoscopic retrograde cholangiography [111,112]. In this section, however, we will focus on overarching challenges that occur when applying soft robotic systems to a majority of endoscopic procedures. Here, fundamental challenges remain to be investigated—some of which require multi- and interdisciplinary approaches—for successful applications in endoscopic surgery:

1. **Maintaining consistent quality in fabrication and manufacturing:** Soft robotic systems have been created using casting and 3D printing fabrication techniques, respectively. Both methods are associated with substantial limitations such as narrow material choices, low durability, and high variation in fabrication quality. In particular, weak points of soft robotic systems might occur at interfaces between soft and rigid components that these robots are made of.

2. **Challenges in miniaturization**: Any medical device for endoscopic interventions must meet size requirements. In particular, constraints in miniaturization occur when embedding sensing, actuation, and variable stiffness mechanisms into a soft robotic system. These are explained by the current manufacturing processes and the limited variety of miniaturized, commercially available components such as micropumps.
3. **Need for next generation of actuators:** A large number of soft robots are fluidically actuated, whereas pressurized fluid and proportional valves are commonly located externally due to size requirements. In cable-driven soft actuators, nonlinear friction might occur which presents challenges in the development of mathematical models and the transmission of forces via tendons to the distal end of a soft manipulator might affect the configuration of the soft body that the tendons are guided through. In addition, a number of smart materials require high electric current resulting in low actuation speed (compared to other fluidic actuation approaches) and complex control methods due to thermomechanical behavior.
4. **New functional materials:** Depending on the application in endoscopic procedures, soft robotic systems need to be sterile and should not suffer from fatigue of any components over the duration of usage. Sterilization of (multiple-use) medical devices can have a negative impact on soft materials [56]. To execute a surgical procedure and manipulate soft tissue requires a level of stability and the availability of on-demand stiffness. Therefore, there is a need for new materials that are unaffected by sterilization and can change their stiffness over a wide range and provide effective stiffness capability even when embedded in miniaturized soft robots.
5. **Soft structured sensing elements:** Creating and integrating transducers for soft robots pose challenges as it requires embedding a conductive element inside the elastomer. As mentioned earlier, the interface between the soft robot body and sensing element might be challenging. Furthermore, the combination of different materials might lead to an increment in the structure's stiffness, artifacts in the acquired data, hysteresis, and risk of damaging soft elements of the device or soft tissue in the environment. These could affect the flexibility, durability, and repeatability of the sensor across several sensing cycles.
6. **Intuitive interfaces for multi-DOFs soft flexible robotic systems:** Soft robots' continuous and elastic body provides infinite DOFs, in theory: they are able to bend, elongate, and squeeze through narrow opening. A new generation of interfaces should enable clinicians to navigate and steer the tip of these high DOFs robotic manipulators intuitively without focusing on how to position the remaining tail, which can result in additional cognitive load and distraction.
7. **Real-time modeling:** To compensate for some of the aforementioned limitations in manufacturing and actuation as well as in the overall performance of soft robots, real-time kinematic and dynamic models that could provide valuable input for model-based controllers are still under development.

It is worth noting that regulatory bodies should establish a standard framework for benchmarking surgical robots as well as endoscopic solutions in general [113]. This framework should consider clinical evidence, economic parameters, and patient benefits in order to scrutinize and compare results among various platforms. One of these key parameters should include training required (learning curve) by surgeons and endoscopists to use these new soft robotic solutions. More intuitive solutions should be preferred as clinicians may need less intensive training to achieve and then deliver current quality, safety, and consistency expected levels across procedures.

4.5 Conclusions and potential opportunities for endoscopic soft robotic systems

Endoscopic procedures have become a preferred method over open surgery due to patients' shorter recovery time, shorter stay in the hospital, lower tissue damage, less blood loss, lower impact in the immune system, and lower requirement of anesthesia. Due to these advantages, endoscopic cameras and instruments are employed in a growing variety of medical examination and treatments.

Medical devices and robotic technology aiming at improving patient's outcome and supporting the clinician during endoscopic procedures have advanced rapidly over the last decades. However, commercially available endoscopic tools such as capsule endoscopes, flexible endoscopes, laparoscopic tools, and other specific endoscopes (for instance, neuroendoscopes or bronchoscopes) still require further development to improve maneuverability, including haptic feedback, enhance control perspectives, and increase endoscopists' and patients' ergonomics. Addressing these challenges might lead to an increase in numbers of patients who are willing to undergo an endoscopic procedure, which again might increase the early detection of illnesses such as cancer.

Soft robotic systems can provide effective solutions to overcome current limitations of endoscopic procedures as these systems are compliant, squeezable, and soft and allow extensive navigation and maneuverability inside the body. These properties, together with variable stiffness mechanisms, can improve the ergonomics of endoscopists and patients. Soft robotic systems can reduce clinicians' unnatural repetitive postures and the need to change a patient's posture during medical procedures.

However, current emerging soft robotic technologies require further advancements focusing on how to maintain consistent high quality in the fabrication and manufacturing process; challenges in miniaturized soft robotic systems; the next generation of actuators, sensors, and functional materials; and real-time modeling approaches. Overcoming shortcomings in these areas will improve the performance of current systems and allow for the development of accurate real-time kinematic and dynamic models. In addition, navigating multi-DOFs, soft flexible robotic manipulators through confined space inside the body can be an additional cognitive load to the clinician. As a consequence, intuitive interfaces enabling clinicians to navigate and steer the tip of these high DOFs robotic manipulators without focusing on how to position the remaining tail should be explored.

It is profoundly important, however, that new developments in the field of soft robotic endoscopic systems concentrate on improving the patient's outcome and health. The focus should be on addressing shortcomings and limitation of current procedures that negatively impact the patient and clinician to make medical procedures safer and less invasive and demanding for patients and clinicians. Hence, interdisciplinary research and collaboration between medical experts and engineers are essential to developing innovative soft robotic systems that deliver the expected positive impact in the field of endoscopic interventions.

Acknowledgment

This work has been supported by the Springboard Award of the Academy of Medical Sciences (grant number: SBF003–1109) and the Engineering and Physical Sciences Research Council (grant numbers: EP/N509577/1 and EP/S014039/1).

References

[1] Li M. Haptic feedback of rigid tool/ soft object interaction in medical training and robot-assisted minimally invasive surgery. PhD thesis, Strand, London WC2R 2LS, UK: King's College London; 2014.

[2] Li M, Liu H, Jiang A, Seneviratne LD, Dasgupta P, Althoefer K, Wurdemann H. Intraoperative tumour localisation in robot- assisted minimally invasive surgery: a review. Proc Inst Mech Eng H J Eng Med 2014;228(5):509–22.

[3] Arezzo A, Mintz ME, Allaix YA, Arolfo S, Bonino M, Gerboni G, Brancadoro M, Cianchetti M, Menciassi A, Wurdemann H, Noh Y, Althoefer K, Fras J, Glowka J, Nawrat Z, Cassidy G. Walker R., and M. Morino. Total mesorectal excision using a soft and flexible robotic arm: a feasibility study in cadaver models. Surg Endosc 2017;31:264–73.

[4] Ciuti G, Skonieczna-Zydecka K, Marlicz W, Iacovacci V, Liu H, Stoyanov D, Arezzo A, Chiurazzi M, Toth E, Thorlacius H, et al. Frontiers of robotic colonoscopy: a comprehensive review of robotic colonoscopes and technologies. J Clin Med 2020;9(6):1648.

[5] Bicchi A, Canepa G, De Rossi D, Iacconi P, Scillingo EP. A sensor-based minimally invasive surgery tool for detecting tissutal elastic properties. In: Proceedings of IEEE international conference on robotics and automation, volume 1. IEEE; 1996. p. 884–8.

[6] Faragasso A. Vision-based sensing mechanism for softtissue stiffness estimation. PhD thesis, Strand, London WC2R 2LS, UK: King's College London; 2016.

[7] Cianchetti M, Ranzani T, Gerboni G, Nanayakkara T, Althoefer K, Dasgupta P, Menciassi A. Soft robotics technologies to address shortcomings in today's minimally invasive surgery: the stiff-flop approach. Soft Rob 2014;1(2):122–31.

[8] Wurdemann H, Stilli A, Althoefer K. An antagonistic actuation technique for simultaneous stiffness and position control. In: Lecture notes in computer science. Intelligent robotics and applications, vol. 9246; 2015. p. 164–74.

[9] Lindenroth L, Back J, Schoisengeier A, Noh Y, Wurdemann H, Althoefer K, Liu H. Stiffness-based modelling of a hydraulically- actuated soft robotics manipulator. In: International Conference on Intelligent Robots and Systems; 2016. p. 2458–63.
[10] Stilli A, Althoefer K, Wurdemann H. Soft robotics. Bio-inspired antagonistic stiffening. In: Developing support technologies. Biosystems & biorobotics, vol. 23; 2018. p. 207–14.
[11] Peters J, Nolan E, Wiese M, Miodownik M, Spurgeon S, Arezzo A, Raatz A, Wurdemann H. Actuation and stiffening in fluid-driven soft robots using low-melting-point material. In: International conference on intelligent robots and systems; 2019. p. 4692–8.
[12] Shafti A, Andorno F, Marchese N, Arolfo S, Aydin A, Elhage O, Noh Y, Wurdemann H, Arezzo A, Dasgupta P, Althoefer K. Comfort and learnability assessment of a new soft robotic manipulator for minimally invasive surgery. In: Annual international conference of the IEEE engineering in medicine and biology society. IEEE; 2015. p. 4861–4.
[13] Nisky I, Huang F, Milstein A, Pugh CM, Mussa-Ivaldi FA, Karniel A. Perception of stiffness in laparoscopy—the fulcrum effect. Stud Health Technol Inform 2012;173:313.
[14] Russo M. Triangulation mechanism for a minimally invasive surgical device; 2014. US Patent 8845517.
[15] Hamilton S, Aaltonen L. Pathology and genetics of tumours of the digestive system. Lyon: IARC Press; 2006.
[16] Barbeiro S, Libânio D, Castro R, Dinis-Ribeiro M, Pimentel-Nunes P. Narrow-band imaging: clinical application in gastrointestinal endoscopy. GE Port J Gastroenterol 2019;26(1):40–53.
[17] Hewett DG, Kaltenbach T, Sano Y, Tanaka S, Saunders BP, Ponchon T, Soetikno R, Rex DK. Validation of a simple classification system for endoscopic diagnosis of small colorectal polyps using narrow-band imaging. Gastroenterology 2012;143(3):599. 607.e1.
[18] Kanao H, Tanaka S, Oka S, Hirata M, Yoshida S, Chayama K. Narrow-band imaging magnification predicts the histology and invasion depth of colorectal tumors. Gastrointest Endosc 2009;69(3 Pt 2):631–6.
[19] NHS. Biopsy, https://www.nhs.uk/conditions/biopsy/. Accessed: 2020-09-30.
[20] Savage N. Another set of eyes for cancer diagnostics. Nature 2020;579(7800):S14–6.
[21] Ai Medical Service Inc. https://ai-ms.com/en/. Accessed: 2020-12-02.
[22] Lunit. https://www.lunit.io/en/. Accessed: 2020-12-02.
[23] Lee JH, Sun HY, Park S, Kim H, Hwang EJ, Goo JM, Park CM. Performance of a deep learning algorithm compared with radiologic interpretation for lung cancer detection on chest radiographs in a health screening population. Radiology 2020;297(3):687–96.
[24] UK National Health Service. Gastroscopy, https://www.nhs.uk/conditions/gastroscopy/. Accessed: 2020-09-15.
[25] Rex DK, Schoenfeld PS, Cohen J, Pike IM, Adler DG, Fennerty MB, Lieb JG, Park WG, Rizk MK, Sawhney MS, et al. Quality indicators for colonoscopy. Gastrointest Endosc 2015;81 (1):31–53.
[26] Senore C, Ederle A, Fantin A, Andreoni B, Bisanti L, Grazzini G, Zappa M, Ferrero F, Marutti A, Giuliani O, et al. Acceptability and side-effects of colonoscopy and sigmoidoscopy in a screening setting. J Med Screen 2011;18(3):128–34.
[27] Baron TH, Petersen BT, Mergener K, Chak A, Cohen J, Deal SE, Hoffman B, Jacobson BC, Petrini JL, Safdi MA, et al. Quality indicators for endoscopic retrograde cholangiopancreatography. Gastrointest Endosc 2006;63(4):S29–34.
[28] Jin F, Mu D, Chu D, Fu E, Xie Y, Liu T. Severe complications of bronchoscopy. Respiration 2008;76(4):429–33.
[29] Stahl D, Richard K, Papadimos T. Complications of bronchoscopy: a concise synopsis. Int J Crit Illn Inj Sci 2015;5(3):189.

[30] Wrightson J. Oxford centre for respiratory medicine: bronchoscopy—information for patients, https://www.ouh.nhs.uk/patient-guide/leaflets/files/12354Pbronchoscopy.pdf; 2015. Accessed: 2020-11-30.

[31] Paul BC, Chen S, Sridharan S, Fang Y, Amin MR, Branski RC. Diagnostic accuracy of history, laryngoscopy, and stroboscopy. Laryngoscope 2013;123(1):215–9.

[32] Silverberg MJ, Li N, Acquah SO, Kory PD. Comparison of video laryngoscopy versus direct laryngoscopy during urgent endotra- cheal intubation: a randomized controlled trial. Crit Care Med 2015;43(3):636–41.

[33] United Kingdom National Health System. Cystoscopy; 2020. https://www.nhs.uk/conditions/cystoscopy/. Accessed: 2020-12-02.

[34] Capmas P, Pourcelot A-G, Giral E, Fedida D, Fernandez H. Office hysteroscopy: a report of 2402 cases. J Gynecol Obstet Biol Reprod 2016;45(5):445–50.

[35] United Kingdom National Health System. Hysteroscopy; 2018. https://www.nhs.uk/conditions/hysteroscopy/. Accessed: 2020-12-02.

[36] American Society for Gastrointestinal Endoscopy. Endoscopic procedures.

[37] Marcus HJ, Seneci CA, Hughes-Hallett A, Cundy TP, Nandi D, Yang G-Z, Darzi A. Comparative performance in single-port versus multiport minimally invasive surgery, and small versus large operative working spaces: a preclinical randomized crossover trial. Surg Innov 2016;23(2):148–55.

[38] Peng C, Ling Y, Ma C, Ma X, Fan W, Niu W, Niu J. Safety outcomes of notes cholecystectomy versus laparoscopic cholecystectomy: a systematic review and meta-analysis. Surg Laparosc Endosc Percutan Tech 2016;26(5):347.

[39] Targarona EM, Maldonado EM, Marzol JA, Marinello F. Natural orifice transluminal endoscopic surgery: the transvaginal route moving forward from cholecystectomy. World J Gastrointest Surg 2010;2(6):179.

[40] Baekelandt JF, De Mulder PA, Le Roy I, Mathieu C, Laenen A, Enzlin P, Weyers S, Mol BWJ, Bosteels JJA. Transvaginal natural orifice transluminal endoscopic surgery (vnotes) adnexectomy for benign pathology compared with laparoscopic excision (notable): a protocol for a randomised controlled trial. BMJ Open 2018;8(1), e018059.

[41] Clark MP, Qayed ES, Kooby DA, Maithel SK, Willingham FF. Natural orifice translumenal endoscopic surgery in humans: a review. Minim Invasive Surg 2012;2012.

[42] Inoue H, Ikeda H, Hosoya T, Yoshida A, Onimaru M, Minami H, Kudo S-E. [per-oral endoscopic myotomy (poem) for esophageal achalasia]. *Nihon Shokakibyo Gakkai zasshi*. Jpn J Gastroenterol 2012;109(5):728–31.

[43] Rahnemai-Azar AA, Rahnemaiazar AA, Naghshizadian R, Kurtz A, Farkas DT. Percutaneous endoscopic gastrostomy: indications, technique, complications and management. World J Gastroenterol 2014;20(24):7739.

[44] Sutcliffe AG, Sebire NJ, Pigott AJ, Taylor B, Edwards PR, Nicolaides KH. Outcome for children born after in utero laser ablation therapy for severe twin-to-twin transfusion syndrome. Br J Obstet Gynaecol 2001;108(12):1246–50.

[45] McAllister WJ, Absalom MJ, Mir K, Shivde S, Anson K, Kirby RS, Lawrence WT, Paterson PJ, Watson GM, Fowler CG. Does endoscopic laser ablation of the prostate stand the test of time? Five-year results from a multicentre randomized controlled trial of endoscopic laser ablation against transurethral resection of the prostate. BJU Int 2000;85(4):437–9.

[46] Mantovani G, Astara G, Manca G, Versace R, Contu P, Carai A. Endoscopic laser ablation as palliative treatment of endobronchial, nonresectable, or recurrent lung cancer: assessment of its impact on quality of life. Clin Lung Cancer 2000;1(4):277–85.

[47] Legrand J, Wenmakers D, Ourak M, Javaux A, Vercauteren T, Deprest J, Ourselin S, Denis K, Vander Poorten E. Handheld active add-on control unit for a cable-driven flexible endoscope. Front Robot AI 2019;6:87.
[48] Feigel A, Sylla P. Transanal endoscopic microsurgery (TEM) and transanal minimally invasive surgery (TAMIS). Tokyo: Springer Japan; 2018. p. 387–435.
[49] Rimonda R, Arezzo A, Arolfo S, Salvai A, Morino M. Transanal minimally invasive surgery (tamis) with Sils port versus transanal endoscopic microsurgery (tem): a comparative experimental study. Surg Endosc 2013;27(10):3762–8.
[50] deBeche Adams T, Nassif G. Transanal minimally invasive surgery. Clin Colon Rectal Surg 2015;28(3):176.
[51] Son D, Gilbert H, Sitti M. Magnetically actuated soft capsule endoscope for fine-needle biopsy. Soft Rob 2020;7(1):10–21.
[52] Olympus Europa SE & CO. KG. Smart and safe endocapsule; 2019. p. 10.
[53] Koulaouzidis A, Iakovidis DK, Karargyris A, Rondonotti E. Wireless endoscopy in 2020: will it still be a capsule? World J Gastroenterol 2015;21(17):5119.
[54] Check-Cap Ltd. C-scan.
[55] Li Y, Wang X, Bao D, Liao Z, Li J, Han X, Wang H, Xu K, Li Z, Stone GW, et al. Optimal antiplatelet therapy for prevention of gastrointestinal injury evaluated by ankon magnetically controlled capsule endoscopy: rationale and design of the opt-peace trial. Am Heart J 2020;228:8–16.
[56] Cianchetti M, Laschi C, Menciassi A, Dario P. Biomedical applications of soft robotics. Nat Rev Mater 2018;3(6):143–53.
[57] Loeve A, Breedveld P, Dankelman J. Scopes too flexible…and too stiff. IEEE Pulse 2010;1 (3):26–41.
[58] United Kingdom National Health System. Laparoscopy (keyhole surgery). https://www.nhs.uk/conditions/laparoscopy/, Accessed: 2021-02-18; 2018.
[59] Dasgupta P, Rose K, Challacombe B. Chapter 1—equipment and technology in robotics. In: Smith JA, Tewari A, editors. Robotics in urologic surgery. Philadelphia: W.B. Saunders; 2008. p. 3–11.
[60] Intuitive. Intuitive for surgeons. https://www.intuitive.com/en-us. Accessed: 2020-11-21.
[61] CMR Surgical Ltd. Versius. https://cmrsurgical.com/versius. Accessed: 2020-12-02.
[62] Senhance. The senhance surgical system; 2020. https://www.senhance.com/us/digital-laparoscopy. Accessed: 2020-10-22.
[63] Inc. Auris Health. Monarch platform. https://www.aurishealth.com/. Accessed: 2020-12-02.
[64] Medrobotics. Flex robotic system technology: how it works; 2020. https://medrobotics.com/gateway/technology-int/. Accessed: 2020-10-22.
[65] Striegel J, Jakobs R, Van Dam J, Weickert U, Riemann JF, Eickhoff A. Determining scope position during colonoscopy without use of ionizing radiation or magnetic imaging: the enhanced mapping ability of the neoguide endoscopy system. Surg Endosc 2011;25 (2):636–40.
[66] Eickhoff A, Van Dam J, Jakobs R, Kudis V, Hartmann D, Damian U, Weickert U, Schilling D, Riemann JF. Computer-assisted colonoscopy (the neoguide endoscopy system): results of the first human clinical trial ("pace study"). Am J Gastroenterol 2007;102(2):261–6.
[67] Rösch T, Adler A, Pohl H, Wettschureck E, Koch M, Wiedenmann B, Hoepffner N. A motor-driven single-use colonoscope controlled with a hand-held device: a feasibility study in volunteers. Gastrointest Endosc Jun 2008;67(7):1139–46.
[68] Yeung C-K, Cheung JLK, Sreedhar B. Emerging next-generation robotic colonoscopy systems towards painless colonoscopy. J Dig Dis 2019;20(4):196–205.

[69] Freehand. http://freehandsurgeon.com. Accessed: 2020-09-30.
[70] Ciftci F. Laparoscopic versus open inguinal hernia repair on patients over 75 years of age. Int J Clin Exp 2015;8(6):10016.
[71] Marcus H, Cundy T, Hughes-Hallett A, Yang G-Z, Darzi A, Nandi D. Endoscopic and keyhole endoscope-assisted neurosurgical approaches: a qualitative survey on technical challenges and technological solutions. Br J Neurosurg 2014;28(5):606–10.
[72] Shergill A, McQuaid K, Rempel D. Ergonomics and gi endoscopy. Gastrointest Endosc 2009;70(1):145–53.
[73] Runciman M, Darzi A, Mylonas G. Soft robotics in minimally invasive surgery. Soft Rob 2019;6(4):423–43.
[74] Wang L, Nurzaman SG, Iida F. Soft-material robotics; 2017.
[75] Laschi C, Cianchetti M. Soft robotics: new perspectives for robot bodyware and control. Front Bioeng Biotechnol 2014;2:3.
[76] Nature. Soft materials. 2020.
[77] Rossiter J, Hauser H. Soft robotics—the next industrial revolution. IEEE Robot Autom Mag 2016;23(3):17–20.
[78] Russo S, Ranzani T, Walsh CJ, Wood RJ. An additive millimeter-scale fabrication method for soft biocompatible actuators and sensors. Adv Mater Technol 2017;2(10):1700135.
[79] Miriyev A, Stack K, Lipson H. Soft material for soft actuators. Nat Commun 2017;8(1):1–8.
[80] Chautems C, Tonazzini A, Boehler Q, Jeong SH, Floreano D, Nelson BJ. Magnetic continuum device with variable stiffness for minimally invasive surgery. Adv Intell Syst 2019;1900086.
[81] Decroly G, Mertens B, Lambert P, Delchambre A. Design, characterization and optimization of a soft fluidic actuator for minimally invasive surgery. Int J Comput Assist Radiol Surg 2020;15(2):333–40.
[82] Zhang B, Fan Y, Yang P, Cao T, Liao H. Worm-like soft robot for complicated tubular environments. Soft Rob 2019;6(3):399–413.
[83] Bernth J, Arezzo A, Liu H. A novel robotic meshworm with segment-bending anchoring for colonoscopy. IEEE Robot Autom Lett 2017;2(3):1718–24.
[84] Ikuta K, Ichikawa H, Suzuki K. Safety-active catheter with multiple-segments driven by micro-hydraulic actuators. In: International conference on medical image computing and computer-assisted intervention. Springer; 2002. p. 182–91.
[85] Gandarias JM, Wang Y, Stilli A, García-Cerezo AJ, Gómez-de Gabriel JM, Wurdemann H. - Open-loop position control in collaborative, modular variable-stiffness-link (vsl) robots. IEEE Robot Autom Lett 2020;5(2):1772–9.
[86] Brancadoro M, Manti M, Grani F, Tognarelli S, Menciassi A, Cianchetti M. Toward a variable stiffness surgical manipulator based on fiber jamming transition. Front Robot AI 2019;6:12.
[87] Shiva A, Stilli A, Noh Y, Faragasso A, De Falco I, Gerboni G, Cianchetti M, Menciassi A, Althoefer K, Wurdemann H. Tendon- based stiffening for a pneumatically actuated soft manipulator. IEEE Robot Autom Lett 2016;1(2):632–7.
[88] Jiang A, Ranzani T, Gerboni G, Lekstutyte L, Althoefer K, Dasgupta P, Nanayakkara T. Robotic granular jamming: does the membrane matter? Soft Rob 2014;1(3):192–201.
[89] Sonar HA, Gerratt AP, Lacour SP, Paik J. Closed-loop haptic feedback control using a self-sensing soft pneumatic actuator skin. Soft Rob 2020;7(1):22–9.
[90] Wurdemann H, Sareh S, Shafti A, Noh Y, Faragasso A, Chathuranga DS, Liu H, Hirai S, Althoefer K. Embedded electro-conductive yarn for shape sensing of soft robotic manipulators. In: International Conference of the IEEE Engineering in Medicine and Biology Society; 2015. p. 8026–9.

[91] Sareh S, Noh Y, Ranzani T, Wurdemann H, Liu H, Althoefer K. A 7.5mm steiner chain fibre-optic system for multi-segment flex sensing. In: IEEE/RSJ international conference on intelligent robots and systems; 2015. p. 2336–41.

[92] Steiger C, Abramson A, Nadeau P, Chandrakasan AP, Langer R, Traverso G. Ingestible electronics for diagnostics and therapy. Nat Rev Mater 2019;4(2):83–98.

[93] Shamsudhin N, Zverev VI, Keller H, Pane S, Egolf PW, Nelson BJ, Tishin AM. Magnetically guided capsule endoscopy. Med Phys 2017;44(8):e91–e111.

[94] Kwack WG, Lim YJ. Current status and research into overcoming limitations of capsule endoscopy. Clin Endosc 2016;49(1):8.

[95] Manfredi L, Capoccia E, Ciuti G, Cuschieri A. A soft pneumatic inchworm double balloon (spid) for colonoscopy. Sci Rep 2019;9(1):1–9.

[96] Bernth JE, Back J, Abrahams G, Lindenroth L, Hayee B, Liu H. Endoscope force generation and intrinsic sensing with environmental scaffolding. In: 2019 International conference on robotics and automation (ICRA). IEEE; 2019. p. 1940–6.

[97] Garbin N, Mamunes A, Sohn D, Hawkins R, Valdastri P, Obstein K. Evaluation of a novel low-cost disposable endoscope for visual assessment of the esophagus and stomach in an ex-vivo phantom model. Endosc Int Open 2019;7(9):E1175.

[98] Garbin N, Stilli A, Shiva A, Fras J, Slawinski P, Obstein K, Althoefer K, Wurdemann H, Valdastri P. Toward a low-cost soft robotic manipulator based on fluid-actuated bellows for gastric cancer screening. In: The Hamlyn symposium on medical robotics; 2017. p. 53.

[99] Ranzani M, Cianchetti T, Gerboni G, Falco ID, Menciassi A. A soft modular manipulator for minimally invasive surgery: design and characterization of a single module. IEEE Trans Robot 2016;32(1):187–200.

[100] Faragasso A, Stilli A, Bimbo J, Noh Y, Liu H, Nanayakkara T, Dasgupta P, Wurdemann H, Althoefer K. Endoscopic add-on stiffness probe for real-time soft surface characterisation in mis. In: International Conference of the IEEE Engineering in Medicine and Biology Society; 2014. p. 6517–20.

[101] Faragasso A, Stilli A, Bimbo J, Wurdemann H, Althoefer K. Multi-axis stiffness sensing device for medical palpation. In: IEEE/RSJ international conference on intelligent robots and systems; 2015. p. 2711–6.

[102] Li J, Li X, Wang J, Xing S, Wang Y, Ren X. Design and evaluation of a variable stiffness manual operating platform for laparoendoscopic single site surgery (less). Int J Med Rob Comput Assisted Surg 2017;13(4), e1797.

[103] Abidi H, Tonazzini A, Cianchetti M, Floreano D, Menciassi A. Low melting point alloy based stiffening of a soft robot. In: International congress of the society for medical innovation and technology; 2017.

[104] Runciman M, Avery J, Zhao M, Darzi A, Mylonas GP. Deployable, variable stiffness, cable driven robot for minimally invasive surgery. Front Robot AI 2020;6:141.

[105] Okamura AM. Haptic feedback in robot-assisted minimally invasive surgery. Curr Opin Urol 2009;19(1):102.

[106] Vallbo B. Properties of cutaneous mechanoreceptors in the human hand-related to touch sensation. Hum Neurobiol 1999;3:3.

[107] Kuebler B, Gruber R, Joppek C, Port J, Passig G, Nagel JH, Hirzinger G. Tactile feedback for artery detection in minimally invasive robotic surgery – preliminary results of a new approach. In: Dössel O, Schlegel WC, editors. World Congress on Medical Physics and Biomedical Engineering, September 7—12 2009, Munich, Germany, pages 299–302, Berlin, Heidelberg. Berlin Heidelberg: Springer; 2009.

[108] Yamamoto T, Abolhassani N, Jung S, Okamura AM, Judkins TN. Augmented reality and haptic interfaces for robot-assisted surgery. Int J Med Rob Comput Assisted Surg 2012;8 (1):45–56.

[109] Koehler M, Usevitch NS, Okamura AM. Model-based design of a soft 3-d haptic shape display. IEEE Trans Robot 2020;36(3):613–28.

[110] Li M, Ranzani T, Sareh S, Seneviratne LD, Dasgupta P, Wurdemann H, Althoefer K. Multi-fingered haptic palpation utilizing granular jamming stiffness feedback actuators. Smart Mater Struct 2014;23(9):095007.

[111] Tenca A, Mustonen H, Lind K, Lantto E, Kolho K-L, Boyd S, Arola J, Jokelainen K, Färkkilä M. The role of magnetic resonance imaging and endoscopic retrograde cholangiography in the evaluation of disease activity and severity in primary sclerosing cholangitis. Liver Int 2018;38(12):2329–39.

[112] North OJ. Medical robotics for use in MRI guided endoscopy; 2013.

[113] Yang G-Z, Bergeless C, Vitiello V. Surgical robotics: the next 25 years successes, challenges, and the road ahead; 2016. https://www. ukras.org/wp-content/uploads/2018/10/UK_RAS_WP_SR25yr_web.pdf Accessed: 2020-09-10.

Chapter 5

Simulators

Andrea Moglia
EndoCAS, Center for Computer Assisted Surgery, University of Pisa, Pisa, Italy

5.1 Introduction

On March 27, 1977, the worst accident in the aviation industry happened when two aircrafts collided in Tenerife island (Spain). As a consequence, 583 persons died. This crash was caused by several concurrent factors, including bad weather conditions, the absence of standards in communication between crews of both aircrafts and staff at the airport, and the nonoptimal team work of the crew of one airplane. There are two lessons learnt from this tragic event: (i) noncrucial errors, if taken singularly, are not fatal; however, they can turn out to be fatal if they happen concurrently; and (ii) the absence of standards to perform effectively and safely routine tasks, even few, can become fatal. The Tenerife accident had a huge impact on improving training in aviation, and the number of victims is small if compared to the number of deaths as a consequence of medical errors, equivalent to 250,000 in the United States alone, as reported by Makary in 2016 [1]. This represents an almost exponential increase over the "To err is human" report, published in 2000 and estimating that each year, 98,000 patients die in the United States [2]. Both of these estimates also include errors of surgical nature. According to an analysis based on a review of 444 cases of surgical malpractices, 52% involved technical errors, defined either as direct manipulative errors or due to bad judgment and inadequate knowledge leading to performance of inappropriate, inadequate, or delayed intervention, e.g., failure to intervene urgently in a patient with a leaking aortic aneurysm [3]. In this study, two-thirds of errors (65%) were manipulative, and the vast majority (84%) involved common routine rather than advanced complex operations. In surgery, patient outcome ultimately reflects on the complex issues related to surgical education. The traditional surgical training approach, known as "See one, do one, teach one" was devised by Dr William Halsted at John Hopkins Hospital at the end of 19th century. This method has an intrinsic limitation, namely the assessment of trainees was subjective. It was challenged by the advent of minimal access surgery (MAS), first laparoscopy and later robot-assisted surgery (RAS).

Endorobotics. https://doi.org/10.1016/B978-0-12-821750-4.00005-0

5.2 Required skills for MAS

MAS requires additional skills to those recognized as essential for conventional open surgery. These include perceptual, visuo-spatial, and hand-eye coordination, which are necessary because the MAS laparoscopic surgeons no longer operate from reality, but from images. Hence, they have to mentally reconstruct 3D anatomy from indirect 2D images displayed on a monitor remote from the actual spatial location of intraabdominal contents of the patients. This imposes a greater perceptual and cognitive processing for valid assessment and interpretation of the spatial relations between the anatomical components of the displayed images, especially as the surgeons cannot see their hands from the monitor when they operate. In addition, precise surgical manipulations by the surgeon are hampered by the reduced tactile feedback transmitted through long (250–300mm) 5-mm in diameter instruments and the greatly reduced degrees of freedom (DOFs) limited to 4 by the invariant point of entry through the abdominal wall. Although it is technically challenging, MAS is a widely recognized and established surgical approach. Laparoscopy has become the gold standard for appendicitis and cholecystitis, while the number of worldwide RAS procedures has increased almost exponentially, exceeding 1 million [4].

5.3 Human factors

Medicine, and in particular surgery, may never have considered human factors, if laparoscopy had not been developed [5]. Human factors is a multidisciplinary field of science studying human-machine interface. Sometimes, it is called ergonomics [5]. However, this is not correct since ergonomics is a subset of human factors [5]. The purpose of human factors is to improve operational performance, safety, and costs [5]. There are five different approaches to integrate human factors: equipment design, task design, environment design, training the individuals, and selecting the individuals [5]. Human factors have to be considered for the design of surgical equipment where there is human-machine interaction, like MAS. Consequently, human factors are relevant also in the teaching and practice of the requested skills in MAS (cognitive, technical, and interpersonal) [5].

5.4 Training: aviation vs surgery

The training paradigm in surgery has been traditionally different from those of other highly qualified fields, like aviation industry, where structured competence-based programs called curricula, and objective assessment of the skills during training are in established use. A key component in aviation training has been the simulator. The first flight simulator was devised by Edwin Link and was capable to simulate a safe landing in a thick fog in 1934 [6]. In surgery, several types of simulators have been devised over the past decades.

5.5 Types of surgical simulators

5.5.1 Animal tissues

Animal tissues such as pieces of chicken and pork have been successfully used for surgical training. They can be used for the training of different tasks, including knot tying, suturing, and vessels dissection. Advantages include the possibility to practice with real surgical tools, their availability, and their low prices. Additionally, they expose trainees to work with real tissue. However, they have some disadvantages. First, the tissue samples can be used for a limited number of times. Second, they are difficult to assess in terms of which metrics to use and also since they need the presence of a tutor. Third, animal tissue models require special facilities for hygiene and safety reasons [5].

5.5.2 Synthetic models

Synthetic models have been increasingly adopted for training after the advent of laparoscopy. As animal tissues, they are readily available, look like the anatomy they simulate, and do not require specific facilities in terms of animal tissues. They share with animal tissues two drawbacks: the finite number of times they can be used and the difficulty to assess them with critical metrics. However, they do not respond as human tissue or animal tissue to needle and thread during tasks like suturing [5].

5.5.3 Cadaver tissues

Human cadavers have always been an important component in medicine education by virtue of their high realism. In particular, in surgery, they allow trainees to understand the three-dimensional organization of anatomical structures as well as their dimensions, densities, and the strength of various tissues. As animal tissues and synthetic models, they allow learners to use actual surgical instruments. However, it is not easy to implement surgical training based on cadavers. In addition to the disadvantages of animal tissues, their use is regulated by laws, thus their use is not allowed in all countries; their availability is limited; they are expensive to embalm, store, maintain, and dispose [5].

5.5.4 Virtual reality simulators

The first virtual reality (VR) surgical simulator was introduced in late 1980s, whereas the first VR surgical simulator for RAS was introduced in 2010 [7]. VR surgical simulators enable to track learners progresses by objectively and automatically measuring a set of metrics, e.g., time of execution and economy of motion. In particular, VR simulators for RAS replicate the ergonomics of the master console of da Vinci surgical system by Intuitive Surgical (Sunnyvale, CA, United States) so that trainees can become familiar with the main controls including camera, clutching, and energy pedals. Currently, there are four VR simulators for RALS used for training: the Robotic Surgical System (Simulated Surgical Systems, San Jose, CA, United States), the dV-Trainer (Mimic, Seattle, WA, United States), the da Vinci Skills Simulator (Intuitive Surgical), and the recently introduced RobotiX Mentor (3D Systems, Simbionix Products, Cleveland, OH, United States) [8]. There is an additional one, the Surgical Education Platform (SEP; SimSurgery, Oslo, Norway), which, however, has a limited adoption. In the next section, the main four platforms are described.

5.6 VR simulators for RALS

5.6.1 da Vinci Skills Simulator

The da Vinci Skills Simulator (dVSS) is a portable hardware unit, which can be mounted on the back of the master console of da Vinci surgical system (Fig. 5.1).

For this reason, it is also known as "Backpack." Currently, there are two versions: one for da Vinci Si and the other for da Vinci Xi. The two versions of dVSS share most exercises. The company recently released SimNow, a software service enabling simulation of surgical procedures, e.g., prostatectomy and lobectomy. The advantage of dVSS over other VR simulators for RAS is that it runs on the original master console of da Vinci surgical system, thus offering the highest fidelity in terms of ergonomics and 3D vision. However, the price of the full system (dVSS and master console) is by far the highest in VR simulation for RAS. Consequently, not all simulation centers can afford its purchase. The typical scenario is to use one master console for dVSS only when it is not used by surgeons for procedures. For this reason, other companies started manufacturing their own standalone VR simulators for RAS, capable to run without the need of the master console. The dVSS Si assesses learners' performances by means of an overall score in percentage form, while the dVSS Xi allows to choose between this way and a proficiency-based system.

5.6.2 dV-Trainer

The dV-Trainer is a standalone VR simulator, whose hardware emulates the master console of da Vinci surgical system (Fig. 5.2).

Since Mimic developed most exercises of early versions of dVSS, dV-Trainer and dVSS share many of them [7]. The dV-Trainer consists of the following units: an upper part integrating the 3D viewer and master control, a foot switch pedal, and a computer [7]. An external screen allows 2D visualization of the 3D virtual scene generated by the simulator. It can be used by a tutor supervising the trainees. The simulator can mimic both Si and Xi versions of da Vinci surgical system. Mimic's Maestro AR (augmented reality) offers simulated procedures enabling interaction between the trainee's virtual robotic instruments and real surgical videos. At present, four procedures are available:

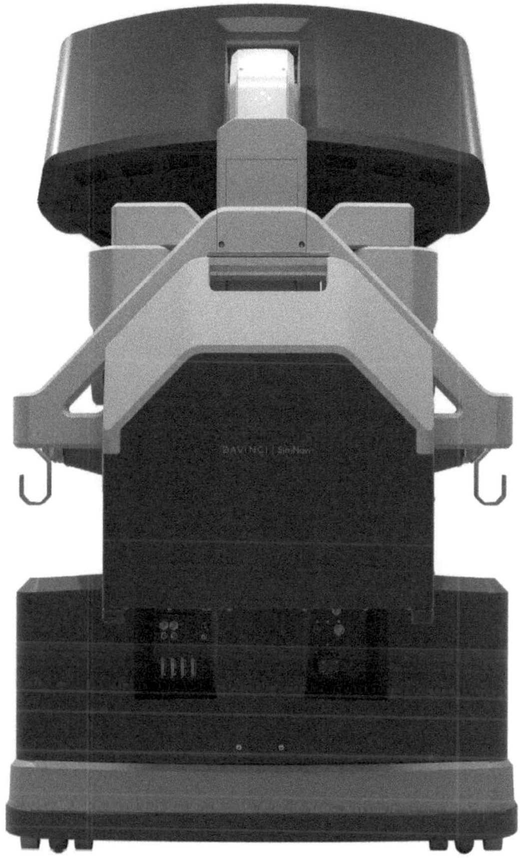

FIG. 5.1 da Vinci Skills Simulator mounted on the back of da Vinci Surgical System. © [2020] Intuitive Surgical, Inc.

partial nephrectomy, hysterectomy, inguinal hernia repair, and prostatectomy [7]. The procedures are split into component tasks which provide learners with instruction and guidance to identify anatomy and improve their technical skills. During the tasks, the trainees' knowledge on the procedural steps is evaluated through multiple choice questions [7].

dV-Trainer can also be connected to a secondary simulator, Xperience Team Trainer (XTT) for team training (Fig. 5.3).

More recently, Mimic has launched a portable simulator called Flex VR (Fig. 5.4).

With dVSS Xi, assessment with dV-trainer can be performed by using a percentage or a proficiency-based system.

FIG. 5.2 dV-Trainer simulator. © [2020] Mimic Simulation Inc.

5.6.3 Robotic Surgical Simulator

Like the dV-Trainer, the Robotic Surgical Simulator (RoSS) is a standalone simulator (Fig. 5.5).

The simulator consists of a 3D viewer, commercial force feedback devices as master controls, foot pedals, and an external monitor for the instructor to view. In addition to exercises on basic skills, RoSS offers Hands-on Surgical Training (HoST), video-based surgical modules, where the learners are guided through the movements necessary to complete a real surgical procedure, e.g., prostatectomy and hysterectomy. HoST exploits the force feedback capabilities of the master controls to push and pull the trainees' hands to follow the simulated instruments [7]. The company recently launched a portable version called RoSS Lite II (Fig. 5.6).

Concerning assessment of trainees, RoSS shows a numeric score and its position relative to a pass threshold [7].

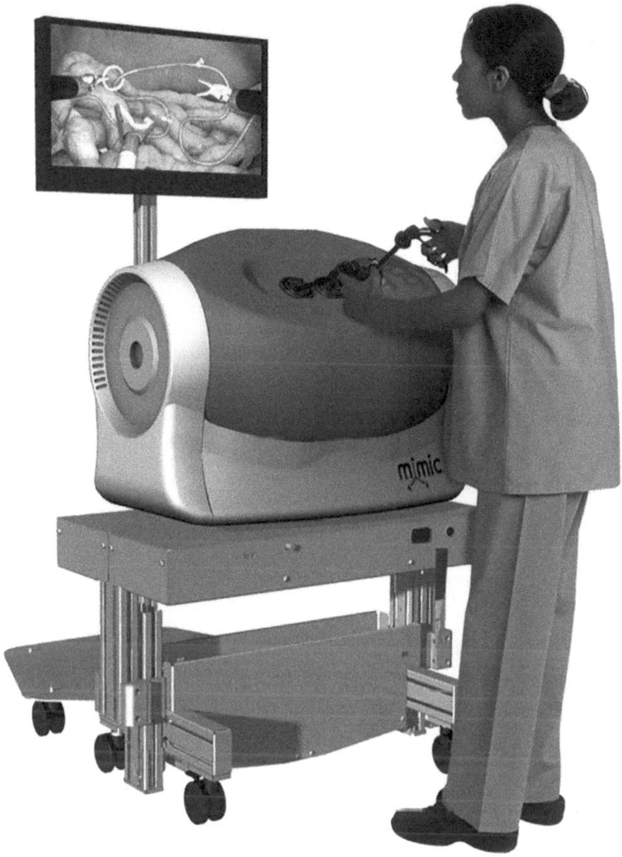

FIG. 5.3 XTT simulator. © [2020] Mimic Simulation Inc.

5.6.4 RobotiX Mentor

RobotiX Mentor is a standalone VR simulator for RAS. It consists of two mobile carts: one replicating the master console of da Vinci surgical system, free-floating master controls and foot pedals; the other hosts the computer and an external monitor for tutoring purposes (Fig. 5.7) [7].

In particular, the master controls are tethered to the armrest by a bundled electronic cable [7]. RobotiX Mentor shares with dVSS some exercises on suturing. It also provides fully simulated VR procedures, including: prostatectomy, hysterectomy, lobectomy, inguinal hernia, and hemicolectomy. RobotiX Mentor can be connected to an external simulator for team training between primary and secondary operators (Fig. 5.8).

RobotiX Mentor measures metrics of learners' performances without benchmark.

FIG. 5.4 Flex VR simulator. © [2020] Mimic Simulation Inc.

5.7 Evidence from the published literature

Most of the conducted studies on VR simulators concern validity. There are six different types of validity. Their definition is reported in Table 5.1.

Several studies assessed face validity on the VR simulators. The realism of the simulators was generally judged good concerning 3D graphics, and instrument controls [9–28]. In terms of content validity, the simulators were rated well by considering their usefulness as training tools and their integration in residency programs [10,11,13–17,19–24]. Face and content validity are assessed subjectively by asking subjects to answer a questionnaire. A criticism to these studies is that some of them also included participants without experience in RAS (novices) to assess the simulator. This adds some degrees of bias to those publications. Another remark concerns the definition of expert surgeons. In some studies, experts are defined in terms of total number of cases, while in others, number of cases per year is used [8]. This is common to all studies,

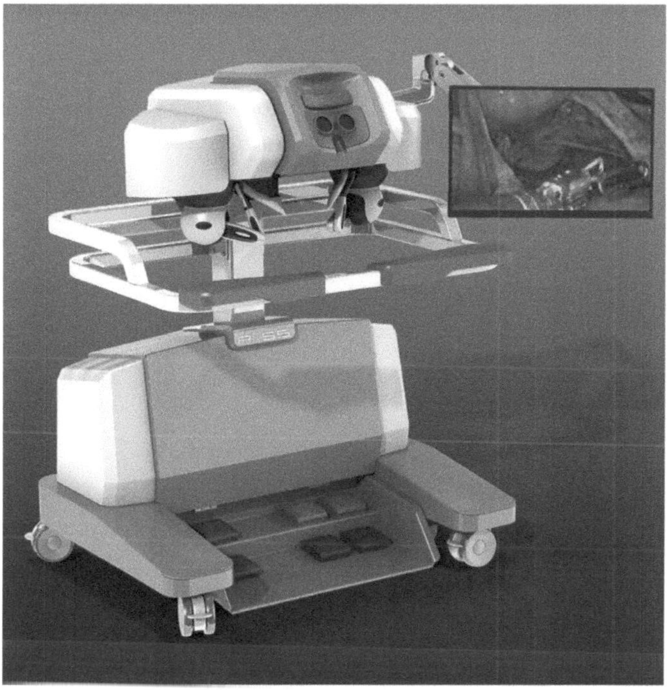

FIG. 5.5 RoSS simulator. © [2020] Simulated Surgical System, LLC.

which makes it difficult to compare reports on the same topic. The other types of validity are instead based on objective assessment since they measure performances of participants through the metrics of the simulators. Several studies were published on construct validity [9–11,13–16,18–23,29–35]. They generally report the capability of the simulator to distinguish between different levels of experiences. However, these studies have a limited number of participants. Moreover, the enrolled cohorts differ among studies, with the majority comparing two groups with different experience, while others comparing three or more groups [8]. The number of tasks assessed in the studies varies substantially for each simulator, thus making it difficult to establish which exercises are more valid to distinguish participants with different levels of experience [8]. The last issue on studies on construct validity is the variable assessment methods: some studies pooled overall score of all tasks, while others analyzed each task separately. Pooling results of the exercises do not give an accurate estimation of the capability of the VR simulator to distinguish between levels of experience, since it inevitably introduces bias by mixing results for tasks of different levels of difficulty [8]. There are fewer reports on concurrent validity since this kind of studies has a more complicated design, which compares performances of participants at the simulator with those performing some tasks (either on synthetic models or on animal tissue) with a real da Vinci surgical system [14,17,18,20,24,36]. In this case, mixed results were reported [8]. There is only

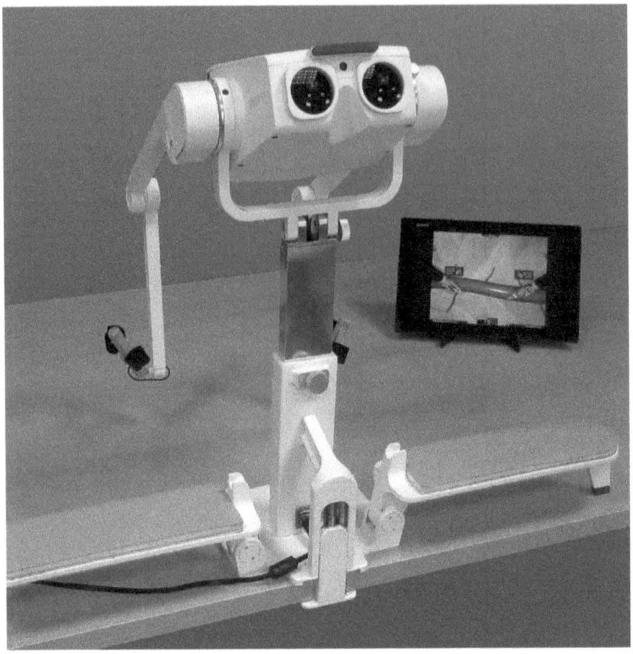

FIG. 5.6 RoSS Lite II simulator. [2020] Simulated Surgical System, LLC.

one study on predictive validity of the VR simulator for RAS. Although it involved a small number of participants, it found a high correlation between the overall score during the initial test on the simulator and the final test on a da Vinci robot [36]. Discriminant validity was reported in only two studies trying to quantify the size of three groups with different levels of innate aptitude for surgery [37,38]. Although each of them used a different simulator, the results are remarkably similar: around 6% of gifted medical students with high level, and around 11.0% of low level of innate ability for surgery [37,38]. Evidence showed that previous exposure to both videogames and playing musical instruments in medical students has more influence on VR simulator scores than experience in only videogames or musical instruments [39].

Besides tasks on basic skills, few validity studies were conducted on simulated procedures. For instance, face, content, construct, and concurrent validity on Maestro AR partial nephrectomy of dV-Trainer were investigated in one study, while construct and face validity on lobectomy procedure of RobotiX Mentor were reported in another study [20,40].

5.8 Curriculum

The aviation industry adopted competency-based curricula requiring pilots to reach defined benchmark performance criteria before moving to the next level

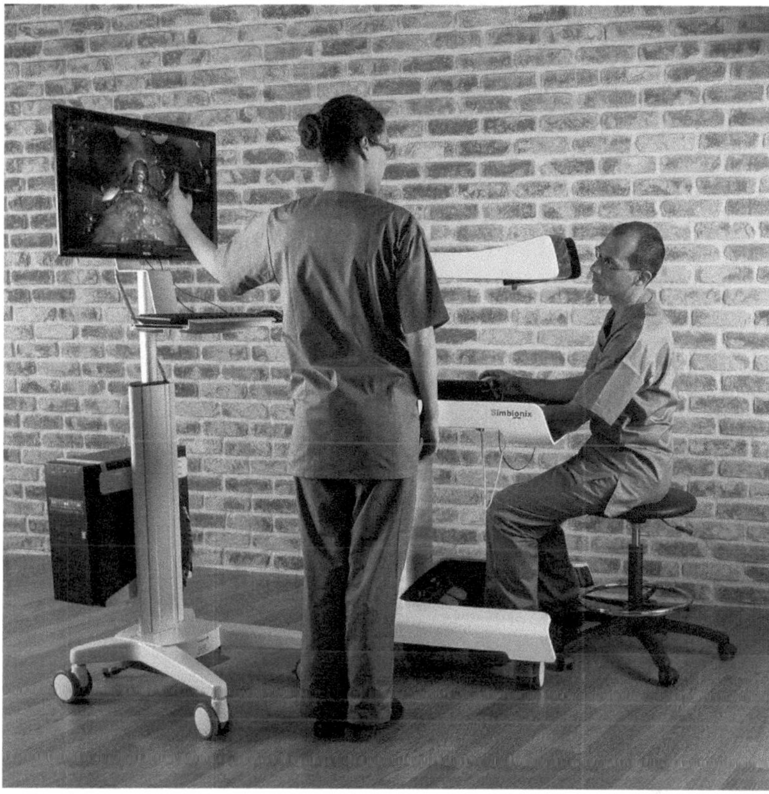

FIG. 5.7 Simbionix RobotiX Mentor simulator. © [2020] 3D Systems.

FIG. 5.8 Simbionix RobotiX Mentor simulator for team training. © [2020] 3D Systems.

TABLE 5.1 Definition of validity applied to surgical simulation.

Validity type	Meaning
Face	Realism
Content	Whether the simulator is useful as a training tool
Construct	Ability to differentiate performance between subjects with different levels of experience, e.g., experts and novices
Concurrent	A measure that reflects whether test scores between different instruments or simulators are in broad agreement
Discriminant	Capability of a simulator to differentiate ability levels within a group with similar experience, e.g., medical students
Predictive	Ability to select those who will not be able to perform surgical operations well, despite training, and vice versa, that is, those who will excel

of training [8]. A simulation-based curriculum should incorporate the three pillars of a comprehensive training program: proficiency, deliberate and distributed practice [41]. First, a curriculum should train learners from novice to proficiency level, according to the Dreyfus and Dreyfus model on surgical skills acquisition [5]. The first time proficiency used as the benchmark on surgical simulation was within the so called "Yale study" (from the name of the institute where it was conducted). A randomized control trial (RCT) involving surgical residents demonstrated that the experimental group trained until reaching proficiency at a VR simulator for laparoscopy outperformed the control group, following traditional training, in terms of time and number of errors during a cholecystectomy procedure on real patient [42]. Second, a well-designed curriculum should also satisfy the five conditions identified by Ericsson for deliberate practice: a defined task to practice, provision of detailed and immediate feedback, motivation to improve, variable level of difficulty for the chosen task, and simple opportunity for repetition and gradual refinement. Third, surgical training sessions should be distributed over time rather than concentrated massively [41]. A curriculum should be validated by a RCT to assess its robustness and skills retention after a follow-up [41].

Additionally, it should deliver training efficiently in view of the limitations on working hours of residents, set at 80 h/week in the United States and 48 h/week in the UK and EU countries. A curriculum may eventually become an integral component of a high-stake examination for certification on competence of surgical skills, following Fundamentals of Laparoscopic Surgery (FLS), and Fundamentals of Endoscopic Surgery (FES), which became training requirements for general surgery certification.

Over the past few years, some initiatives have been carried out to establish competence-based curricula for RAS, in most cases with the endorsement of surgical societies. These curricula concern basic skills, a single specialty, or a specific procedure.

5.8.1 Fundamentals of Robotic Surgery

Fundamentals of Robotic Surgery (FRS) is a multispecialty curriculum on fundamental cognitive and technical skills for RALS. It was designed during four consensus conferences by 66 subject matter experts from several surgical societies [43]. At the end of the conferences, seven tasks addressing 25 technical skills, the contents of online course on cognitive skills, the development of the simulator model for skills training, and the metrics for performance assessment were defined [43]. A multicenter RCT, involving 12 simulation centers accredited by the American College of Surgeons Accredited Education Institutes, assessed three experimental groups (trained at dVSS, dV-Trainer, or FRS physical dome (Fig. 5.9)) vs a control group in a test on avian tissue model.

Results showed that residents and fellows performed tasks faster (dVSS and FRS dome groups) and with fewer errors than control one [43].

5.8.2 Robotic Training Network

The Robotic Training Network curriculum involves over 50 centers in the United States and aims to standardize the robotic surgical curriculum and education for residents in gynecology and general surgery [8].

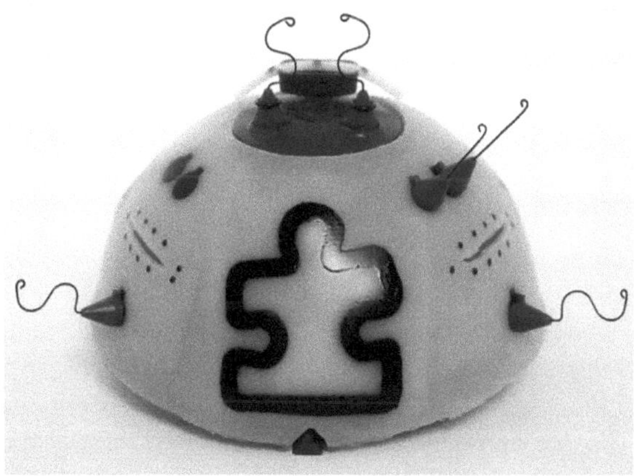

FIG. 5.9 FRS Dome. © [2020] AdventHealth Nicholson Center.

5.8.3 Society of European Robotic Gynaecological Surgery

The curriculum by the Society of European Robotic Gynaecological Surgery (SERGS) consists of three stages: (i) a didactic introduction into correct docking, use of instruments, and solving technical problems; (ii) VR training, and wet-lab; and (iii) procedural training where trainees perform moderate to complex gynecological procedures [44]. A pilot study on four fellows reported good or acceptable technical quality [44].

5.8.4 European Society of Thoracic Surgeons and European Association for Cardio-Thoracic Surgery

Concerning thoracic surgery, the result of a Delphi survey involving members of European Society of Thoracic Surgeons (ESTS) and European Association for Cardio-Thoracic Surgery (EACTS) showed consensus on a basic curriculum, and an advanced one on the procedure [45]. The basic one should consist of VR simulation, dry-lab, wet lab, and case observation. The advanced one should include e-learning on index procedure, simulation, modular procedure training at the console before training of the full procedure [45].

5.8.5 European Association of Urology Robotic Urology Section

The European Association of Urology Robotic Urology Section (ERUS) developed a curriculum on radical prostatectomy. It consists of cognitive skills, technical skills (dry lab, wet lab, and VR simulation), case observation, and a fellowship program on performing tasks of a procedure before undertaking a whole procedure (modular training using the dual console) [46]. A pilot study reported that eight out of 10 fellows were judged by their mentors able to perform a RAS radical prostatectomy independently, safely, and efficiently on completion of the ERUS curriculum, and three were considered able to perform a complex RAS radical prostatectomy independently, safely, and effectively [46].

5.8.6 Morristown protocol

The Morristown protocol is a structured training program including cognitive skills, technical skills at a VR simulator for RAS and use of a da Vinci surgical system on pig models, and assessment during a RAS procedure on patient. Evidence showed that surgeons following Morristown protocol performed better than a control group during a hysterectomy in terms of time and volume of blood loss [47]. Although the results showed skills transfer of this curriculum from simulation setting to operating room, the study is not randomized and is imbalanced (10 surgeons in the experimental group vs four in the control one).

5.8.7 Curriculum on pancreaticoduodenoctomy

A modular curriculum on pancreaticoduodenoctomy was developed at the University of Pittsburgh Medical Center. It includes VR simulation, inanimate model, and operative curriculum [48]. A study on the first step showed improved performances of 17 fellows on tasks at VR simulator and during dry-lab with actual da Vinci surgical systems [48].

5.9 Future research

Although VR simulators for RAS measure several metrics on learner's performances, they do not assess other important parameters, e.g., bimanual dexterity. Global Evaluative Assessment of Robotic Skills (GEARS) is a global rating scale evaluating six skills for RAS using a 5-point Likert scale: depth perception, bimanual dexterity, efficiency, force sensitivity, autonomy, and robotic control [49]. Although assessment by means of GEARS is subjective since it is used by expert reviewers to rate videos, it can complement assessment through the VR simulator. However, analysis by GEARS takes long time. Crowdsourcing is a problem-solving approach in which a specific task is completed more efficiently in terms of time by a large cohort of nonspecialized individuals compared to a small group of experts [50]. Amazon Mechanical Turk is an example of an online crowdsourcing platform. A study showed that performance scores on GEARS of a crowd of 30 nonexperts agreed with those of three experts on two tasks for RAS [51]. More recently, a linear predictive model on the GEARS score was developed using overall score and metrics for two exercises of dVSS and dV-Trainer [52].

Surgery stands to gain from the explosion of artificial intelligence (AI), and RAS is no exception [53]. Machine learning, in particular deep neural networks, may be helpful to automate analysis of surgical videos. However, programming neural networks for video analysis requires advanced competencies in computer science. To address this issue, major companies including Amazon, IBM, Google, and Microsoft offer AI cloud services. A preliminary study showed feasibility of using services by Google and IBM to apply convolutional neural networks on the analysis of videos of tasks of the dV-Trainer simulator to classify performances into novices and experts [54].

VR simulators for RAS are expensive tools, starting from approximately 100,000 dollars, excluding additional software modules, and contracts for long-term support and maintenance. The price of dVSS is even higher since it requires a dedicated console of da Vinci surgical system. Additionally, for each VR simulator, a tutor is required to instruct trainees to understand how to use its hardware and software. A sort of virtual tutor would allow trainers to utilize the VR simulator without a proctor. An intelligent tutor system to train the cognitive process and procedural steps on camera targeting and suture FRS tasks was developed with the aim to integrate it into one of the existing VR simulators for RAS [55].

Several simulation-based curricula on RAS have been proposed and the results of their validation have been reported above. However, there is still the need of a proof in a RCT that trainees who reached proficiency at a VR simulator perform better than their colleagues, following traditional training, during intraoperative assessment on patients, as demonstrated for laparoscopy [42]. There is also room to assess maintenance of acquired skills after follow-ups.

Finally, since VR simulators for RAS are expensive, it would be essential to assess their cost-effectiveness, as done for flight simulators, to justify their purchase [8]. Transfer effectiveness ratio was employed by aviation industry to compute the saved time, and consequently money, when using a flight simulator rather than an actual aircraft to train pilots to become competent. In aviation, 1 h on a flight simulator saves approximately 40–60 min of real flying time, while a RCT showed that one hour at a VR simulator for laparoscopy saved more than 2 h of traditional training [56,57]. However, at present, there is no published report on transfer effectiveness ratio for the VR simulator for RAS.

References

[1] Makary MA, Daniel M. Medical error—the third leading cause of death in the US. BMJ 2016;353:i2139.

[2] Institute of Medicine (US) Committee on Quality of Health Care in America. To err is human: building a safer health system. Washington, DC: National Academies Press (US); 2000.

[3] Regenbogen SE, Greenberg CC, Studdert DM, Lipsitz SR, Zinner MJ, Gawande AA. Patterns of technical error among surgical malpractice claims: an analysis of strategies to prevent injury to surgical patients. Ann Surg 2007;246:705–11.

[4] Soper NJ, Stockmann PT, Dunnegan DL, Ashley SW. Laparoscopic cholecystectomy. The new 'gold standard'? Arch Surg 1992;127:917–21.

[5] Gallagher AG, O'Sullivan GC. Fundamentals of surgical simulation. London, UK: Springer Verlag; 2011.

[6] Levine AI, De Maria S, Schwartz AD, Sim AJ. The comprehensive textbook of healthcare simulation. Springer; 2014.

[7] Julian D, Tanaka A, Mattingly P, Truong M, Perez M, Smith R. A comparative analysis and guide to virtual reality robotic surgical simulators. Int J Med Robot 2018;14, e1874.

[8] Moglia A, Ferrari V, Morelli L, Ferrari M, Mosca F, Cuschieri A. A systematic review of virtual reality simulators for robot-assisted surgery. Eur Urol 2016;69:1065–80.

[9] Lendvay TS, Casale P, Sweet R. Initial validation of a virtual-reality robotic simulator. J Robot Surg 2008;2:145–9.

[10] Kenney PA, Wszolek MF, Gould JJ, et al. Face, content, and construct validity of dV-trainer, a novel virtual reality simulator for robotic surgery. Urology 2009;73:1288–92.

[11] Sethi AS, Peine WJ, Mohammadi Y, et al. Validation of a novel virtual reality robotic simulator. J Endourol 2009;23:503–8.

[12] Korets R, Mues AC, Graversen JA, et al. Validating the use of the Mimic dV-trainer for robotic surgery skill acquisition among urology residents. Urology 2011;78:1326–30.

[13] Hung AJ, Zehnder P, Patil MB, et al. Face, content and construct validity of a novel robotic surgery simulator. J Urol 2011;186:1019–24.

[14] Lee JY, Mucksavage P, Kerbl DC, et al. Validation study of a virtual reality robotic simulator—role as an assessment tool? J Urol 2012;187:998–1002.
[15] Liss MA, Abdelshehid C, Quach S, et al. Validation, correlation, and comparison of the da Vinci trainer(™) and the daVinci surgical skills simulator(™) using the Mimic(™) software for urologic robotic surgical education. J Endourol 2012;26:1629–34.
[16] Perrenot C, Perez M, Tran N, et al. The virtual reality simulator dV-Trainer(®) is a valid assessment tool for robotic surgical skills. Surg Endosc 2012;26:2587–93.
[17] Egi H, Hattori M, Tokunaga M, et al. Face, content and concurrent validity of the Mimic® dV-Trainer for robot-assisted endoscopic surgery: a prospective study. Eur Surg Res 2013;50:292–300.
[18] Schreuder HW, Persson JE, Wolswijk RG, et al. Validation of a novel virtual reality simulator for robotic surgery. ScientificWorldJournal 2014;2014:507076. https://doi.org/10.1155/2014/507076.
[19] Kang SG, Cho S, Kang SH, et al. The Tube 3 module designed for practicing vesicourethral anastomosis in a virtual reality robotic simulator: determination of face, content, and construct validity. Urology 2014;84(2):345–50.
[20] Hung AJ, Shah SH, Dalag L, Shin D, Gill IS. Development and validation of a novel robotic procedure specific simulation platform: partial nephrectomy. J Urol 2015;194(2):520–6. https://doi.org/10.1016/j.juro.2015.02.2949.
[21] Alzahrani T, Haddad R, Alkhayal A, et al. Validation of the da Vinci Surgical Skill Simulator across three surgical disciplines: a pilot study. Can Urol Assoc J 2013;7:E520–9.
[22] Kelly DC, Margules AC, Kundavaram CR, et al. Face, content, and construct validation of the da Vinci Skills Simulator. Urology 2012;79:1068–72.
[23] Lyons C, Goldfarb D, Jones SL, et al. Which skills really matter? Proving face, content, and construct validity for a commercial robotic simulator. Surg Endosc 2013;27:2020–30.
[24] Ramos P, Montez J, Tripp A, et al. Face, content, construct and concurrent validity of dry laboratory exercises for robotic training using a global assessment tool. BJU Int 2014;113:836–42.
[25] Seixas-Mikelus SA, Kesavadas T, Srimathveeravalli G, et al. Face validation of a novel robotic surgical simulator. Urology 2010;76:357–60.
[26] Chowriappa A, Raza SJ, Fazili A, et al. Augmented-reality-based skills training for robot-assisted urethrovesical anastomosis: a multi-institutional randomised controlled trial. BJU Int 2015;115:336–45.
[27] Whittaker G, Aydin A, Raison N, et al. Validation of the RobotiX Mentor robotic surgery simulator. J Endourol 2016;30(3):338–46. https://doi.org/10.1089/end.2015.0620.
[28] Alshuaibi M, Perrenot C, Hubert J, Perez M. Concurrent, face, content, and construct validity of the RobotiX Mentor simulator for robotic basic skills. Int J Med Robot 2020;16(3). https://doi.org/10.1002/rcs.2100, e2100.
[29] Seixas-Mikelus SA, Stegemann AP, Kesavadas T, et al. Content validation of a novel robotic surgical simulator. BJU Int 2011;107:1130–5.
[30] Finnegan KT, Meraney AM, Staff I, et al. da Vinci Skills Simulator construct validation study: correlation of prior robotic experience with overall score and time score simulator performance. Urology 2012;80:330–5.
[31] Hung AJ, Jayaratna IS, Teruya K, et al. Comparative assessment of three standardized robotic surgery training methods. BJU Int 2013;112:864–71.
[32] Connolly M, Seligman J, Kastenmeier A, et al. Validation of a virtual reality-based robotic surgical skills curriculum. Surg Endosc 2014;28:1691–4.
[33] Chowriappa AJ, Shi Y, Raza SJ, et al. Development and validation of a composite scoring system for robot-assisted surgical training—the Robotic Skills Assessment Score. J Surg Res 2013;185:561–9.

[34] Raza SJ, Froghi S, Chowriappa A, et al. Construct validation of the key components of Fundamental Skills of Robotic Surgery (FSRS) curriculum—a multi-institution prospective study. J Surg Educ 2014;71:316–24.

[35] Balasundaram I, Aggarwal R, Darzi A. Short-phase training on a virtual reality simulator improves technical performance in tele-robotic surgery. Int J Med Robot 2008;4:139–45.

[36] Hung AJ, Patil MB, Zehnder P, et al. Concurrent and predictive validation of a novel robotic surgery simulator: a prospective, randomized study. J Urol 2012;187:630–7.

[37] Moglia A, Ferrari V, Morelli L, et al. Distribution of innate ability for surgery amongst medical students assessed by an advanced virtual reality surgical simulator. Surg Endosc 2014;28:1830–7.

[38] Moglia A, Morelli L, Ferrari V, Ferrari M, Mosca F, Cuschieri A. Distribution of innate psychomotor skills recognized as important for surgical specialization in unconditioned medical undergraduates. Surg Endosc 2018;32(10):4087–95.

[39] Moglia A, Perrone V, Ferrari V, Morelli L, Boggi U, Ferrari M, Mosca F, Cuschieri A. Influence of videogames and musical instruments on performances at a simulator for robotic surgery. Minim Invasive Ther Allied Technol. 2017 Jun;26(3):129–134.

[40] Whittaker G, Aydin A, Raveendran S, Dar F, Dasgupta P, Ahmed K. Validity assessment of a simulation module for robot-assisted thoracic lobectomy. Asian Cardiovasc Thorac Ann 2019;27(1):23–9.

[41] Zevin B, Levy JS, Satava RM, Grantcharov TP. A consensus-based framework for design, validation, and implementation of simulation-based training curricula in surgery. J Am Coll Surg 2012;215:580–6.

[42] Seymour NE, Gallagher AG, Roman SA, et al. Virtual reality training improves operating room performance: results of a randomized, double-blinded study. Ann Surg 2002;236(4):458–64.

[43] Satava RM, Stefanidis D, Levy JS, et al. Proving the effectiveness of the fundamentals of robotic surgery (FRS) skills curriculum: a single-blinded, multispecialty, multi-institutional randomized control trial. Ann Surg 2020;272(2):384–92. https://doi.org/10.1097/SLA.0000000000003220.

[44] Rusch P, Kimmig R, Lecuru F, et al. The Society of European Robotic Gynaecological Surgery (SERGS) Pilot Curriculum for robot assisted gynecological surgery. Arch Gynecol Obstet 2018;297(2):415–20. https://doi.org/10.1007/s00404-017-4612-5.

[45] Veronesi G, Dorn P, Dunning J, et al. Outcomes from the Delphi process of the Thoracic Robotic Curriculum Development Committee. Eur J Cardiothorac Surg 2018;53(6):1173–9. https://doi.org/10.1093/ejcts/ezx466.

[46] Volpe A, Ahmed K, Dasgupta P, et al. Pilot validation study of the European Association of Urology robotic training curriculum. Eur Urol 2015;68:292–9.

[47] Culligan P, Gurshumov E, Lewis C, Priestley J, Komar J, Salamon C. Predictive validity of a training protocol using a robotic surgery simulator. Female Pelvic Med Reconstr Surg 2014;20(1):48–51. https://doi.org/10.1097/SPV.0000000000000045.

[48] Hogg ME, Tam V, Zenati M, et al. Mastery-based virtual reality robotic simulation curriculum: the first step toward operative robotic proficiency. J Surg Educ 2016. https://doi.org/10.1016/j.jsurg.2016.10.015.

[49] Goh AC, Goldfarb DW, Sander JC, Miles BJ, Dunkin BJ. Global evaluative assessment of robotic skills: validation of a clinical assessment tool to measure robotic surgical skills. J Urol 2012;187(1):247–52. https://doi.org/10.1016/j.juro.2011.09.032.

[50] Dai JC, Lendvay TS, Sorensen MD. Crowdsourcing in surgical skills acquisition: a developing technology in surgical education. J Grad Med Educ 2017;9(6):697–705. https://doi.org/10.4300/JGME-D-17-00322.1.

[51] White LW, Kowalewski TM, Dockter RL, Comstock B, Hannaford B, Lendvay TS. Crowd-sourced assessment of technical skill: a valid method for discriminating basic robotic surgery skills. J Endourol 2015;29(11):1295–301. https://doi.org/10.1089/end.2015.0191.

[52] Dubin AK, Julian D, Tanaka A, Mattingly P, Smith R. A model for predicting the GEARS score from virtual reality surgical simulator metrics. Surg Endosc 2018;32:3576–81.

[53] Moglia A, Georgiou K, Georgiou E, Satava RM, Cuschieri A. A systematic review on artificial intelligence in robot-assisted surgery. Int J Surg. 2021 Oct 22;95:106151. https://doi.org/10.1016/j.ijsu.2021.106151. Epub ahead of print.

[54] Smith R, Julian D. Proceedings of interservice/industry training, simulation, and education conference, December 2–6, 2019, Orlando, FL, United States; 2019.

[55] Julian D, Smith R. Developing an intelligent tutoring system for robotic-assisted surgery instruction. Int J Med Robot 2019;15(6). https://doi.org/10.1002/rcs.2037, e2037.

[56] Fletcher JD, Orlansky J. Recent studies on the cost-effectiveness of military training in TTCP countries. IDA Paper P-1896. Alexandria, VA: Institute for Defense Analyses; 1986.

[57] Orzech N, Palter VN, Reznick RK, Aggarwal R, Grantcharov TP. A comparison of 2 ex vivo training curricula for advanced laparoscopic skills: a randomized controlled trial. Ann Surg 2012;255:833–9.

Part 2

Materials and engineering design

Chapter 6

Smart materials for mini-actuators

Gianluca Rizzello[a] and Paul Motzki[a,b]
[a]*Department of Systems Engineering, Saarland University, Saarbrücken, Germany,* [b]*Center for Mechatronics and Automation Technology (ZeMA gGmbH), Saarbrücken, Germany*

6.1 Introduction

The increasing need for actuator systems which are more energy-efficient, lightweight, and less cumbersome has recently generated a growing interest in the field of smart materials. This term refers to those transducers such as shape memory materials, electroactive polymers, piezoelectric ceramics, or magnetorheological fluids, which are able to change their mechanical properties (e.g., geometry, force, stiffness, damping) in response to an external stimulus (e.g., electrical, magnetic, thermal, chemical, optical). What makes smart materials particularly attractive is their unique combination of features, which often include high-energy efficiency, low power consumption, lightweight, high flexibility, silent operations, and the possibility of working as actuators and sensors at the same time (this operation mode is often referred to as self-sensing). Smart materials allow to design mechatronic devices which are, in many cases, capable of outperforming conventional actuator systems based on electromagnetic or fluidic transducers. A number of smart material systems have been presented in the recent literature, with examples in various application fields such as industry [1,2], automotive [3], biomedical systems [4,5], aerospace [6], robotics [7,8], and MEMS [9,10], to mention a few. At the same time, some successful applications of market-ready smart material systems have also been commercialized, e.g., pneumatic valves for seat comfort [11,12], thermostatic valves [13], and flexible strain sensors [14,15]. Despite many successful examples, there are still a number of factors that limit the widespread of smart materials in real-life applications, such as the lifetime and fatigue behavior, the need for special driving electronics, and the strong nonlinearities which significantly complicate design and control. Nevertheless, by considering the increasing amount of research conducted in the field of smart materials and their

Endorobotics. https://doi.org/10.1016/B978-0-12-821750-4.00006-2

applications, it can be foreseen that smart actuators will undergo remarkable improvements in the next few years.

Among the many types of smart materials, two particularly attractive ones are represented by thermal shape memory alloys (SMAs) and dielectric elastomers (DEs). An SMA consists of metal alloys, typically arranged as thin wires or coils working as a muscle fiber, which undergoes a reversible contraction when subject to heat. DEs, on the other hand, are thin elastomeric films that react to an applied high voltage with a large area of expansion. Despite both materials generating relatively large actuation strains (4%–8% for SMAs, and more than one order of magnitude higher for DEs), their complementary characteristics make them suited for different application requirements, e.g., high-energy density and flexibility for SMA, fast and large deformation with low-energy consumption for DEs. Both materials have found successful applications in several areas of engineering, which range from conventional valves and pumps to robotic and biometrical systems. Some relevant examples of existing smart devices which fall in the latter categories include endoscopes, stents, artificial muscles, cooperative robots, and prostheses. Because of the high degree of integration, miniaturization, and compliance required in those applications, the actuator components play a key role in the achievement of the target performance. While a number of such systems have been successfully developed based on either electromagnetic [16] or fluidic [17] actuators, those technologies present some drawbacks which may appear as limiting factors in specific applications. On the one hand, fluidic actuators require compressors which, in turn, increase the construction space, weight, and energy losses. On the other hand, electromagnetic motors are heavy, limited in terms of design flexibility, subject to acoustic noise emission, and often require expensive materials such as copper and rare earths. In contrast, smart actuators based on SMAs and DEs permit to directly address the natural requirements of those applications thanks to their high compliance, large flexibility, lightweight, smooth controllability, biocompatibility, and ease of miniaturization. Clearly, the effective development of highly integrated, intrinsically compliant, and self-sensing robotic/biomedical devices requires the complex operating principle of those smart transducers to be properly understood first.

The aim of this chapter is to introduce some of the most representative applications of SMA and DE technologies in the fields of biorobotics and biomedical engineering. The chapter is structured into two main sections, each one targeted to a specific material technology. After introducing the basic operating principle and the actuation mode of the material, a wide spectrum of applications from the literature will be presented. While the main focus is on endorobotic applications, a number of related examples will also be discussed, ranging from prostheses and rehabilitation robots to artificial muscles and soft grippers. Finally, future perspectives of both technologies are discussed.

6.2 Shape memory alloy (SMA)

This section discusses a number of biomedical applications of SMA actuator technology. First, the operating principle of the material itself will be presented. Then, the main SMA actuator configurations which are relevant for the target applications, namely SMA wires and SMA coils, will be discussed. Based on the gained knowledge, a number of SMA actuator applications in the biomedical and robotic fields will be covered in detail. The major emphasis will be on slender soft robots for endoscopic applications, such as catheters and stents. Additionally, the use of SMAs in the development of prostheses, artificial muscles, and rehabilitation robots will also be discussed.

6.2.1 SMA material operating principle

Thermal SMAs consist of metal alloys that, after being mechanically deformed, are able to recover their original shapes upon heating. This phenomenon, named *shape memory effect*, was first discovered by Ölander in 1932, based on an AuCd alloy [18]. Over the years, also other types of alloys have proved to exhibit a similar behavior, such as CuZn [19] and AuCd [20]. The NiTi alloys, whose shape memory properties were first discovered in 1963 [21], represent up to date the most common type of SMA material adopted in commercial applications. The large success of NiTi alloys is due to their cheaper production cost, easier and safer handling, and better mechanical properties in comparison to alternative SMA materials [22].

Despite the different atomic compositions, all types of SMA materials exhibit the same basic behavior. The operating principle of SMAs can be effectively understood by looking at the material microstructure [23]. At the crystal lattice level, an SMA can exist into two main variants, or phases, denoted as austenite and martensite, respectively. The austenite is characterized by a cubic unit cell and is stable at high temperatures, while the martensite is described by a tetragonal shape and is stable at room temperature. In general, a mix of austenitic and martensitic variants coexist in an SMA material, each one characterized by an arbitrary orientation in space. However, under certain external loading conditions, some variants tend to be more stable from a thermodynamic standpoint, and thus they are more likely to appear in the material microstructure. From a practical point of view, we can consider the following three relevant cases, also as shown in Fig. 6.1:

- The SMA rests at thermal equilibrium at a sufficiently low temperature, and no external stress is applied. In this setting, the crystal lattice is made of a mix of martensitic variants with various orientations in space. This lattice configuration is commonly referred to as *twinned martensite*.
- The SMA rests at thermal equilibrium at a sufficiently low temperature, and it is subject at the same time to a sufficiently large tensile stress. In this case,

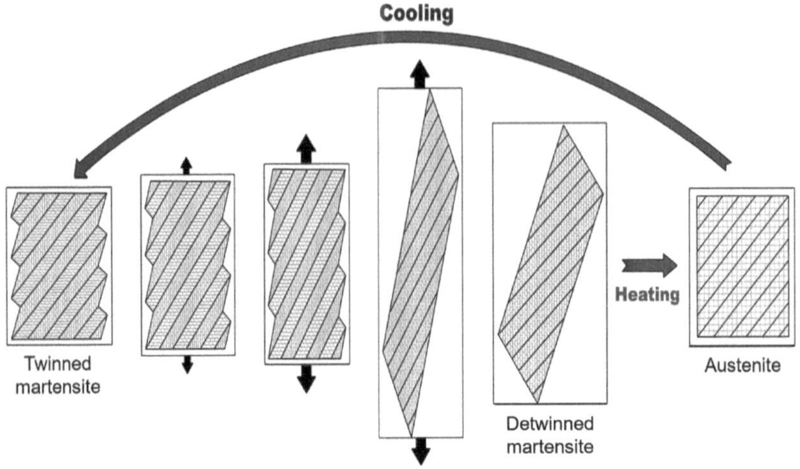

FIG. 6.1 Different phase configurations of an SMA material and transformations among them: *twinned martensite*, stable in case of low temperature and no external stress; *detwinned martensite*, stable in case an external tensile stress is applied; *austenite*, stable in case the material rests at a sufficiently high temperature.

no austenite exists, while all the martensitic variants are aligned with the axis of the applied load. The resulting lattice configuration is commonly denoted as *detwinned martensite*. From Fig. 6.1, it can also be noted how the material remains in detwinned martensite even after the external load is removed.

- The SMA is heated up to a specific transformation temperature. In this case, the martensitic phase is no longer stable from a thermodynamic viewpoint, and the material transforms entirely to austenite. The resulting lattice configuration is simply denoted as *austenite*. After cooling down, the material transforms back to twinned martensite, as illustrated in Fig. 6.1.

The specific values of temperature and stress which induce a crystal lattice transformation are not fixed, but rather depend on the specific alloy composition as well as on the thermomechanical training that the material has undergone. The microstructural configuration described above has also some relevant implications on the macroscopic behavior of the alloy. In particular, a random arrangement of twinned martensitic variants has an overall size that is similar to the one of a fully austenitic material. Macroscopically, no significant difference in geometry is observed among the two configurations. Conversely, the detwinned martensite is characterized by larger size (up to 4%–8%) compared to the other two variants. This effect is qualitatively illustrated in Fig. 6.1, as well.

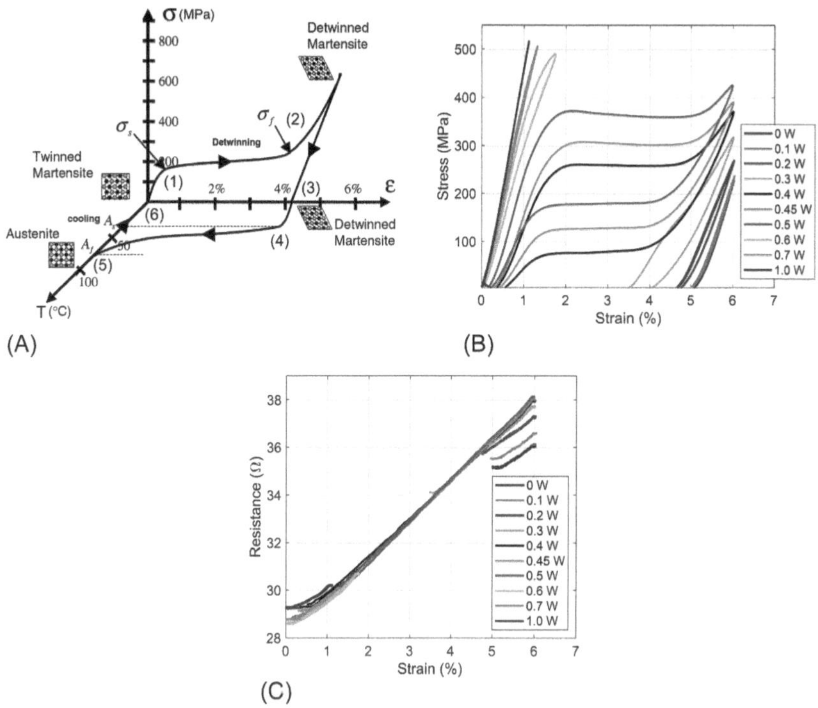

FIG. 6.2 Typical SMA stress-strain-temperature characteristics under a thermomechanical loading history, adapted from [24] (A); typical stress-strain hysteresis of an SMA wire exposed to different constant Joule heating values (B); corresponding electrical resistance-strain curves (C).

The macroscopic implications of the SMA phase transformation can be readily understood by inspecting the material thermomechanical characteristic curves. A typical SMA thermomechanical loading curve is shown in Fig. 6.2A. Different behaviors can be identified:

- Starting from an unloaded material at zero strain and at room temperature, an external tensile stress is applied and the material undergoes an elastic deformation while remaining in twinned martensite (1).
- The stress is increased above σ_s and a quasi-plastic deformation occurs; the material progressively transforms into detwinned martensite. The phase transformation is complete when the stress reaches the value σ_f (2).
- The external stress is removed, the material relaxes while remaining in the detwinned martensite phase, and the strain is larger than the one at the beginning of the experiment (3).
- The material is heated up, and when the temperature rises above the value A_s, a transformation in austenite begins (4). The transformation is complete when the temperature reaches the value A_f (5). A reduction in strain is observed.

- The material is cooled down to room temperature, the austenite transforms back to twinned martensite while the strain remains overall unchanged (6).

By evaluating the cyclic stress-strain curve of an SMA at different constant temperatures, a family of hysteresis loops is obtained. The hysteretic behavior is representative of energy dissipation phenomena, which are due to the irreversibility of the phase transformation. Examples of stress-strain hysteresis curves measured at different constant temperatures are shown in Fig. 6.2B [25]. Isothermal conditions are obtained by cycling the material quasistatically (in order to neglect the self-heating effects due to the latent heat of the phase transformation) while applying a constant heat flow. From here, it can be noted how the material transitions between two linear characteristics as the temperature increases from low to high, and shows a hysteretic behavior for intermediate values. In Fig. 6.2C, the corresponding electrical resistance-strain curves are reported for the same experiment. From here, a clear relationship between resistance and mechanical deformation can readily be established.

When discussing the transition between austenitic and martensitic phases, we have always referred to as low and high temperature in general terms. Different nomenclatures are normally adopted to describe the SMA, depending on the actual shape of the hysteresis under room temperature regimes. More specifically, if at room temperature and under load-free conditions the SMA is in twinned martensite phase, the material is usually referred to as *quasi-plastic*. This naming comes from the fact that, if a specimen in twinned martensite is stretched and then released, it will end up in full detwinned martensite. Therefore, at the end of the experiment, the material will exhibit a longer size compared to the start, despite being subjected in both cases to a zero stress load. From the point of view of the external force, the material has then undergone an apparent plastic transformation. On the other hand, alloys that exhibit a full austenitic composition at room temperature are commonly denoted as *superelastic*. Given a zero stress loading condition, such types of alloys always exhibit the same length, since they are in a full austenitic phase. At the same time, they can undergo a large deformation (compared to standard steel) when subject to an external load, due to the shape of their mechanical characteristic curve. Both types of materials can be usually found on the market, each one serving for different types of applications [26,27].

6.2.2 SMA actuators

As discussed in the previous section, SMA materials undergo transformations in their crystal lattice when subject to specific temperature and stress. By properly triggering the phase transformation in a controlled way, the SMA can be exploited to generate a stroke. To use an SMA as an actuator, the material needs to be preloaded with an external biasing force, e.g., provided by a pretensioned spring. In this way, the alloy naturally rests in a detwinned martensite configuration. Once heated, the material transforms into austenite and, in turn, it contracts in shape. Once the heat source is removed, the material cools down to room temperature and, due to the restoring force provided by the biasing element, it recovers its initial configuration (detwinned martensite). By performing many heating/cooling cycles, a repeatable stroke can be generated. Despite in principle any SMA can work as an actuator, quasi-plastic alloys are normally preferred over superelastic ones, since for the former the detwinned martensite can be generated by means of a lower mechanical stress (which, in turn, favors the material lifetime). To ensure effective control of the material temperature, SMA actuators are commonly shaped as thin wires or coils. In this way, the SMA can be effectively heated via Joule heating which results from the application of an electric current. Major advantages of SMA wire actuators include strains up to 4%–8%, in combination with a remarkably large-energy density (up to $10\,J/cm^3$), high flexibility, and silent operations [28]. Such a combination of features allows in designing highly compact and lightweight devices [29]. Moreover, the electrical resistance of a wire can be easily measured during actuation and used to extract information on the SMA elongation (cf. Fig. 6.2C). In this way, a closed-loop control architecture can be effectively implemented without explicitly introducing an external sensor [30]. Such an operating mode is usually referred to as self-sensing, and permits to further reduce the size, weight, and cost of SMA actuator systems. Despite their many advantages, SMA actuators are characterized by a number of drawbacks, such as the hysteretic response which significantly complicates their modeling and control, as well as the thermal activation which unavoidably leads to low-energy conversion efficiency (on the order of few percent points) and, in some cases, to slow-driving frequency. These factors need to be properly considered when designing SMA-based applications and devices.

As discussed above, the operating principle of SMA actuators strongly relies on the use of a biasing system. Different types of actuator configurations can be obtained, depending on the SMA material topology as well as the type of mechanical bias. Some of the most popular solutions adopted in the literature are summarized in the following.

6.2.2.1 SMA wire actuators

The simplest possible type of actuator can be obtained by combining a SMA wire with a passive mechanical element, which provides the required restoring force. Such a biasing mechanism usually consists of a mass or a pretensioned spring. Springs are normally preferred over masses, due to the possibility of working in any direction independently of gravity, as well as to their reduced inertial load which results in less oscillations. A sketch of a simple spring-biased SMA wire actuator is shown in Fig. 6.3A, in both unactivated (blue) and activated (red) states, respectively. As it can be observed, the wire activation allows the SMA to shrink and, in turn, the spring elongates to a more

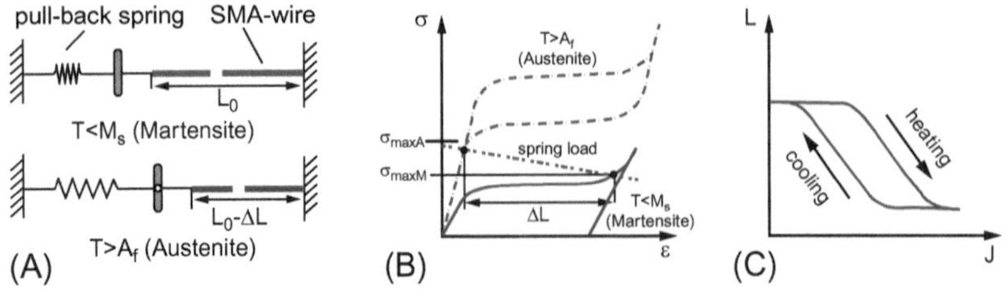

FIG. 6.3 Sketch of spring-loaded SMA wire actuator (A), corresponding stress-strain hysteresis (B), overall actuator input-output hysteresis (C).

relaxed state. In this way, a linear actuation is obtained. Alternatively, both SMA and spring can be pivoted around a rotational joint in such a way to transfer the SMA contraction into a rotational motion. The stroke of the spring-biased SMA wire strongly depends on the choice of the biasing spring. In particular, the role of the biasing element in determining the actuation performance can readily be understood by inspecting the hysteretic curve of the material in Fig. 6.3B, which shows the SMA force-displacement hysteresis for the unactivated (solid blue line) and activated (dashed red line) cases, respectively. The plots also report the characteristics of the spring (dash-dot blue line). Note that the spring exhibits a negative slope in this diagram, since its force characteristics are plotted with respect to the SMA strain. At equilibrium, the forces of the SMA and the bias must balance each other. This condition can be graphically expressed in terms of an intersection between SMA hysteresis and bias characteristics. As the SMA is activated, the shape of its hysteresis changes and, in turn, the equilibrium shifts to a lower displacement value. Once an SMA wire is given, the spring characteristic needs to be properly designed (e.g., by modulating the stiffness and pretension) in such a way to match the shape of the SMA hystereses and, in turn, optimize the achievable stroke. Fig. 6.3C shows the resulting curve obtained by plotting the SMA length as a function of the applied Joule heating. This curve is characterized by a hysteretic shape as well, thus reflecting the fact that multiple intersections between SMA and spring curves may exist for some intermediate Joule heating values. In this case, the true value of SMA equilibrium is determined by the previous loading history.

The main advantage of a simple SMA wire actuator consists of its design simplicity, which ensures low manufacturing effort, compactness, and lightweight. However, the natural cooling time of the wire represents the bottleneck for the actuator bandwidth. It is remarked how the cooling time can be reduced by increasing the surface-to-volume ratio of the wire, i.e., by reducing the wire radius. For instance, commercially available wires can be found with diameters ranging from 25 up to 500 μm [26]. In this way, actuation frequencies up to 20 Hz can be also potentially achieved based on 25 μm diameter wires. On the other hand, the reduced diameter is also accompanied by a reduction in force. This problem can be circumvented by bundling together several wires, even though this would result in an increase in the overall size which might be undesirable in some applications.

6.2.2.2 SMA coil actuators

Simple SMA wires exploit the standard transformation between austenite and martensite to generate a stroke, which corresponds to a macroscopic strain of the order of 4%. SMA coils represent a popular alternative to simple SMA wires, which allow achieving a remarkable increase in the stroke for the same actuator length. An SMA coil is simply obtained by winding an SMA wire into a threadlike spring [31]. The operating principle of the obtained SMA spring is essentially the same as a simple SMA wire. Fig. 6.4A and B provide a sketch of a spring-loaded SMA coil and its corresponding characteristic curves, respectively [26]. The overall actuator performance can be deduced with plots that are qualitatively similar to the ones in Fig. 6.3B and C.

Compared to a spring-loaded SMA wire, a spring-loaded SMA coil exhibits a remarkably larger stroke, which can be on the order of 10%–20% up to 50% of the actuator size [26,31]. The increase in stroke, however, is accompanied by a reduction in force. The different load distribution within the coil also implies that only a small portion of the material volume undergoes a phase transformation, in contrast to SMA wires which exploit the phase change in a more efficient way. Additionally, SMA coils occupy a larger radial space compared to a wire (typically on the order of several millimeters), thus preventing from obtaining a very thin tendon-like actuator. The higher compactness also worsens the overall thermal conduction properties, thus further slowing down the actuation performance. These differences have to be accounted for when choosing between an SMA wire and an SMA coil.

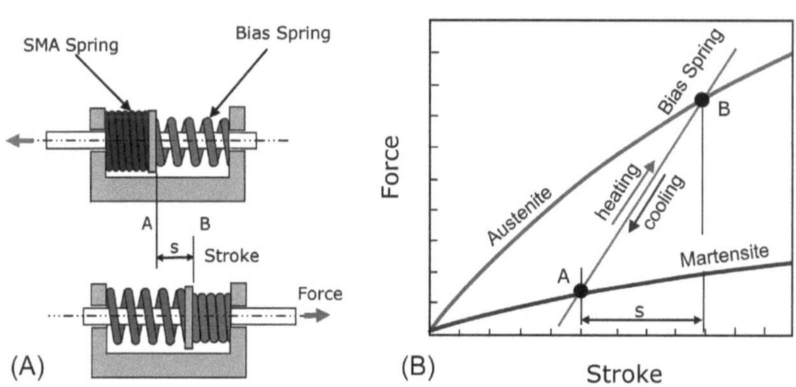

FIG. 6.4 Sketch (A) and characteristic curves (B) of spring-loaded SMA coil actuator [26].

6.2.2.3 Agonist-antagonist SMA wire actuators

All the actuator configurations considered so far are based on an active SMA element working against a passive load. One of the major drawbacks of this type of actuator is the limited active control during the cooling phase. In fact, while one can always try to increase the activation speed by injecting a higher amount of heat, the deactivation time is usually dictated by the natural speed of convective cooling. A potential means to enhance this aspect consists of biasing an SMA wire with a second wire actuator. In this way, a bidirectional displacement can be achieved depending on which wire is activated. The active displacement control in both directions also permits to further increase the actuation speed. An exemplary sketch of the agonist-antagonist SMA is shown in Fig. 6.5A. When the left hand side wire (SMA wire 1) is activated, a corresponding stroke to the left is generated. The converse is true when the right SMA wire (SMA wire 2) is actuated. The graphical criterion based on the intersection between mechanical characteristics, already discussed for the simple SMA wire actuator, can be used here as well, as shown in Fig. 6.5B. The only difference with the previous case is that we need to replace the spring characteristics with another hysteresis, which represents the antagonist wire. The hystereses of both agonist and antagonist wires can be modified independently of each other, by controlling the corresponding Joule heating. Based on such a setting, the resulting displacement-Joule heating hysteresis is reported in Fig. 6.5C. Here, the positive and negative abscissa semiaxes represent the Joule heating applied to SMA wire 1 and SMA wire 2, denoted as J_1 and J_2, respectively. Note how the input-output response is symmetric with respect to the actuated wire. Moreover, it can be observed how the overall actuator exhibits a bistable behavior, i.e., a non-zero position is maintained in case none of the wires is activated. Such a result highlights an additional interesting feature of agonist-antagonist SMA actuators, which can be exploited for energy-saving purpose. Despite the numerous advantages discussed above, agonist-antagonist SMA wires are generally more complex than standard spring-biased actuators. This fact has clearly important implications in terms of design, modeling, and control.

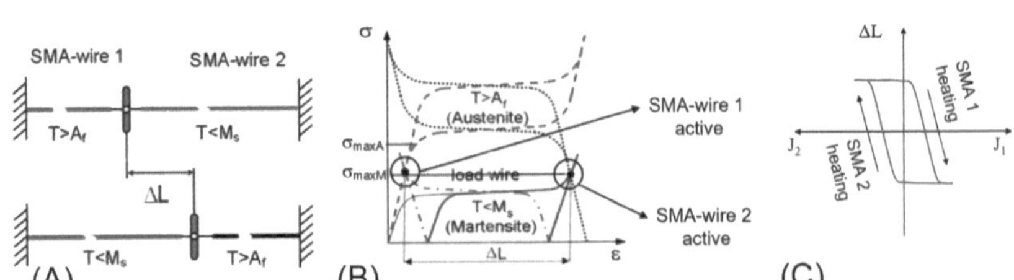

FIG. 6.5 Sketch of agonist-antagonist SMA wire actuator (A), corresponding stress-strain hysteresis (B), overall actuator input-output hysteresis (C).

6.2.3 SMA applications in the biomedical field

Due to the unique combination of features, SMA actuators have found applications in a wide number of areas of engineering. Some relevant examples of SMA application fields include aerospace, automotive, adaptive structures, micropositioning actuators, bioinspired robots, continuum robots, and many more [22]. By means of SMA technology, one can either replace conventional actuators in exchange for improved performance, or eventually develop completely novel actuation concepts that are not possible with standard transducers. Undoubtedly, some of the most successful applications of SMA actuators have been presented in the field of biomedicine [32]. Since the 1980s, a number of SMA devices capable of assisting both human patients and doctors have been developed. Those range from novel robots such as flexible endoscopes and stents to crawling robots for colonoscopy, from artificial muscles to prosthetic hand actuators, from assistive devices for muscular rehabilitation to reconfigurable braille displays. Part of this great success is due to the fact that NiTi SMA wires exhibit excellent properties for biomedical applications. These include biocompatibility, high corrosion resistance, magnetic-free activation, compliance similar to human tissues, and operating range compatible with the human body temperature regime [22]. This combination of features, together with the high flexibility and the ease for miniaturization, has allowed researchers and technicians to develop a large number of smart SMA solutions.

In the following, a number of SMA applications in the field of biomedicine and biorobotics will be discussed. The list is not intended to be exhaustive, but rather to showcase some of the most successful examples of biomedical SMA devices. A strong emphasis will be put on endoscopic applications, since they represent one of the most attractive fields for SMAs. Nevertheless, other types of related applications will be also discussed, such as artificial muscles, prostheses, and assistive robotics.

6.2.3.1 *Slender soft robots for endoscopy*

Due to many similarities with the biological tendons, it is no surprise that SMA technology has been widely exploited to design novel actuators for robotics [33,34]. Other than working as tendon actuators for standard rigid robots, SMA wires have also found natural applications in the more recent field of soft robotics. Differently from conventional rigid robots, soft robots exhibit no rigid structures or joints, but exploit instead highly compliant structures to create complex motion patterns. In many cases, soft robots are designed to resemble slender and flexible arms, which are inspired by biological trunks, snakes, and tentacle arms. This allows to achieve improved dexterity and flexibility, together with ease to miniaturization and better safety during interactions with humans. Due to their mechanical properties, SMA wire actuators can be naturally integrated into slender soft robot structures without requiring large driving units and cumbersome transmission

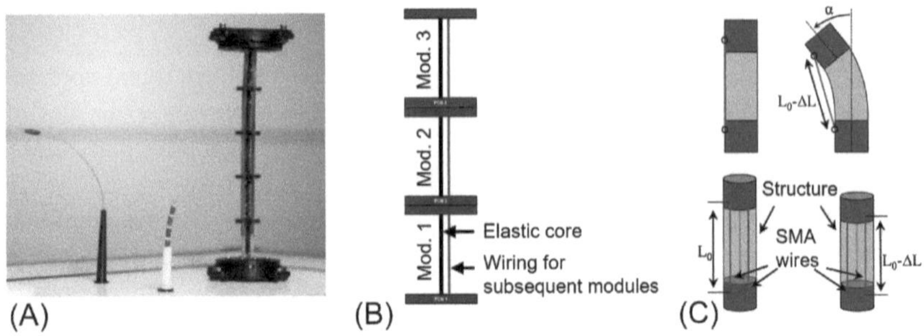

FIG. 6.6 Different prototypes of SMA-driven slender soft robots (A), modular design concept (B), and SMA-based driving principle of the individual module (C) [35,36].

mechanisms. Some examples of slender SMA-driven soft robots are shown in Fig. 6.6A, based on the work presented by Goergen et al. [35,36]. Conceptually, such soft robots consist of a serial arrangement of modules, see Fig. 6.6B. The different modules can either be actuated simultaneously, or eventually in an independent way, by activating different groups of wires. The operating principle of each individual module is illustrated in Fig. 6.6C. Three pretensioned SMA wires are located around the circumference of the module flexible backbone, spaced by 120 degrees. When a wire is activated, it contracts in length, and the module bends accordingly along the corresponding direction. Of course, by properly optimizing the control strategy, bending along every possible direction in space can be made possible. Note how the basic principle of such soft robots is essentially grounded in the agonist-antagonist SMA actuators discussed in Section 6.2.2, which is now taken to a higher complexity level due to the increased dimensionality.

The general principle of SMA-driven soft robots has been exploited since the 1980s to develop flexible and steerable endoscopes. One of the earliest examples in the literature is the work presented in Ikuta et al. [37]. The research focused on an active endoscope depicted in Fig. 6.7A, whose mechanical structure is based on the basic architecture illustrated in Fig. 6.6. The active endoscope consists of five segments of 13-mm diameter, each one 40-mm long. A fiberscope is inserted in the center core of the structure, used for monitoring purposes. Antagonistic SMA coil actuators are exploited to activate the multimodule flexible structure, thus allowing to guide the fiberscope. The authors highlight how SMA technology is highly suitable for the considered applications due to its ease for miniaturization, as well as due to the noncritical requirements in terms of the activation speed and energy efficiency. By means of a resistance-based feedback control, the authors were also able to control the bending angle and improve the robustness against material hysteresis and thermal disturbances. Based on similar principles, a number of SMA endoscopes and catheters have been presented over the subsequent years, some of which are discussed in the following. In Fukuda et al.

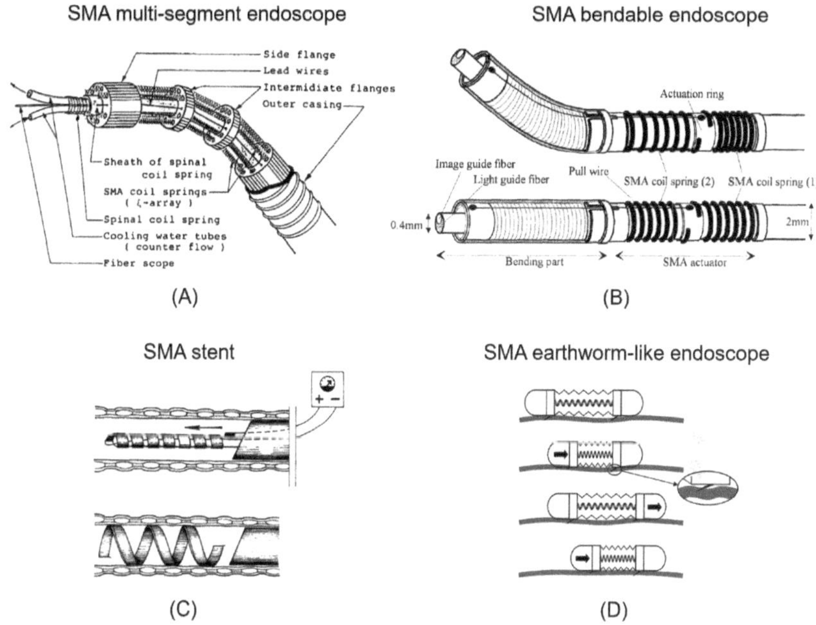

FIG. 6.7 Illustrations of different SMA endoscopic robots concepts: SMA multisegment endoscope [37] (A), SMA bendable endoscope [43] (B), SMA stent [44] (C), SMA earthworm-like endoscope [45] (D).

[38], an active microcatheter consisting of three SMA-driven units was presented, capable of a maximum bending angle of 30 degrees. A multimodular snake-like catheter-used to navigate blood vessels was developed by Lim et al. [39]. The system is actuated via SMA coils, has an outer diameter of 2.8 mm, and is capable of bending angles up to 30 degrees. Mineta et al. [40] presented a multimodular active catheter with a 3 degrees of freedom (DOFs) structure, as shown in Fig. 6.8A. The activation is governed by batch-fabricated SMA coils. The total outer diameter is 0.8 mm, while the achieved bending angle is 35 degrees. In Makishi et al. [5], an endoscope driven by SMA coils was developed. The system has an outer diameter of 5.5 mm and is capable of a 90 degrees bending. A multimodule robotic catheter capable of in-plane bending was presented by Crews and Buckner [41]. Each module exerts 2 DOFs by means of an agonist-antagonist SMA wire actuator mechanism, achieving bending angles up to 90 degrees in both directions. Kadir et al. [42] presented a three-segment endoscope, also shown in Fig. 6.8B. In this system, every module consists of a flexible backbone activated by three independent SMA actuators. A maximum bending angle of 30 degrees is achieved.

Despite a wide number of SMA catheters exploiting a multimodular structure as described above, a number of alternative design concepts have also been presented. One relevant example is the work by Maeda et al. [43], whose

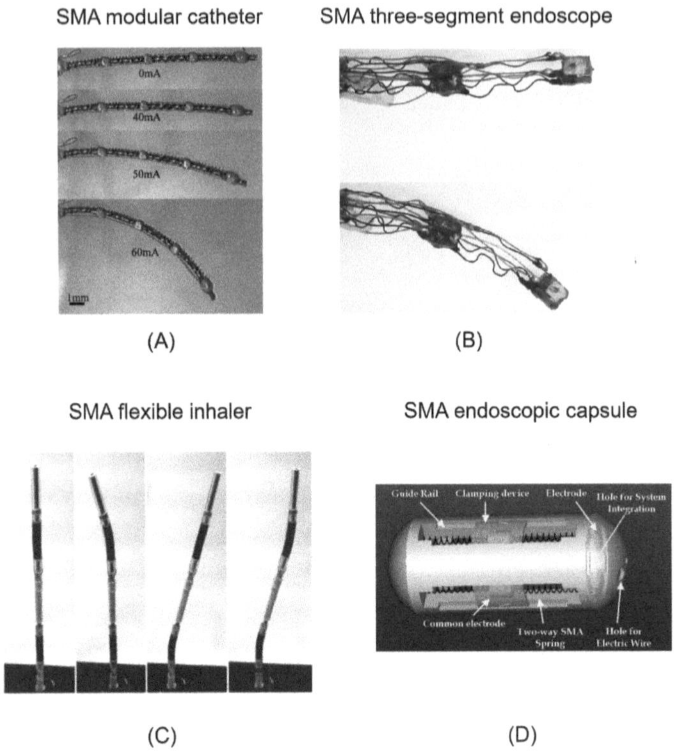

FIG. 6.8 Several possible realizations of SMA endoscopic robots: SMA modular catheter [40] (A), SMA three-segment endoscope [42] (B), SMA flexible inhaler [47] (C), SMA endoscopic capsule [48] (D).

functional sketch is shown in Fig. 6.7B. Here, the active endoscope (2-mm outer diameter) consists of a bending part connected in series with two SMA coil actuators. The SMA coils are connected to an actuation ring, which is linked to the endoscope tip via a passive wire. By heating the SMA coils, the actuation rings move along the structure, allowing the structure to bend due to the pulling force exerted by the passive wire. The SMA actuation permits to achieve bending angles up to 60 degrees and forces up to 1 N. A similar actuation concept was presented by Reynaerts et al. [49], targeted at a gastrointestinal intervention application. The outer diameter and length of the robot are 12 and 95 mm, respectively. Vertebra-like segments are considered, whose control is ensured via separate SMA-actuated modules, i.e., an extension-contraction module, a locking module, and a 2 DOFs bending module. A further alternative design concept exploiting SMA coils winded around tubes was also presented by Haga et al. [50]. The soft tube has an outer diameter of 6 mm and a length of 40 mm and can exert a maximum bending angle of 110 degrees.

6.2.3.2 Slender soft robots for stents

The compliance matching between SMA and human tissues makes the technology highly suitable for the realization of stents. Differently from a soft robotic endoscope, in which the main goal of SMA is to provide a flexible actuation, a stent is generally used to remove occlusions inside human vessels. The first example of an SMA-driven stent was presented by Dotter et al. [51]. Here, an SMA coil actuator is placed at the end of an inner coaxial catheter. By properly guiding the stent at the right location inside a vessel, it can be expanded via an external heat source, thus clearing the obstruction. Note how, here, the SMA actuator is exploited in an expansion mode, rather than in elongation. A further example of an SMA-based stent for tracheal treatment was presented by Vinograd et al. [44]. The stent is made of a flat sheet of NiTi SMA (about 1.8mm thick and 0.18mm wide), arranged in a coil configuration. The operating principle of such device is illustrated in the sketch provided in Fig. 6.7C. By heating the stent up to 40°C, the wire diameter expands from 2mm up to 7.2mm. The effectiveness of the device was tested on 20 young rabbits. A self-expanding SMA stent, based on a similar principle,was also used by Yanagihara et al. [52] to treat the tracheal stenosis of two patients. The stent measures 100mm in length, and it can expand up to 18mm in diameter.

6.2.3.3 Slender soft robots for drug delivery

While most of the endoscopic devices described above are used for inside-body examinations, slender SMA soft robots can be also used for delivering medicaments or drugs inside the body in a localized way. One example is provided by Ayvali et al. [53], where a multi-DOFs steerable cannula based on SMA actuators is presented. The device can be used to deliver diagnostic as well as therapeutic devices to a target location, by means of a hollow inner core. The cannula itself consists of three straight segments having a 3mm outer diameter, and is able to steer up to 30 degrees. A 2 DOFs actuation is made possible via SMA wire actuator elements connected in each module, while temperature feedback is used to achieve accurate control of the bending angle. In Furst and Seelecke [47], fabrication and characterization of a smart inhaler nozzle for aerosol drug delivery were presented. The adaptive nozzle consists of three rigid sections separated by two compliant joints. The overall nozzle has an outer diameter of about 3mm and a length of about 70mm. Three sets of independent SMA wires (50μm diameter) are used to induce a 3 DOFs actuation on the flexible joints. The nozzle tip is shown to exhibit displacements up to 6mm in each spatial direction. A picture of such a nozzle is shown in Fig. 6.8C.

6.2.3.4 Slender soft robots for minimally invasive surgery

Despite many popular applications of flexible SMA robots are related to endoscopic applications, the technology has also been considered in a number of minimally invasive surgery applications. In Nakamura et al. [54], the investigation of SMA actuators in laparoscopic applications was considered. In this type of intervention, forceps need to be inserted into the abdominal cavity. Conventional forceps are difficult to operate, due to their straight shapes. This problem is addressed by the proposed SMA forceps, which can undergo a bending up to 30 degrees. The accuracy of the prototype is then furthered by means of a PID (proportional, integrative, derivative) control algorithm, which allows for SMA hysteresis compensation. Kode and Çavuşoğlu [55] presented a millimeter-scale actuator used to activate the end-effector of a robot, to be used in minimum invasive surgery applications. The end-effector consists of a 5mm diameter endoscopic needle. The actuation mechanism consists of a serial combination of a DC micromotor and a bundle of SMA wires. While the DC motor is used for opening and closing the gripper jaw, the SMA produces high forces required to grip and hold the needle.

6.2.3.5 *Locomoting robots for endoscopy*

The insertion of SMA actuators inside human bodies can be alternatively realized by means of a locomoting principle. An example of an earthworm-like SMA microrobot capable of locomotion functionalities was presented by Kim et al. [45]. The body of the robot consists of an SMA coil actuator embedded in a silicone bellow, which acts as an elastic biasing element. The overall length and diameter of the robot are 50mm and 9.5 mm, respectively. When the SMA is activated, needles placed on the front part of the robot clamp to the walking surface, while the rear part of the robot slides forward. When the SMA is cooled down, the front part of the robot elongates while the rear stays still, clamped to the surface. By activating the SMA coil in a periodic way, a locomotion pattern is achieved. A conceptual sketch of such a locomotion principle is reported in Fig. 6.7D. The fabricated microrobot can reach a stroke of 2 mm per activation cycle, with an average speed of 10 mm/min. A similar concept was also exploited by Yuk et al. [56] to realize a microrobot capable of mimicking the crawling motion of the *Caenorhabditis elegans*. The inchworm robot is 300mm long and 2mm wide. It is made of 12 separate driving units, whose activation is based on an SMA coil element working against an elastic bias. By properly controlling the SMAs, a sinusoidal undulatory motion at 0.17 Hz is produced, obtaining a locomotion comparable to that of a real worm.

These SMA-based locomotion concepts have been extensively exploited in biomedical applications, to develop endoscopic devices capable of exploring complex environments such as the human colon. Some of the earlier works implemented this principle by combining SMA technology with pneumatic actuation. In Mi et al. [46], a colonoscopy inchworm robot mechanism based on a mixed actuation system was proposed. The robot body consists of three units, namely two grippers and one extensor. While the grippers provide adherence to the colon walls via expansion, the extensor provides the extensibility needed to advance through the colon. By alternating activations of the grippers and the extensor, locomotion through the colon can be effectively implemented. All of such components are inflatable, and are activated pneumatically. A cylindrical bendable mechanism based on three SMA components is used to implement a 3 DOFs turning motion. A similar locomoting principle based on grippers and extensors was presented by Menciassi et al. [57]. Also, in this case, pneumatic actuation is used for advancing through the colon, while SMA actuators implement a steering mechanism capable of bending up to 90 degrees along all three spatial directions. An alternative concept was also presented by Carrozza et al. [58], in which SMA elements are used to drive pneumatic microvalves. The principle of the robot is based on an alternance of expansion and elongation motions, similarly to the cases discussed above.

An example of a fully SMA-driven robot for peristaltic locomotion was presented by Kim et al. [48], see Fig. 6.8D. Pairs of SMA coil actuators are used to implement the locomotion, while a simple clamping device is developed based on a biomimetic approach. The full prototype consists of four pairs of SMA coils and four clampers, having a total length of 33 mm and a diameter of 13 mm. By means of comparative studies with alternative locomotion principles, based on both electric motors and ionic metal polymer composites, the SMA colonoscopic system is shown to provide the best performance. In Yan et al. [59], a colonoscopy robot made of several 2 DOFs joints was considered. Both directions of movement and angle of imaging are controlled via SMA coil actuators. The locomotion principle is implemented by replicating the squirm of inchworms, by properly alternating the activation of several serial modules. In Cotta et al. [60], the gripping-and-expansion principle described above was implemented by means of a multimodular architecture, in which each segment is driven by three SMA coil actuators.

6.2.3.6 *Artificial muscles and prostheses*

Due to their natural resemblance to human tendons, SMA wire actuators have been extensively used for the realization of artificial muscles, organs, and prostheses. An artificial urethral valve, based on a dual-SMA plate actuation mechanism, was developed by Chonan et al. [61]. Experiments conducted on male dogs allowed the authors to verify the correct functioning of the artificial valve. In Luo et al. [62], a pair of all-round SMA plates were used to design an anal sphincter actuation mechanism. Since the element requires an increase of 20°C to be activated, thermal insulation with the surrounding tissues is also implemented. The effectiveness of the prototype is demonstrated both *in vitro* and *in vivo*, based on animal models. In Miki et al. [63], an actuator for artificial peristaltic motion was presented. The operating principle of such actuator is illustrated in Fig. 6.9A. In here, a pair of SMA coil actuators is connected to two plates. By activating the SMA elements, the resulting shrinkage introduces a bending that changes the distance among the plates. In this way, food can be

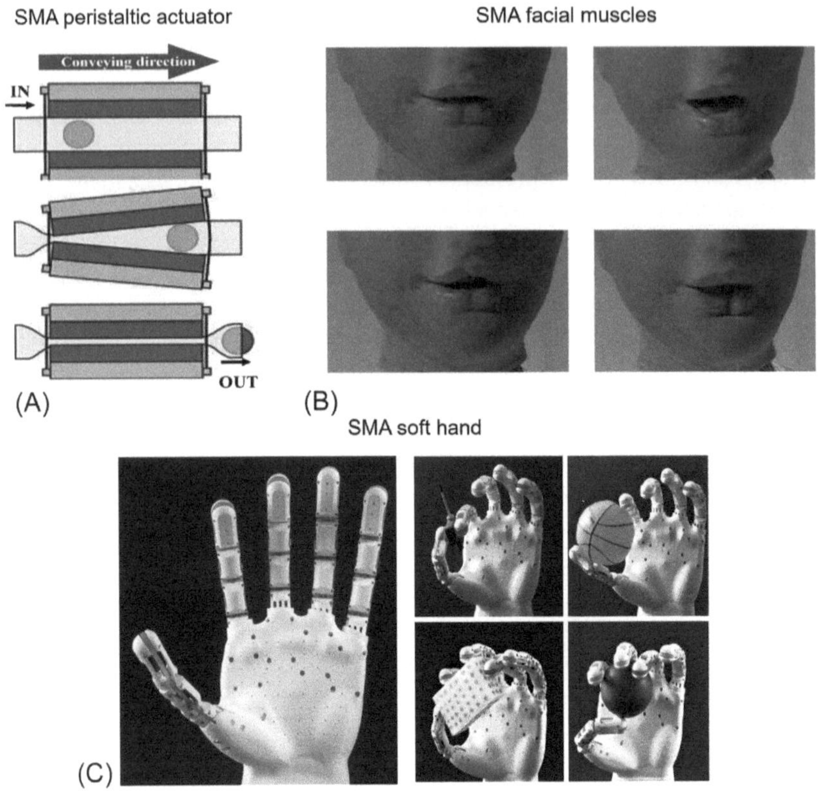

FIG. 6.9 Examples of artificial muscles and prostheses based on SMA technology: artificial SMA peristaltic actuator [63] (A), SMA-driven artificial facial muscles [64] (B), five-finger SMA soft hand [65] (C).

transferred through the esophagus. The overall artificial esophagus is obtained as a serial combination of many of those elements. *In vitro* experiments showed how a gel-like material can be conveyed at a speed of 15 mm/s, while solid materials can be transported at a higher velocity of about 23–34 mm/s.

SMA-based facial muscles capable of reproducing realistic expressions in an anthropomorphic robot were studied by Hara et al. [66]. Thirteen independent driving units, controlled via SMA wire actuators, are connected to the specific location of an artificial face. The proposed prototype is able to produce six different kinds of facial expressions, i.e., surprise, fear, anger, disgust, sad, and happy. A special feature of SMAs in this particular application is represented by their significantly reduced weight, which allows to design a functional human-sized head (210 mm) weighing 2.1 kg only. A similar problem was also investigated by Tadesse et al. [64]. In here, numerous SMA actuators are combined with 35 routine pulleys to generate facial expressions in a biped robot to be used for children education. The resulting humanoid face has a size of 140 mm × 90 mm × 110 mm and a total number of 12 DOFs. Some of the robot facial expressions are illustrated in Fig. 6.9B. An SMA-driven artificial muscle for the control of a robotic eyeball prosthesis was presented by Wolfe et al. [67]. In this work, a pair of agonist-antagonist SMA coil actuators coupled with an elastic bias allows to steer an eyeball actuator by ±30 degrees.

The tendon-like behavior of SMA wires has attracted a number of researchers interested in the development of artificial hands. An example of a miniature five-fingered hand was developed by Maeno and Hino [68]. The prototype consists of 20 DOFs, whose fingers exhibit motion ranges of 30 degrees and a tip force of 0.75 N. Lee et al. [69] developed a lightweight and compact hand prototype, whose actuation is generated by spring-biased SMA wires embedded in the palm. Each finger is characterized by 2 DOFs, while the thumb is completely passive. To ensure compactness and sufficient stroke, the SMA wires are wrapped around pulleys. The hand is capable of grasping various objects weighing up to 500 g. Andrianesis and Tzes [70] proposed an underactuated prosthetic hand, in which only 11 of the total 15 DOFs can be actively controlled. The fingers are driven by a spring-biased SMA actuation mechanism. The resulting system can perform an almost-complete hand closure with a fingertip force up to 11 N. Simone et al. [8] presented a hand concept in which bundles of agonist-antagonist SMA wires are used to activate three of the five fingers. Due to the benefits of the agonist-antagonist configuration, a full 90 degrees rotation can be obtained with settling times on the order of 1–1.5 s, with corresponding tip forces of 1.4 N for each phalanx. Self-sensing features based on resistance measurements are also evaluated experimentally. A second version of such SMA hand was later developed by the same authors [65]. The improved design is characterized by a higher grasping force (45 N), five fully actuated fingers, and compliant joint mechanisms which allow soft manipulation features. Fig. 6.9C shows the described hand while holding a number of objects having different shapes and sizes.

6.2.3.7 Rehabilitation and assistive devices

Thanks to the possibility of modulating their force-displacement behavior by means of electric stimuli, SMA technology has found a number of applications in the field of active neuromuscular rehabilitation. For example, Stirling et al. [71] analyzed the possibility of using SMA coils to develop active soft orthoses for knee rehabilitation. Such a system is shown in Fig. 6.10A. Active control of the SMA coils provides a variable assistance to the patient, depending on several factors and needs. Compared to a passive brace, the developed active system can provide target-specific control, muscular assistance, and reeducation of the neuromotor system. The effectiveness of the prototype is validated by means of a robotic leg. An exoskeleton system used to assist hand rehabilitation was proposed by Tang et al. [72]. In here, SMA wires are used to drive a differential

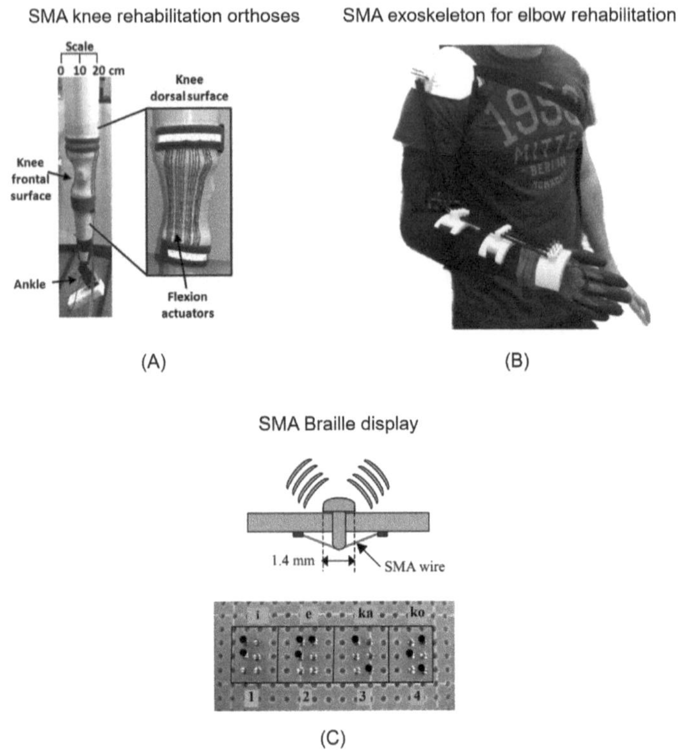

FIG. 6.10 Example of rehabilitative actuators and devices based on SMA: soft orthoses for knee rehabilitation [71] (A), SMA wearable exoskeleton for elbow rehabilitation [74] (B), SMA Braille display [75] (C).

rotational actuator. Due to the low weight of SMA transducers, the system portability can be enhanced, and its structure can be simplified as well. The use of SMA wires for the design of a guide and a limb mobilizer for neuromuscular rehabilitation was also investigated by Pittaccio et al. [73]. A wearable exoskeleton for elbow rehabilitation was presented by Copaci et al. [74], also reported in Fig. 6.10B. The exoskeleton is driven by SMA wire actuators embedded in a 3D-printed soft structure, which ensure an overall lightweight (0.6kg) and low-noise system. Preliminary experimental results showed that the proposed device can provide a suitable task performance for medical rehabilitation, despite the presence of some issues related to the cooling process slowness.

While all of the above devices provide assistance from the outside of patient body, SMA technology also lends itself to the realization of smart implant devices. In Kha et al. [76], an electrode array embedded with SMA wire actuators was designed to assist patients with auditory issues. The device can be located beneath the basilar membrane inside the *cochlear scala tympani*, in order to deliver growth factors into the cochlea and stimulate auditory neurons.

The feasibility of the concept is proved by means of a finite element model. The use of SMA technology for the realization of stiffness-alterable osteosynthesis plates was investigated by Pfeifer et al. [77]. By modulating the stiffness of the implant over time, the duration of the bone healing process can be reduced. By triggering the SMA phase transformation with an induction device, the stiffness of the implant can be altered at an arbitrary date after the intervention. The paper focuses on the feasibility of the design, without performing experiments on patients.

It is remarked how SMA technology has also been used to develop active braille displays, targeted at assisting blind patients as well as in providing haptic feedback functionalities. An example of a dynamic Braille display was presented by Haga et al. [78]. In here, 16 individual pins arranged in an array configuration are controlled by means of SMA coil actuators. The SMA actuators are combined with a magnetic latch mechanism, which permits keeping the pins in a desired position even when the wires are deactivated, thus furthering the overall system energy efficiency. Velázquez et al. [79] presented a compact and low-cost tactile display interface based on SMA technology. The actuation mechanism is based on agonist-antagonist pairs of SMA coils. The actuator has a diameter of 1.5 mm and a weight of 150 mg, and can deliver a pull force of 320 mN with a bandwidth of 1.5Hz using simple forced-air convection. A microvibration actuator based on SMA wires was developed by Zhao et al. [75]. Features of the device include compactness, low energy consumption, and quick response time, which enhance its use for mobile applications. The effectiveness of the display was evaluated by means of several experiments, conducted in a school for blind people. A picture of the considered actuator is reported in Fig. 6.10C.

6.3 Dielectric elastomer (DE)

The focus of this section is on DE actuator technology. After summarizing the material operating principle, the main DE actuator configurations will be discussed. A number of recent and promising applications of DE technology in the biomedical and biorobotic fields will finally be discussed.

6.3.1 DE material operating principle

A DE consists of a thin film of dielectric material (thickness on the order of 50–100 μm), coated with compliant electrodes on both surfaces. Typical materials used as dielectric are acrylics (e.g., VHB 4905 by 3M) [80], silicones (e.g., Wacker Elastosil 2030 by Wacker Chemie) [81], natural rubbers [82], or synthetic rubbers [83], while the electrodes are usually manufactured via carbon-based compounds [84] or thin metal films [85]. The physical effect governing the activation of a DE is based on an electrostatic transduction principle, which is discussed in the following. A conceptual sketch of an unactuated DE

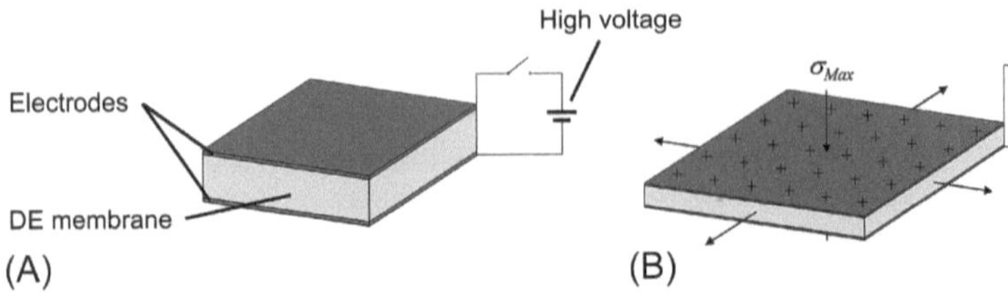

FIG. 6.11 Operating principle of a DE membrane, voltage off (A) and voltage on (B).

membrane is reported in Fig. 6.11A. When a voltage is applied to the DE, electric charges of opposite signs are stored on the two electrodes. If the applied voltage is sufficiently high (on the order of several kV), the attractive electrostatic forces among the charges on the two electrodes become so large that induce a squeezing in the membrane. Due to the incompressibility of the elastomer, such a reduction in thickness is followed by an expansion in the area, see Fig. 6.11B. Both thickness contraction and area expansion can be effectively exploited to generate an actuation stroke. The compressive stress which arises due to an applied voltage is usually referred to as Maxwell stress, denoted as σ_{Max}, and can be computed as follows:

$$\sigma_{Max} = -\epsilon_0 \epsilon_r E^2, \tag{6.1}$$

where E is the electric field inside the material, while ϵ_0 and ϵ_r represent the vacuum and DE relative permittivities, respectively [86]. The minus sign in Eq. (6.1) is representative of the fact that the Maxwell stress always operates in compression.

The first documented analysis of a DE dates back to the ninetieth century, when Röntgen demonstrated their operating principle by spraying electric charges on a piece of natural rubber [87]. The effect was then rediscovered more than one century afterward by Pelrine et al. [88]. Since then, a significant amount of basic research has been done on DE transducers, with the first standards having also been published in 2015 [89].

In order to properly design high-performance DE applications, their behavior needs to be properly understood first. An electromechanical characterization of a DE membrane is normally based on tests conducted on strip-shaped specimens, as the one reported in Fig. 6.12A. By performing tensile experiments at various voltage levels, the material stress-strain curves can be obtained. An example of typical stress-strain behavior observed in DE materials is reported in Fig. 6.12B. For the sake of representation, Fig. 6.12B reports the results in terms of engineering stress, i.e., force normalized over the initial area. As it can be seen, the membrane exhibits three main phenomena:

- A strong nonlinearity in the stress-strain response, which is a direct consequence of the large deformation, and is usually referred to as *hyperelasticity*.

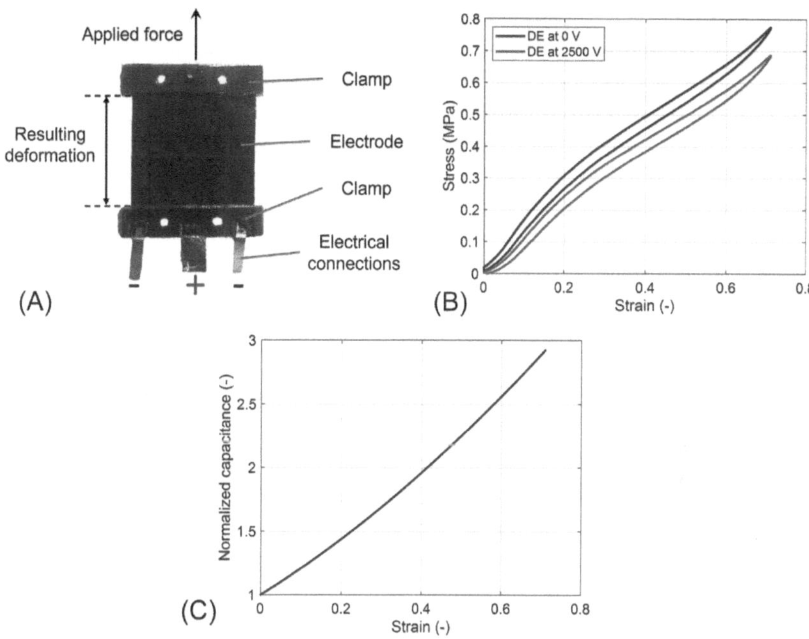

FIG. 6.12 Example of DE specimen for material properties characterization (A), typical stress-strain response of a DE for different applied voltages (B), typical normalized capacitance-strain response (C).

- A dependency of the characteristics on the applied voltage. This is a direct consequence of the Maxwell stress, which results in a modification of the in-plane force. This effect is mainly responsible for the *electromechanical coupling* existing in DEs. Due to the nonlinear dependency of the Maxwell on the applied electric field (cf. Eq. 6.1), the electromechanical coupling introduces further nonlinearities in the material behavior.
- A small hysteresis, which is a manifestation of the *viscoelasticity* within the material response. This hysteresis is rate-dependent, and tends to vanish as the loading rate tends to zero since it is an intrinsically dynamic phenomenon. Note how this is conceptually different from the hysteresis observed in SMAs, which is mainly due to the material phase transformation and does not disappear as the loading rate tends to zero. Nevertheless, both hystereses reflect dissipative mechanisms taking place inside the material.

Note how all of these phenomena contribute to increasing the nonlinearity and complexity of the DE response. The change in geometry that a DE undergoes when stretched also produces a change in electrical capacitance. For a thin DE membrane, the electrical capacitance C can be well approximated by the parallel-plate capacitor equation, given as follows:

$$C = \epsilon_0 \epsilon_r \frac{A}{z}, \tag{6.2}$$

where A is the electrode surface area and z is the dielectric thickness. When a DE membrane is stretched in-plane, A increases while z decreases as a result of the incompressibility (the total DE volume zA remains constant). According to Eq. (6.2), both effects contribute to the increase in electrical capacitance. A qualitative plot showing the normalized capacitance-strain characteristics of a DE is reported in Fig. 6.12C. A monotonic relationship between capacitance and mechanical deformation can readily be observed.

6.3.2 DE actuators

The Maxwell stress principle discussed in Section 6.3.1 relates an applied electric field to a mechanical stress. This fact immediately suggests the use of a DE membrane as an actuator, capable of converting an applied voltage into a displacement. The main advantage of DE actuators is represented by a large amount of deformation, which can be up to 100% for silicone-based DEs and even up to 400% for other types of elastomers, such as acrylics [90]. Further attractive features of DE technology include high flexibility, high energy efficiency (on the order of 60%–90%), low power consumption (on the order of mW), lightweight, and silent operations. The specific type of DE material also plays a key role in the overall system performance. For instance, while acrylics can undergo much larger strains than silicone DEs, they are affected by low energy efficiency due to viscoelastic losses, and also exhibit a strong sensitivity to environmental conditions. In contrast, the moderate viscoelastic losses of silicone DEs make them more suitable for energy-efficient and dynamic applications. The high flexibility and compliance similar to human muscles also make DEs highly attractive for the realization of bioinspired and soft robotic solutions [7]. In addition, DEs also offer the possibility of working as self-sensing actuators. More specifically, if the electrical capacitance is acquired during actuation, it can be used to reconstruct the amount of membrane deformation [91,92] or force [93]. Such information can be eventually used to implement sensor-free position control architectures [94]. All these features make DEs highly attractive for the realization of novel, highly flexible, and intelligent artificial muscles. Despite many attractive features, DE technology still suffers from a number of limitations. Up to date, the most critical aspects concern the high amount of voltage required to activate the material (on the order of several kV, which makes it necessary to develop special driving electronics), the dependency of the material response from environmental conditions and fatigue, and the strong nonlinearities which complicate system design and control.

The high flexibility and scalability of DE material, together with the existence of two main actuation modes (area expansion and thickness reduction), makes it possible to design actor systems in a wide variety of configurations. In addition, in some (but not all) cases, the performance of DE actuators is strongly dependent on the biasing system used to preload the membrane. In the following, a number of relevant types of DE actuator configurations will be discussed. The list does not aim at being exhaustive but rather focuses on those types of actuators that are commonly encountered in bioinspired, robotic, and biomedical applications.

6.3.2.1 In-plane membrane actuators

A standard membrane actuator exploits the area expansion of a simple DE film to generate an in-plane stroke. In order to operate as an actuator, a membrane DE needs to be preloaded with a mechanical bias (similarly to an SMA wire). In the case of membrane DEs, the biasing system plays a crucial role in determining the actuation performance, as clarified in the sequel. A prediction of the stroke of a membrane DE actuator can be simply performed by considering the force equilibrium between elastomer and bias. Such a force equilibrium can be solved graphically, by evaluating the intersection between bias and DE force-displacement characteristics for low voltage (minimum position) and high voltage (maximum position). An example of performance prediction based on a linear biasing spring is shown in Fig. 6.13A. Similar to the SMA case, the spring exhibits a negative slope due to the fact that it is being plotted in the DE coordinate system. As it can be observed, the stroke resulting from a linear spring bias turns out to be small in comparison to the achievable membrane deformation. A possible means to increase the actuation stroke consists of using special types of biasing systems with an *inverted slope*, such that they match the gap between the DE curves and, in turn, compensate the elastomer stiffness. An example of such an ideal bias behavior is shown in Fig. 6.13B. Here, it can be clearly seen how the use of such type of mechanism permits to achieve a remarkable increase in stroke, based on the same DE membrane and applied voltage from Fig. 6.13A. Note that a change in slope in the force-displacement diagram can be physically interpreted as a negative stiffness, i.e., a spring which exhibits some sort of mechanical instability. Possible means of physically realizing those types of bias behavior can be based on precompressed buckled beams [95], similar to the one shown in Fig. 6.13C, attracting permanent magnets [96], soft polymeric domes [97], or complex mechanisms [98,99].

Based on the concept described above, a suitable mechanical biasing system can be designed and combined with a given membrane DE. A conceptual sketch showing the operating principle of a spring-biased DE membrane is shown in Fig. 6.14A, while a picture of the same actuator is reported in Fig. 6.14B [95]. Once a voltage is applied to such an actuator, it undergoes an in-plane expansion. Advantages of this specific configuration include the flatness of the resulting design, which might be desirable for some applications, together with the reduced amount of nonlinearities due to the simple kinematic configuration. However, this type of membrane may exhibit a necking during the actuation, as also shown in Fig. 6.14A. Such a necking may result in losses of actuation performance and increase, at the same time, the modeling and simulation complexity [100].

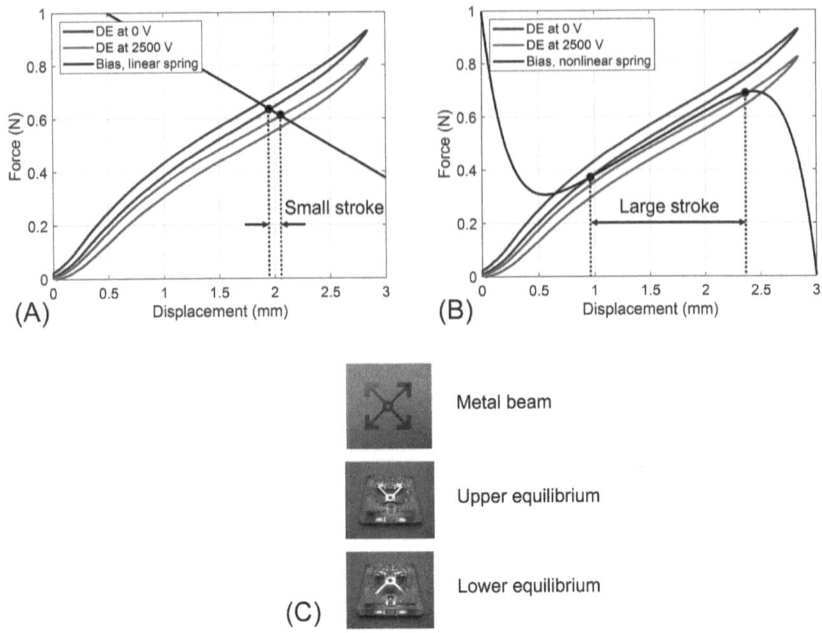

FIG. 6.13 Force displacement equilibrium analysis for DE preloaded with a linear (A) and a nonlinear (B) biasing element. Possible physical realization of a nonlinear-biasing spring, based on a precompressed buckled beam (C).

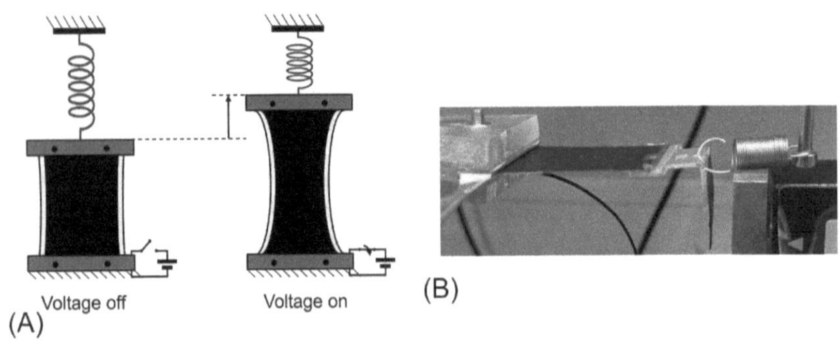

FIG. 6.14 Sketch (A) and picture (B) of an in-plane membrane DE actuator [95].

6.3.2.2 *Out-of-plane membrane actuators*

Out-of-plane membrane actuators are essentially a variation of in-plane membrane DEs, in which the direction of the external force does not act parallelly to the plane on which the membrane lies. A simple example of such an actuator is shown in Fig. 6.15. Such a configuration, normally denoted as cone DE, consists of an annular DE membrane with a central rigid inclusion and fixed within a

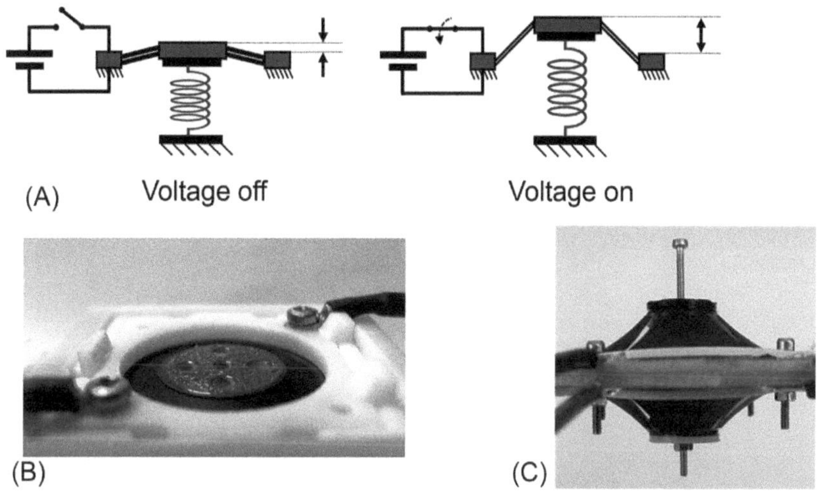

FIG. 6.15 Sketch (A) and picture (B) of an out-of-plane DE cone actuator. Picture of a double-cone DE actuator with partitioned electrodes, which allows for a multi-DOFs actuation [101] (C).

rigid frame. Similar to in-plane membranes, also those types of cone DEs require a biasing force, which can be provided, e.g., by means of a precompressed spring connected to the central rigid inclusion, see Fig. 6.15A. When a voltage is applied to the DE, the elastomer relaxes and the inner spacer moves upward. In this way, a stroke is generated out of the plane of the DE membrane itself. The graphical method illustrated in Fig. 6.13 can be used also in this case to optimize the actuation performance.

One of the main advantages of this type of DE is the compactness, which allows to design actuators with remarkably high work density [102]. In addition, the basic cone DE configuration can be easily modified to achieve more complex motion patterns. One example of such configuration, denoted as double-cone DE, was presented by Nalbach et al. [101]. This system consists of two cone DE membranes rigidly connected through an inner rigid spacer, and precompressed against each other. The electrodes of both membranes are partitioned into several sections which can be activated separately. In this way, the double-cone DE can generate a multi-DOFs actuation, e.g., translations and rotations with respect to different axes. Such a design has been proven to represent an effective actuator configuration for robotic applications [103].

6.3.2.3 *Roll actuators*

Roll DEs represent yet another variation of membrane configuration. In this case, the membrane is rolled along the side orthogonal to the one which undergoes a stroke, thus resulting in a very thin muscle-like fiber. An example

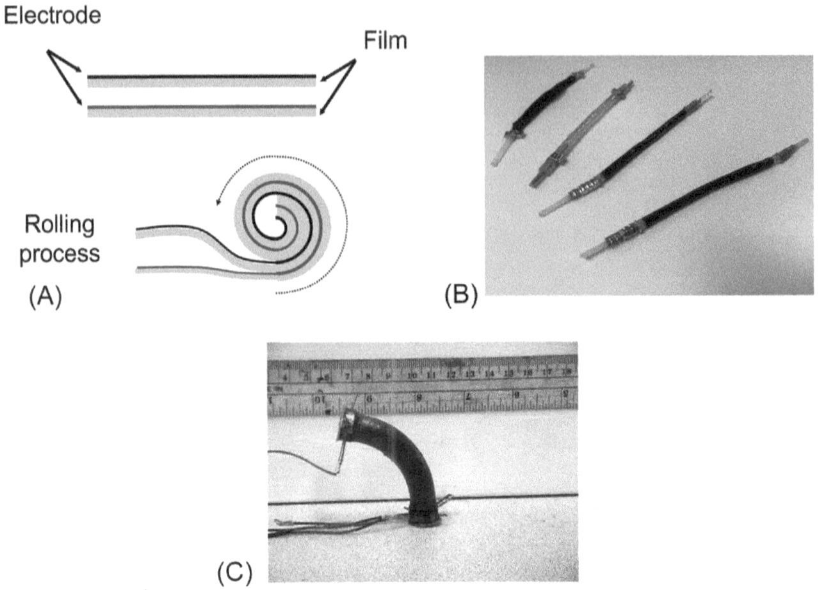

FIG. 6.16 Sketch of the rolling process of a DE membrane (A). Picture of roll DE used for axial [104] (B) and bending actuation [105] (C).

of such a concept is illustrated in Fig. 6.16A [104]. In here, two different DE membranes with a one-sided electrode are rolled together, in order to avoid shortcuts and ensure proper electrical insulation. Note that the two electrodes are colored in black and red for illustrative purposes only. By activating the roll, it undergoes an axial expansion as a consequence of the Maxwell stress. Eventually, a biasing system can be used also in this case to improve the actuation stroke. A further element that affects the manufacturing and behavior of a roll DE is the inner core of the actuator. In general, the core may consist of a pretensioning element, a hollow volume, or either it can be absent in case the DE film is tightly rolled as a spiral.

Different types of roll DEs have been proposed in the literature. The most basic ones simply operate by providing an axial deformation, as the ones shown in Fig. 6.16A [104]. Such types of actuators are highly suitable for providing a one-dimensional actuation within an extremely thin design. In addition, many of such fiber-like actuators can be bundled together and used to build high-energy-dense artificial muscles. A number of authors have also proposed alternative roll DE configurations, which are capable of undergoing bending upon actuation. One example is shown in Fig. 6.16B based on the work by Pei et al. [105], in which a 2 DOFs actuation (bending and elongation) was obtained by means of a special patterned electrode. Similar systems were also presented by Wang et al. [106] and Li et al. [107]. By means of a special type of design, complex bioinspired robots capable of walking or crawling like snakes can be effectively designed.

6.3.2.4 Stack actuators

While all the DE configurations discussed so far make use of the membrane area expansion, stack actuators exploit the thickness reduction to implement a contractile actuation. An example of a DE stack actuator is shown in Fig. 6.17, based on the work presented by Kovacs et al. [108]. It essentially consists of a serial arrangement of several prestrained DE membranes, which contract all at the same time when subject to high voltage. Efficient processes have been developed to ensure the automatic and repeatable fabrication of a multilayer stack, based on a single prepatterned elastomer film [109]. The actuator force is transmitted to an external load by means of rigid supports at the end-points of the stack. Note that, in general, these types of actuators do not make use of biasing mechanisms to amplify the stroke. As a result, the individual layer undergoes a thickness contraction few of micrometers only. The stacking of several layers which contract at the same time permits to magnify the absolute displacement to a meaningful value.

The stroke and the force of a stack DE can be easily tuned by modifying the number of DE layers and the electrode area of a single layer, respectively. The major features of stack actuators include the high compactness and energy density of the resulting design, as well as the possibility of generating a contractile actuation, in contrast to most DE actuators which work in expansion mode. A stack DE is generally preferred to a membrane actuator whenever high forces need to be prioritized over large displacements.

6.3.2.5 Adaptive structures

All the configurations examined so far permit to achieve relatively simple actuation patterns, such as a linear deformation, or a bending in the case of roll DEs. Adaptive DE structures permit to achieve a complex actuation pattern within a simple and lightweight design. One example of such adaptive systems, which has been effectively designed based on DE technology, is represented by minimum energy structures [110,111]. Essentially, a DE minimum energy structure consists of a prestretched elastomer combined with an elastic frame, see Fig. 6.18A. When the encapsulated elastomer is cut out of the frame, it gives rise to a complex structure due to the transfer of elastic energy from the DE to the frame. Based on an energetic equilibrium between DE and frame, a

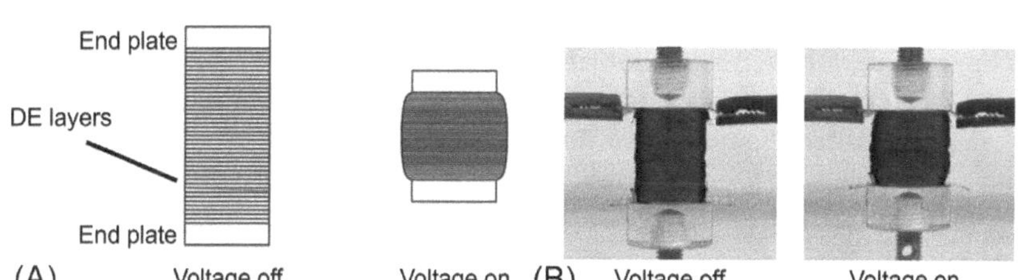

FIG. 6.17 Sketch (A) and picture (B) of a stack DE actuator [108].

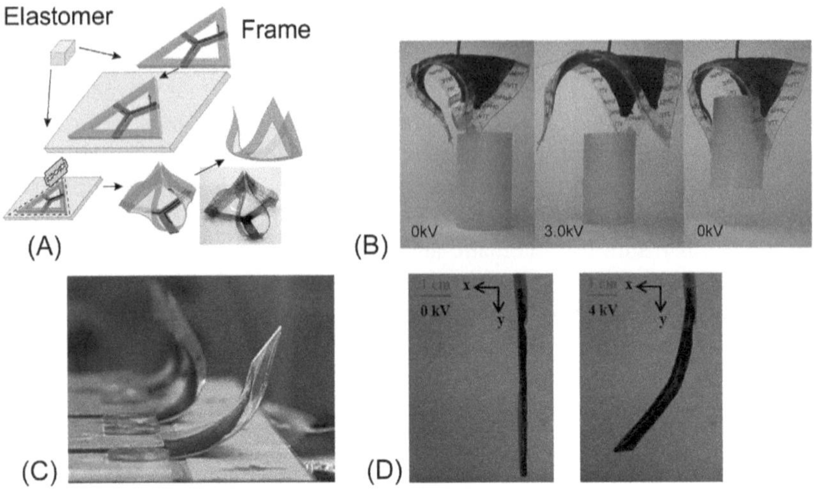

FIG. 6.18 Manufacturing of a DE minimum energy structure [110] (A). Various examples of DE adaptive structures used to achieve a bending actuation: minimum energy structures from [110] (B) and [111] (C), origami structure from [112] (D).

complex pattern can be achieved. Additionally, the application of a high voltage permits to further modify this energetic equilibrium, thus resulting in a structural reconfiguration. Examples of DE minimum energy structures used to generate gripping [110] and bending [111] are reported in Fig. 6.18B and C, respectively. A further type of adaptive DE structure is shown in Fig. 6.18C, in which a bending motion is obtained based on the origami principle [112].

The main advantages of such adaptive structures consist of the possibility of generating an out-of-plane actuation pattern based on an in-plane fabricated membrane, as well as the elimination of bulky biasing mechanisms. In this way, high compact and lightweight actuators capable of performing complex actuation patterns and structural reconfigurations can be developed.

6.3.3 DE applications in the biomedical field

The unique combination of large deformation, lightweight, energy efficiency, and high compliance has led DE technology to gain significant popularity since the early 2000s. Examples of successful DE applications presented in the literature include valves, pumps, soft and bioinspired robots, artificial muscles, and loudspeakers. Despite most of the current DE devices operating at the macroscale, some first attempts toward the development of meso- and microscale actuators have also been proposed [10,113,114].

Due to the characteristics of the material, it should be no surprise that numerous applications of DE technology have also been presented in the field of biomedicine. The compliance similar to human muscles and the nonmagnetic behavior certainly play a major role in this regard, since they allow safe interactions with humans. In principle, the high voltage required to activate the material (on the order of kV) could represent a critical point for safety. Such a high voltage, however, is associated with low-energy levels (on the order of mJ), and thus it can be managed via proper insulation layers. We point out that, up to date, DEs have been around only for 20 years, in contrast to SMA wires which have been known for more than twice this period. As a result, despite the great potential of the technology being well acknowledged, the actual number of devices that have been produced is unavoidably affected by the shorter lifespan. Up to date, most of the current DE applications in the biomedical field tend toward artificial muscles, prostheses, and rehabilitation devices. Despite a number of DE flexible mechanisms and soft robots have being presented, they are mostly targeted at bioinspired robotic applications rather than at endoscopy. Nevertheless, current efforts toward miniaturization and the safety of interaction will foreseeably lead to novel breakthrough solutions over the next few years.

In the following, an overview of DE applications with a special focus on the biomedical field will be provided. Different from the analogous discussion carried out for SMAs, in which the focus was explicitly on endoscopic applications, this section has a more general scope toward the recently established applications for the biomedical field, as well as future-promising solutions, e.g., for endoscopy based on slender soft robots.

6.3.3.1 Slender soft robots

Multi-DOFs flexible robots can be effectively achieved based on the DE technology, by combining multiple actuators in agonist-antagonist configurations. This concept is similar to the one already presented when discussing SMA soft robots in Section 6.2.3. The existence of a wide variety of possible DE actuator configurations (cf. Section 6.3.2), however, opens up a much larger number of potential mechanical configurations.

As stated above, up to date, most DE flexible arm applications are designed for industrial or bioinspired robotic solutions [7]. One of the few examples explicitly dealing with endoscopy is the work presented by Kim et al. [115]. In this paper, the authors introduce a smart colonoscope based on an inchworm-like locomotion principle. SMA actuators mounted at the top of the endoscope are used to implement camera steering. A double-cone DE actuator with partitioned electrodes is used as an external controller for camera steering. The eight independent active areas of the double-cone DE permit to achieve a 5 DOFs control of the camera steering motion. A tracking mechanism is also used to synchronize the eye movement of the doctor with the DE actuator, thus allowing to automatize the implementation of the steering motion. Note how in this design the DE is used to provide an external control of the device, rather than direct actuation for the endoscope bending and locomotion.

It can be foreseen that this is mainly due to high-voltage insulation issues. Nevertheless, the authors acknowledge the high compactness and work capabilities of the DE technology.

The research work mentioned above dates back to the early years of DE technology, thus, it represents one of the earliest attempts to integrate the DE technology with soft robotic applications. Over subsequent years, a number of authors have presented several soft robotic solutions in which DE elements are used as main actuator components. For example, Pei et al. [105] proposed a serpentine robot made of a serial connection of four roll DEs, 2 DOFs each, based on acrylic material. By alternating activations of different groups of actuators, the resulting bending of the roll can be used to generate a wavelike snake motion. A picture of such a concept is shown in Fig. 6.19A. Another type of DE

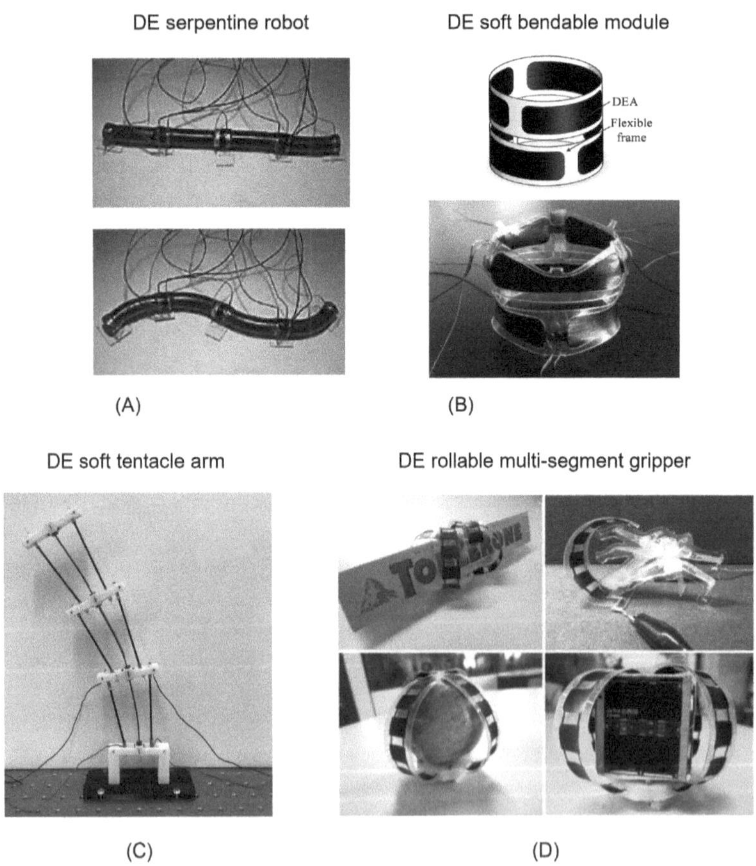

FIG. 6.19 Examples of slender soft robots based on DE technology: DE serpentine robot [105] (A), DE soft bendable module [121] (B), DE soft tentacle arm [119] (C), DE rollable multisegment gripper [120] (D).

serpentine robot was presented by Plante [116]. The robotic actuator is developed for needle insertion in intra-MRI interventions. In this application, DEs appear suitable due to their insensitivity to magnetic disturbances as well as to their smooth controllability. The proposed design is based on diamond actuator modules, consisting of in-plane acrylic DE membranes combined with a bistable mechanism. Based on the same type of actuator concept, two different prototypes are proposed. The former consists of 24 independent DE actuators, occupies a volume of 600 mm × 300 mm × 300 mm, and is capable of a tip angular displacement of ±15 degrees along every direction. The latter is a simplified version of the same robot consisting of 7 independent DEs, has a total weight smaller than 200 g, and allows for a tip angular displacement between −35 degrees and +20 degrees. According to the author, the improved performance of the smaller actuator is amenable to a reduced mass. Petralia and Wood [117] presented a snake-like robot made up of a serial connection of five DE minimum energy structures. Each DE actuator is made of acrylic material, and is capable of a remarkable bending actuation when actuated with 1.2 kV only (this value is remarkably small in comparison to typical DE applications). The authors demonstrated experimentally how the robot allows for an increase in length of 1.7 times and a decrease in height of 1.4 times, upon simultaneous activation of all segments. An overall energy efficiency of about 30% is also estimated.

Some recent works have also developed trunk-like robots, which are capable of moving within a three-dimensional space. Zhao et al. [121] proposed a novel concept for a 3 DOFs bendable module, shown in Fig. 6.19B. Many of these modules can be potentially connected in series, thus allowing to obtain a system resembling a biomimetic elephant trunk. Each module is based on a prestretched acrylic fixed to a flexible but inextensible frame. Three separate active areas are present in each module, whose activation results in a bending along a desired spatial direction. For a single 3 DOFs module, the measured output force is 875 mN, which is about 30 times higher than the robot's own weight, while the angular displacement of the single module is 9 degrees. A similar type of multimodular flexible robot is also proposed by Xing et al. [118]. Each DE module is based on a concept similar to the one of [121], with the only difference that a 2 DOFs rather than a 3 DOFs actuation is considered. One of the most attractive features of this design is the remarkably small weight of 18 g, for an actuator length of about 320 mm, together with a bending angle per module larger than 50 degrees. Duduta et al. [122] presented a further similar concept, based on a serial connection of three modules having 3 DOFs each. Both extensive roll DEs and contractile stack DEs are considered as actuators, whose activation is based on ultrathin carbon nanotube electrodes. The resulting design allows to achieve an energy density up to 20 J/kg, which is comparable to the theoretical upper limit for natural muscles, i.e., 40 J/kg. Prostheses and surgical robots are pointed out as possible target applications for the device. Note how most of the solutions presented in the literature are based on acrylic elastomers, which provide large deformations but also low efficiency and bandwidth. An alternative modular design concept, based on silicone DE actuators, was presented by Prechtl et al. [119]. In this work, roll DEs are used to drive a 2 DOFs bendable module with a flexible backbone. Many of those modules can be connected in series and used to design a soft robotic tentacle arm, as shown in Fig. 6.19C. Once a roll DE is activated, it expands and causes a bending of the structure along the opposite direction. Theoretical predictions based on available material data show potential bending angles up to 20 degrees per module, and are also confirmed experimentally by the picture in Fig. 6.19C. This result shows remarkable improvements in terms of both angular performance and design scalability, compared to previous attempts of modular silicone DE robots based on double-cone actuators [101]. The considered type of actuators clearly represent an important step toward the development of future flexible endoscopic robots.

Despite the wide attention received by bendable soft robotic arms, some authors also investigated the use of DEs for the realization of compliant grippers. In Araromi et al. [120], a rollable gripper based on a multisegment DE minimum energy structure was presented. The high aspect ratio (size of 105 mm × 22 mm) allows the gripper to wrap around objects of various sizes, as shown in Fig. 6.19D, by means of a lightweight design (0.65 g). Despite the relatively low grasping force generated by the actuators, the authors acknowledge the design as a milestone for the development of DE grippers technology. A similar type of actuator was also presented by Lau et al. [123]. This design is characterized by an area of 103 mm × 40 mm, and allows to generate high grasping forces up to 8–9 times the system weight (7.5 g). The capability of the gripper in lifting practical-sized objects such as a mandarin orange is highlighted.

6.3.3.2 Artificial muscles and prostheses

The mechanical characteristics of DEs make them the perfect candidates for the realization of artificial muscles and prostheses. One of the first works demonstrating the capabilities of the technology in the field of artificial muscles was presented by Jung et al. [124]. In here, fabrication and characterization of a three-finger robotic hand driven by DEs are presented. The main actuator element consists of a silicone-based DE roll with an elastic core. Such actuators are arranged in an agonist-antagonist configuration, and permit to achieve up to 15–30 DOFs for each finger. In Kovacs et al. [125], a wrestling arm robotic mechanism actuated by 250 roll DEs was presented. Bundles of rolls are arranged in order to reproduce an agonist-antagonist muscle configurations, to achieve an arm-like bidirectional rotation. Despite the promising results, it was remarked how further advances in terms of material properties, control algorithms, and efficient design solutions are still required to properly exploit DEs in limb prosthesis applications [126].

One relevant category which has received a significant interest is represented by facial muscles. In Kwak et al. [127], several types of DE actuators were used to actuate different parts of an artificial robot face. Segmented double-cone DEs are used to actuate eye rotation, while in-plane membrane DEs in agonist-antagonist configurations are used to drive the muscles regulating eyebrows and cheeks. Actuation of neck, jaw, and eyelids is implemented via servomotors, due to the higher force requirement. Based on the proposed hybrid facial actuation, six fundamental expressions are effectively reproduced, i.e., surprise, fear, angry, disgust, sadness, and happiness. The use of DE membranes for jaw movement actuation was investigated by Wang and Zhu [128]. To mimic the real jaw motion, two in-plane membrane DEs made of acrylic material were installed on a robotic skull, at a position similar to one of human masseters, see Fig. 6.20A. Experiments demonstrated how the maximum achieved vertical displacement is compatible with the maximum motion of the natural human jaw. The use of an alternative DE concept for implementing the lateral eyeball motion in a robotic face was investigated by Carpi and De Rossi [129]. The actuator concept, shown in Fig. 6.20B, is based on pairs of agonist-antagonist stacks of silicone DE. A maximum rotation of ± 25 degrees is obtained by means of the proposed solution, which is compatible with the ± 20 degrees occurring in human eyes. A similar concept for eyeball actuation was presented by Li et al. [130], in which three roll DEs based on an acrylic material are used to implement horizontal, vertical, and circular motions. A further contribution in the field of artificial eye actuation was made by Carpi et al. [90]. Differently from the previous contributions, in here the authors focused on using DEs for the implementation of tunable lenses. The concept is based on a fluid-filled elastomeric lens integrated with an annular acrylic DE, which operates as an in-plane membrane actuator. By activating the DE, the fluid chamber is compressed and the focus of the lens changes accordingly, as shown in Fig. 6.20C. The proposed actuator is capable of responding to step voltage inputs with time constants on the order of 60 ms.

One further application of DE technology for prosthetics concerns bioinspired fluidic pumps. As an example, Goulbourne et al. [131] performed a theoretical study aiming at designing a prosthetic blood pump based on an acrylic DE membrane. The target performance is a swept volume increase of 70 cc into a systolic pressure of 120 mmHg. The change in shape of a DE, designed in such a way to resemble a heart chamber, permits to pump blood similarly to the ventricular walls in the human heart. Based on the numerical study, it is concluded that the target swept volume can be obtained based on a realistic DE design configuration, thus showcasing the feasibility of the technology. Carpi et al. [132] proposed a concept for a lightweight DE peristaltic pump. The system is based on a series of radially expanding flexible tubes, based on hollow DE rolls. Relative radial deformations up to 12% were experimentally demonstrated. A bioinspired soft pumping unit based on a similar concept was also presented by Bowers et al. [133]. The overall actuator occupies a space of

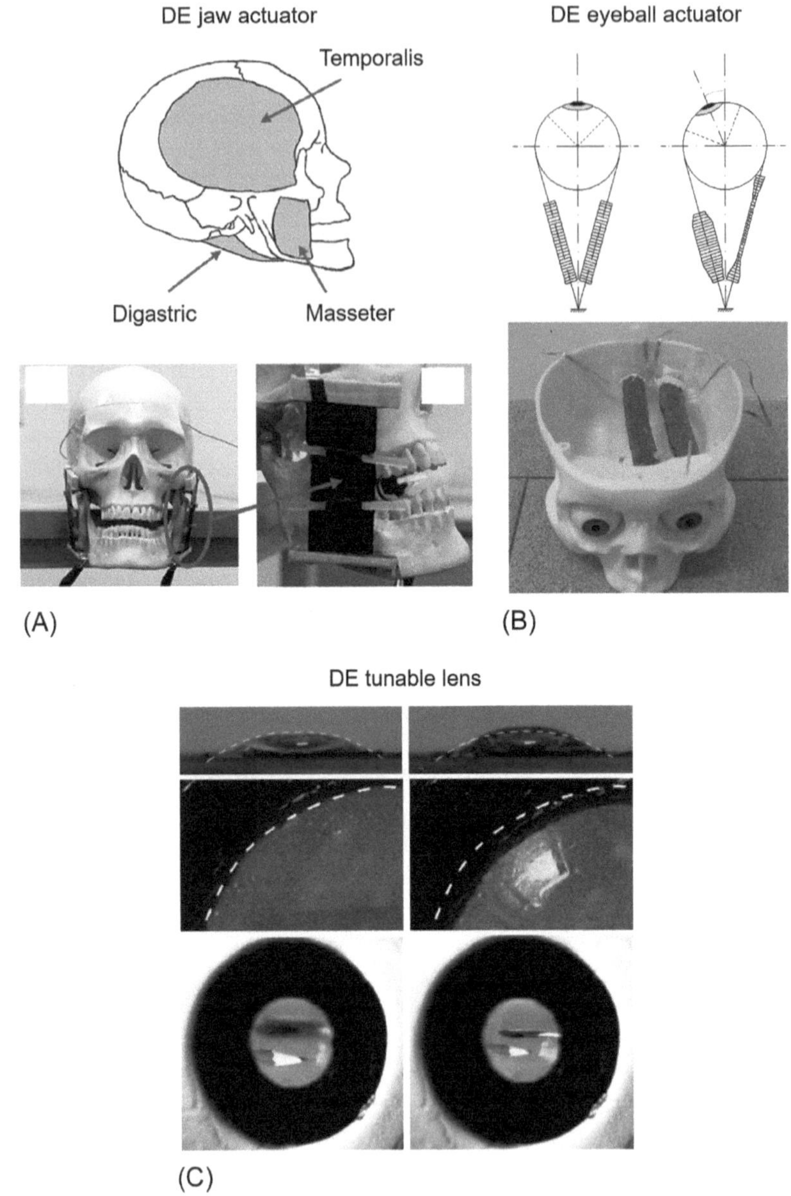

FIG. 6.20 Examples of artificial muscles and prostheses based on DE technology: DE jaw actuator [128] (A), DE eyeball actuator [129] (B), DE tunable lens [90] (C).

40 mm × mm 25 × 2 mm, and exploits silicone-based roll DEs to achieve flow rates up to 40 μL/s.

6.3.3.3 Rehabilitation and assistive devices

The possibility of modulating the material stiffness via an applied voltage makes it possible to exploit DE transducers to implement devices for neuromuscular rehabilitation. An active DE compression bandage for treating patients with disorders in the lower extremity venous system was presented by Pourazadi et al. [134]. The system consists of a silicone-based DE membrane that needs to be wrapped around the patient's ankle and knee, also shown in Fig. 6.21A. By controlling the DE membranes electrically, the pressure applied by the bandage on the patient's leg can be effectively modulated. The conducted

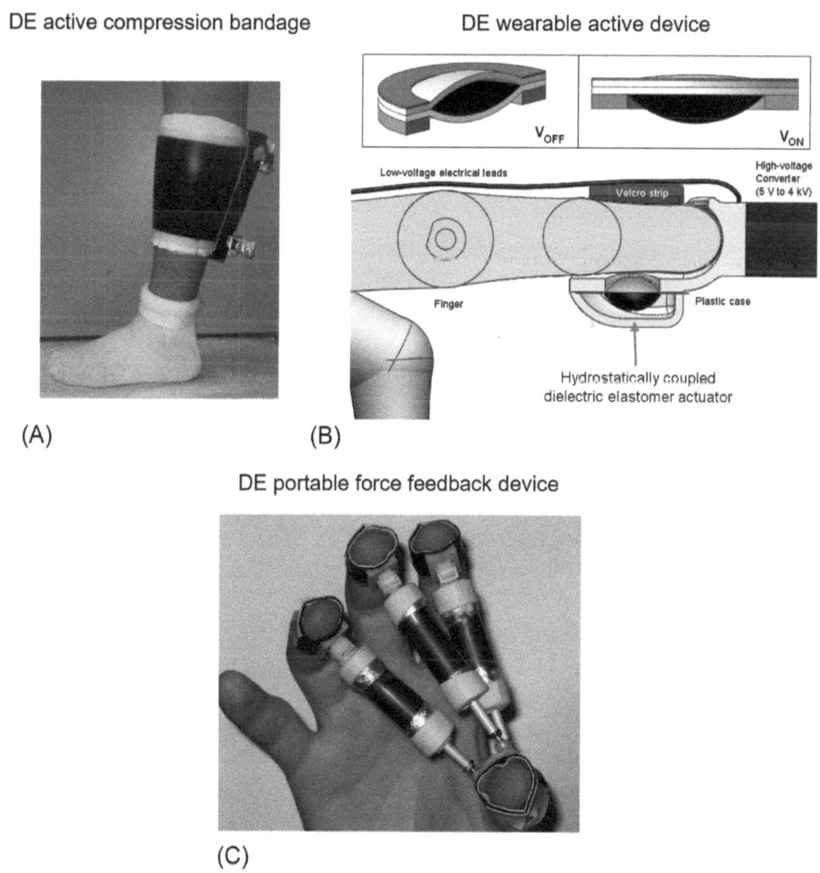

FIG. 6.21 Examples of rehabilitation and assistive devices based on DE technology, DE active compression bandage [134] (A), DE wearable haptic device [136] (B), DE portable force feedback device [137] (C).

investigation shows how the proposed DE active bandage permits to modulate the pressure level to values similar to those of compression stockings. A variable stiffness DE orthosis for hand rehabilitation was presented by Carpi et al. [4]. Multilayer in-plane DE membranes based on acrylic material are used to achieve active modulation of the stiffness perceived by the patient's hand. The effectiveness of the proposed solution was proved based on a psychophysical testing campaign, conducted on human test subjects. The lightweight, low power consumption, and silent operations of DE transducers permit to enhance the portability of the system. A further hand rehabilitation device was presented by Kelley and Kauffman [135], with the specific goal of suppressing hand tremors in patients affected by Parkinson's disease. People suffering from Parkinson's are subjected to undesired motions at faster frequencies (>3 Hz) than voluntary motions (>2 Hz). The solution proposed by the authors is based on canceling such undesired tremors by generating an opposite force via a DE membrane attached to the patient's wrist. The effectiveness of the solution was demonstrated by means of numerical simulations.

DE actuators have also been proven to provide effective support to patients affected by blindness. In Choi et al. [138], a dynamic Braille display driven by DEs was presented. The different pins of the display are driven by precompressed silicone DE membranes which buckle out of plane upon activation. Compared to alternative existing Braille displays, the proposed solution has advantages in terms of intrinsic softness, enhanced spatial density of the pins, ease of fabrication, low cost, and ease of miniaturization. The effectiveness of the device is confirmed by means of psychophysical experiments. An alternative design for tactile DE displays was proposed by Carpi et al. [136]. The device consists of bubble-like hydrostatically coupled DEs. The actuator configuration is based on out-of-plane DE membranes, made of acrylic material. In this specific configuration, an incompressible fluid acts as a biasing force needed for the out-of-plane actuation, while providing at the same time electrical insulation. A schematic depiction of the actuator is reported in Fig. 6.21B. The proposed configuration was shown to be effective for dynamic Braille displays and finger rehabilitation, as well as for providing haptic feedback during navigation in virtual environments. An improved version of the same type of actuator, capable of exerting a force of 0.7 N under a 4 kV actuation, was subsequently presented by Boys et al. [139].

The use of DE haptic feedback devices has also been proposed to support surgeons. Zhang et al. [137] presented a portable force feedback device for surgery simulations. The core actuation components consist of roll DE actuators based on an acrylic elastomer, capable of exerting a maximum force of 7.2 N and an elongation of 31%. A picture of the overall device is shown in Fig. 6.21C. It has an overall weight of 38 g for a three-finger actuation, thus it is a suitable solution for mobile applications. A haptic display for mimicking the compliance of organs and tissues was presented by Carpi et al. [140]. The concept is once again based on the hydrostatically coupled DE transducers already discussed before. Prototypes of haptic displays for cardiac contractility, pulsating blood pressure, and compliance of soft tissues are manufactured based on both silicone and acrylic DEs. By means of preliminary physical and psychophysical tests, the authors assessed the feasibility of the proposed solution.

6.4 Conclusions

In this chapter, two different types of smart material actuators were discussed, namely thermal SMAs and DEs, with a focus on their applications in biomedical and biorobotic fields. Features and limitations of each technology were highlighted, and their most relevant configuration in actuator applications were pointed out. A comparative summary with different figures of merit for both materials is presented in Table 6.1, based on the data available in Fu et al. [28], Carpi et al. [141], and Kornbluh et al. [142]. From a direct comparison, it can be seen how none of the considered material excels in every field, but rather shows some clear advantages and drawbacks. As an example, SMAs are normally preferred in applications where a high amount of work needs to be performed with an overall compact space, and in which the actuation speed is not a critical parameter. This is also reflected in the applications discussed in

TABLE 6.1 Performance comparison between SMA and DE materials [28,141,142].

Parameter	Unit	SMA	DE
Maximum strain	(%)	4÷8	100 silicone, ~400 acrylic
Young modulus	(MPa)	~30,000 martensite, ~70,000 austenite	0.1÷10
Volumetric mass density	(g/cm^3)	6.45	1
Energy density	(J/cm^3)	Up to 10	0.4
Energy conversion efficiency	(%)	3÷5	70÷90
Operating frequency	(Hz)	1 typical, up to 10÷20 for thin wires	>1000, up to 50,000
Activation	(–)	Thermal	Electrostatic
Self-sensing principle	(–)	Resistive	Capacitive

Section 6.2.3, which are mostly related to endoscopy and prostheses. On the other hand, DEs appear more attractive in terms of deformation range, response speed, and energy efficiency. This fact also emerges from the examples presented in Section 6.3.3, in which portable devices and artificial muscles occur frequently. The two technologies appear then as highly complementary, yet they are both highly suitable for the design of compact and highly integrated smart actuators.

Despite the many attractive features and the high potential of both materials, SMA and DE actuators are still largely unexploited in many real-life application areas. This may be partly due to the fact that such materials are still relatively young, compared to competing actuator technologies such as fluidics and electromagnetics. Therefore, further progresses in material science are still required in order to improve their functionalities up to a level compatible with the requirements of real-life applications (i.e., robustness, lifetime, and production cost). Specific issues also need to be addressed for each material. In the case of SMAs, for instance, the research is moving toward the development of thinner wires, capable of exerting a faster and more efficient actuation. For DEs, reduction of activation voltage (nowadays still on the order of several kV), together with an increase of material lifetime, still represents the main technological challenges. Other than improvements in the material itself, it is also expected that further advances in systems engineering will lead to the development of more efficient applications. Up to now, a significant amount of work has been put into showing how smart materials can be used to replace existing technologies. On the other hand, the development of novel applications entirely tailored around the operating principles of SMAs and DEs represents the key step that has to be carried out, in order to promote their spread in real-life applications. To this end, the development of numerically efficient simulation tools, the design of fast and accurate control architectures, and the implementation of estimation algorithms for practical exploitation of self-sensing in complex systems represent some of the most relevant research challenges for both SMAs and DEs. Hopefully, joint efforts in system development, material science, modeling, and control will allow this vision to become reality in the near future.

References

[1] Hill M, Rizzello G, Seelecke S. Development and experimental characterization of a pneumatic valve actuated by a dielectric elastomer membrane. Smart Mater Struct 2017;26(8), 085023.

[2] Motzki P, Seelecke S. Industrial applications for shape memory alloys. In: Reference module in materials science and materials engineering. Amsterdam, Netherlands: Elsevier; 2019. p. 1–13.

[3] Williams E, Elahinia MH. An automotive SMA mirror actuator: modeling, design, and experimental evaluation. J Intell Mater Syst Struct 2008;19(12):1425–34.

[4] Carpi F, Frediani G, Gerboni C, Gemignani J, De Rossi D. Enabling variable-stiffness hand rehabilitation orthoses with dielectric elastomer transducers. Med Eng Phys 2014; 36(2):205–11.

[5] Makishi W, Matunaga T, Haga Y, Esashi M. Active bending electric endoscope using shape memory alloy coil actuators. In: The first IEEE/RAS-EMBS international conference on biomedical robotics and biomechatronics, BioRob 2006. IEEE; 2006. p. 217–9.

[6] Hartl DJ, Lagoudas DC. Aerospace applications of shape memory alloys. Proc Inst Mech Eng G: J Aerospace Eng 2007;221(4):535–52.

[7] Gu G-Y, Zhu J, Zhu L-M, Zhu X. A survey on dielectric elastomer actuators for soft robots. Bioinspir Biomimet 2017;12(1), 011003.

[8] Simone F, Rizzello G, Seelecke S. Metal muscles and nerves—a self-sensing SMA-actuated hand concept. Smart Mater Struct 2017;26(9), 095007.

[9] Kohl M. Shape memory microactuators. Berlin Heidelberg: Springer Verlag; 2004.

[10] Kornbluh RD, Pelrine R, Prahlad H, Heydt R. Electroactive polymers: an emerging technology for MEMS. In: MEMS/MOEMS components and their applications, 5344. SPIE; 2004. p. 13–27.

[11] Actuator Solutions GmbH; 2021. Available from: https://www.actuatorsolutions.de/english/products/.

[12] Köpfer M. Industrialisierung der FGL-Technologie in hochvolumigen Serienprodukt. Karlsruhe: VDI-Expertenforum: Smart Materials—aus der Forschung in die industrielle Anwendung; 2017.

[13] Ingpuls GmbH; 2021. Available from: https://ingpuls.de/en/products-services/systems.html.

[14] Parker Hannifin Corporation; 2021. Available from: https://promo.parker.com/promotionsite/flexsense/us/en/home.

[15] StretchSense; 2021. Available from: https://stretchsense.com/solution/gloves/.

[16] Medynski C, Rattray B. Bebionic prosthetic design. In: Myoelectric symposium. Duke University Libraries; 2011. p. 1–4.

[17] Zhang H, Wang MY, Chen F, Wang Y, Kumar AS, Fuh JY. Design and development of a soft gripper with topology optimization. In: 2017 IEEE/RSJ international conference on intelligent robots and systems (IROS). IEEE/RSJ; 2017. p. 6239–44.

[18] Ölander A. An electrochemical investigation of solid cadmium-gold alloys. J Am Chem Soc 1932;54(10):3819–33.

[19] Greninger A, Mooradian V. Strain transformation in metastable beta copper-zinc and beta copper-tin alloys; 1938. Available from: https://www.bcin.ca/bcin/detail.app?id=431236.

[20] Kurdjumov GV, Khandros LG. Dokl Akad Nauk SSSR 1949;66:211–3.

[21] Buehler WJ, Gilfrich JV, Wiley RC. Effect of low-temperature phase changes on the mechanical properties of alloys near composition TiNi. J Appl Phys 1963;34(5):1475–7.

[22] Jani JM, Leary M, Subic A, Gibson MA. A review of shape memory alloy research, applications and opportunities. Mater Des 2014;1078–113.

[23] Otsuka K, Wayman CM. Shape memory materials. Cambridge University Press; 1999.

[24] Lagoudas DC. Shape memory alloys: modeling and engineering applications. Springer; 2008.

[25] Lewis N, York A, Seelecke S. Experimental characterization of self-sensing SMA actuators under controlled convective cooling. Smart Mater Struct 2013;22(9), 094012.

[26] SAES Getters. Smartflex Brochure; 2021. Available from: https://www.saesgetters.com/sites/default/files/SmartFlex%20Brochure_2.pdf.

[27] Dynalloy. Technical characteristics of FLEXINOL ® actuator wires. Available from: http://www.dynalloy.com/pdfs/TCF1140.pdf; 2020.

[28] Fu YQ, Luo JK, Flewitt AJ, Huang WM, Zhang S, Du HJ. Thin film shape memory alloys and microactuators. Int J Comput Mater Sci Surf Eng 2009;2(3–4):208–26.

[29] Furst SJ, Crews JH, Seelecke S. Stress, strain, and resistance behavior of two opposing shape memory alloy actuator wires for resistance-based self-sensing applications. J Intell Mater Syst Struct 2013;24(16):1951–68.

[30] Prechtl J, Seelecke S, Motzki P, Rizzello G. Self-sensing control of antagonistic SMA actuators based on resistance-displacement hysteresis compensation. In: Smart materials, adaptive structures and intelligent systems, 84027. ASME; 2020, V001T03A001.

[31] Kim S, Hawkes E, Choy K, Joldaz M, Foleyz J, Wood R. Micro artificial muscle fiber using NiTi spring for soft robotics. In: 2009 IEEE/RSJ international conference on intelligent robots and systems. IEEE; 2009. p. 2228–34.

[32] Morgan NB. Medical shape memory alloy applications—the market and its products. Mater Sci Eng, A 2004;378(1–2):16–23.

[33] Furuya Y, Shimada H. Shape memory actuators for robotic applications. Mater Des 1991; 12(1):21–8.

[34] Sreekumar M, Nagarajan T, Singaperumal M, Zoppi M, Molfino R. Critical review of current trends in shape memory alloy actuators for intelligent robots. Ind Robot 2007;34(4):285–94.

[35] Goergen Y, Chadda R, Britz R, Scholtes D, Koev N, Motzki P, et al. Shape memory alloys in continuum and soft robotic applications. In: Smart materials, adaptive structures and intelligent systems, 59131. ASME; 2019. p. 1–6. V001T04A014.

[36] Goergen Y, Rizzello G, Seelecke S, Motzki P. Modular design of an SMA driven continuum robot. In: Smart materials, adaptive structures and intelligent systems, 84027. ASME; 2020, V001T04A007.

[37] Ikuta K, Tsukamoto M, Hirose S. Shape memory alloy servo actuator system with electric resistance\nfeedback and application for active endoscope. In: Proceedings. 1988 IEEE international conference on robotics and automation. IEEE; 1988. p. 427–30.

[38] Fukuda T, Guo S, Kosuge K, Arai F, Negoro M, Nakabayashi K. Micro active catheter system with multi degrees of freedom. In: Proceedings of the 1994 IEEE international conference on robotics and automation. IEEE; 1994. p. 2290–5.

[39] Lim G, Minami K, Yamamoto K, Sugihara M, Uchiyama M, Esashi M. Multi-link active catheter snake-like motion. Robotica 1996;14(5):499–506.

[40] Mineta T, Mitsui T, Watanabe Y, Kobayashi S, Haga Y, Esashi M. Batch fabricated flat meandering shape memory alloy actuator for active catheter. Sens Actuators, A 2001; 88(2):112–20.

[41] Crews JH, Buckner GD. Design optimization of a shape memory alloy-actuated robotic catheter. J Intell Mater Syst Struct 2012;23(5):545–62.

[42] Kadir MRA, Dewi DEO, Jamaludin MN, Nafea M, Ali MSM. A multi-segmented shape memory alloy-based actuator system for endoscopic applications. Sens Actuators, A 2019; 296:92–100.

[43] Maeda S, Abe K, Yamamoto K, Tohyama O, Ito H. Active endoscope with SMA (shape memory alloy) coil springs. In: Proceedings of ninth international workshop on micro electromechanical systems. IEEE; 1996. p. 290–5.

[44] Vinograd I, Klin B, Brosh T, Weinberg M, Flomenblit Y, Nevo Z. A new intratracheal stent made from nitinol, an alloy with "shape memory effect". J Thoracic Cardiovasc Surg 1994;107(5):1255–61.

[45] Kim B, Lee MG, Lee YP, Kim YI, Lee GH. An earthworm-like micro robot using shape memory alloy actuator. Sens Actuators, A 2006;125(2):429–37.

[46] Mi Z-N, Gong Z-B, Qian J-W, Mi Z-W, Shen L-Y. Study on moving principle of colonoscopic robot. J Shanghai Univ 2001;5(2):143–6.

[47] Furst SJ, Seelecke S. Fabrication and characterization of a dual-joint smart inhaler nozzle actuated by embedded SMA wires. Smart Mater Struct 2014;23(3), 035008.

[48] Kim B, Lee S, Park JH, Park J-O. Design and fabrication of a locomotive mechanism for capsule-type endoscopes using shape memory alloys (SMAs). IEEE/ASME Trans Mechatron 2005;10(1):77–86.

[49] Reynaerts D, Peirs J, Van Brussel H. Shape memory micro-actuation for a gastro-intestinal intervention system. Sens Actuators, A 1999;77(2):157–66.

[50] Haga Y, Mizushima M, Matsunaga T, Esashi M. Medical and welfare applications of shape memory alloy microcoil actuators. Smart Mater Struct 2005;14:S266.

[51] Dotter CT, Buschmann RW, McKinney MK, Rösch J. Transluminal expandable nitinol coil stent grafting: preliminary report. Radiology 1983;147(1):259–60.

[52] Yanagihara K, Mizuno H, Wada H, Hitomi S. Tracheal stenosis treated with self-expanding nitinol stent. Ann Thoracic Surg 1997;63(6):1786–9.

[53] Ayvali E, Liang C-P, Ho M, Chen Y, Desai JP. Towards a discretely actuated steerable Cannula for diagnostic and therapeutic procedures. Int J Robot Res 2012;31(5):588–603.

[54] Nakamura Y, Matsui A, Saito T, Yoshimoto K. Shape-memory-alloy active forceps for laparoscopic surgery. In: Proceedings of 1995 IEEE international conference on robotics and automation. IEEE; 1995. p. 2320–7.

[55] Kode VRC, Çavuşoğlu MC. Design and characterization of a novel hybrid actuator using shape memory alloy and DC micromotor for minimally invasive surgery applications. IEEE/ASME Trans Mechatron 2007;12(4):455–64.

[56] Yuk H, Kim D, Lee H, Jo S, Shin JH. Shape memory alloy-based small crawling robots inspired by *C. elegans*. Bioinsp Biomim 2011;6(4), 046002.

[57] Menciassi A, Park JH, Lee S, Gorini S, Dario P, Park J-O. Robotic solutions and mechanisms for a semi-autonomous endoscope. In: IEEE/RSJ international conference on intelligent robots and systems. IEEE; 2002. p. 1379–84.

[58] Carrozza MC, Lencioni L, Magnani B, Dario P, Reynaerts D, Trivella MG, et al. A microrobot for colonoscopy. MHS'96 proceedings of the seventh international symposium on micro machine and human science. IEEE; 2002. p. 223–8.

[59] Yan G, Wang K, Shi J. Research on micro robot for colonoscopy. In: 2005 IEEE engineering in medicine and biology 27th annual conference. IEEE; 2005. p. 5050–3.

[60] Cotta F, Icardi F, Zurlo GT, Molfino RM. Peristaltic locomotion: application to a worm-like robot. In: Climbing and walking robots. Berlin Heidelberg: Springer; 2006. p. 501–8.

[61] Chonan S, Jiang ZW, Tani J, Orikasa S, Tanahashi Y, Takagi T, et al. Development of an artificial urethral valve using SMA actuators. Smart Mater Struct 1997;6(4):410.

[62] Luo Y, Okuyama T, Takagi T, Kamiyama T, Nishi K, Yambe T. Thermal control of shape memory alloy artificial anal sphincters for complete implantation related content thermal responses of SMA artificial anal sphincters. Smart Mater Struct 2004;14(1):29.

[63] Miki H, Okuyama T, Kodaira S, Luo Y, Takagi T, Yambe T, et al. Artificial-esophagus with peristaltic motion using shape memory allo. Int J Appl Electromagnet Mech 2010; 33(1–2):705–11.

[64] Tadesse Y, Hong D, Priya S. Twelve degree of freedom baby humanoid head using shape memory alloy actuators. J Mech Robot 2011;3(1), 011008.

[65] Simone F, Rizzello G, Seelecke S, Motzki P. A soft five-fingered hand actuated by shape memory alloy wires: design, manufacturing, and evaluation. Front Robot AI 2020;7, 608841.

[66] Hara F, Akazawa H, Kobayashi H. Realistic facial expressions by SMA driven face robot. In: Proceedings 10th IEEE international workshop on robot and human interactive communication. ROMAN 2001. IEEE; 2001. p. 504–11.

[67] Wolfe TB, Faulkner MG, Wolfaardt J. Development of a shape memory alloy actuator for a robotic eye prosthesis. Smart Mater Struct 2005;14(4):759.

[68] Maeno T, Hino T. Miniature five-fingered robot hand driven by shape memory alloy actuators. Proceedings of the 12th IASTED international conference, robotics and applications. IASTED; 2006. p. 174–9.

[69] Lee JH, Okamoto S, Matsubara S. Development of multi-fingered prosthetic hand using shape memory alloy type artificial muscle. Comput Technol Appl 2012;3(7):477–84.

[70] Andrianesis K, Tzes A. Development and control of a multifunctional prosthetic hand with shape memory alloy actuators. J Intell Robot Syst 2014;78(2):257–89.

[71] Stirling L, Yu C-H, Miller J, Hawkes E, Wood R, Goldfield E, et al. Applicability of shape memory alloy wire for an active, soft orthotic. J Mater Eng Perform 2011;20(4):658–62.

[72] Tang T, Zhang D, Xie T, Zhu X. An exoskeleton system for hand rehabilitation driven by shape memory alloy. In: 2013 IEEE international conference on robotics and biomimetics (ROBIO). IEEE; 2013. p. 756–61.

[73] Pittaccio S, Garavaglia L, Ceriotti C, Passaretti F. Applications of shape memory alloys for neurology and neuromuscular rehabilitation. J Funct Biomater 2015;6(2):328–44.

[74] Copaci D, Cano E, Moreno L, Blanco D. New design of a soft robotics wearable elbow exoskeleton based on shape memory alloy wire actuators. Appl Bionics Biomech 2017;2017:1605101. https://doi.org/10.1155/2017/1605101.

[75] Zhao F, Fukuyama K, Sawada H. Compact Braille display using SMA wire array. In: RO-MAN 2009—the 18th IEEE international symposium on robot and human interactive communication. IEEE; 2009. p. 28–33.

[76] Kha H, Chen B, Clark G, Jones R. Finite element modeling of final placement and insertion depth of new cochlear implant electrode array embedded with nitinol shape memory alloy actuators. In: Proceedings of the 2006 international conference on modeling, simulation & visualization methods, MSV 2006. Citeseer; 2006. p. 27–32.

[77] Pfeifer R, Müller CW, Hurschler C, Kaierle S, Wesling V, Haferkamp H. Adaptable orthopedic shape memory implants. Procedia CIRP 2013;5:253–8.

[78] Haga Y, Makishi W, Iwami K, Totsu K, Nakamura K, Esashi M. Dynamic Braille display using SMA coil actuator and magnetic latch. Sens Actuators, A 2005;119(2):316–22.

[79] Velázquez R, Pissaloux E, Szewczyk J, Hafez M. Miniature shape memory alloy actuator for tactile binary information display. In: Proceedings of the 2005 IEEE international conference on robotics and automation. IEEE; 2005. p. 1344–9.

[80] Patrick L, Gabor K, Silvain M. Characterization of dielectric elastomer actuators based on a hyperelastic film model. Sens Actuators, A 2007;135(2):748–57.

[81] York A, Dunn J, Seelecke S. Experimental characterization of the hysteretic and rate-dependent electromechanical behavior of dielectric electro-active polymer actuators. Smart Mater Struct 2010;19(9):094014. 094014.

[82] Kaltseis R, Keplinger C, Koh SJA, Baumgartner R, Goh YF, Ng WH, et al. Natural rubber for sustainable high-power electrical energy generation. RSC Adv 2014;4(53):27905–13.

[83] Chen Y, Agostini L, Moretti G, Fontana M, Vertechy R. Dielectric elastomer materials for large-strain actuation and energy harvesting: a comparison between styrenic rubber, natural rubber and acrylic elastomer. Smart Mater Struct 2019;28(11), 114001.

[84] Fasolt B, Hodgins M, Rizzello G, Seelecke S. Effect of screen printing parameters on sensor and actuator performance of dielectric elastomer (DE) membranes. Sens Actuators, A 2017; 265:10–9.

[85] Rosset S, Shea HR. Flexible and stretchable electrodes for dielectric elastomer actuators. Appl Phys A 2013;110(2):281–307.

[86] Pelrine R, Kornbluh R, Pei Q, Joseph J. High-speed electrically actuated elastomers with strain greater than 100%. Science 2000;287(5454):836–9.
[87] Röntgen WC. Ueber die durch Electricität bewirkten Form- und Volumenänderungen von dielectrischen Körpern. Ann Phys Chem 1880;247(13):771–86.
[88] Pelrine R, Kornbluh R, Joseph J. Electrostriction of polymer dielectrics with compliant electrodes as a means of actuation. Sens Actuators, A 1998;64(1):77–85.
[89] Carpi F, Anderson I, Bauer S, Frediani G, Gallone G, Gei M, et al. Standards for dielectric elastomer transducers. Smart Mater Struct 2015;24(10), 105025.
[90] Carpi F, Frediani G, Turco S, De Rossi D. Bioinspired tunable lens with muscle-like electroactive elastomers. Adv Funct Mater 2011;21(21):4152–8.
[91] Hoffstadt T, Griese M, Maas J. Online identification algorithms for integrated dielectric electroactive polymer sensors and self-sensing concepts. Smart Mater Struct 2014;23(10), 104007.
[92] Rizzello G, Naso D, York A, Seelecke S. A self-sensing approach for dielectric elastomer actuators based on online estimation algorithms. IEEE/ASME Trans Mechatron 2017; 22(2):728–38.
[93] Rizzello G, Serafino P, Naso D, Seelecke S. Towards sensorless soft robotics: self-sensing stiffness control of dielectric elastomer actuators. IEEE Trans Robot 2020;36(1):174–88.
[94] Rizzello G, Naso D, York A, Seelecke S. Closed loop control of dielectric elastomer actuators based on self-sensing displacement feedback. Smart Mater Struct 2016;25(3), 035034.
[95] Hau S, Bruch D, Rizzello G, Motzki P, Seelecke S. Silicone based dielectric elastomer strip actuators coupled with nonlinear biasing elements for large actuation strains. Smart Mater Struct 2018;27(7), 074003.
[96] Loew P, Rizzello G, Seelecke S. A novel biasing mechanism for circular out-of-plane dielectric actuators based on permanent magnets. Mechatronics 2018;56:48–57.
[97] Neu J, Hubertus J, Croce S, Schultes G, Seelecke S, Rizzello G. Fully polymeric domes as high-stroke biasing system for soft dielectric elastomer actuators. Front Robot AI 2021;8:171.
[98] Berselli G, Vertechy R, Vassura G, Parenti-Castelli V. Optimal synthesis of conically shaped dielectric elastomer linear actuators: design methodology and experimental validation. IEEE/ASME Trans Mechatron 2011;16(1):67–79.
[99] Moretti G, Sarina L, Agostini L, Vertechy R, Berselli G, Fontana M. Styrenic-rubber dielectric elastomer actuator with inherent stiffness compensation. Actuators 2020;9(2):44.
[100] Rizzello G, Loew P, Agostini L, Fontana M, Seelecke S. A lumped parameter model for strip-shaped dielectric elastomer membrane transducers with arbitrary aspect ratio. Smart Mater Struct 2020;29(11), 115030.
[101] Nalbach S, Banda RM, Croce S, Rizzello G, Naso D, Seelecke S. Modeling and design optimization of a rotational soft robotic system driven by double cone dielectric elastomer actuators. Front Robot AI 2020;6:150.
[102] Hau S, Rizzello G, Seelecke S. A novel dielectric elastomer membrane actuator concept for high-force applications. Extreme Mech Lett 2018;23:24–8.
[103] Nguyen CT, Phung H, Nguyen TD, Jung H, Choi HR. Multiple-degrees-of-freedom dielectric elastomer actuators for soft printable hexapod robot. Sens Actuators, A 2017;267: 505–16.
[104] Kunze J, Prechtl J, Bruch D, Fasolt B, Nalbach S, Motzki P, et al. Design, manufacturing, and characterization of thin, core-free, rolled dielectric elastomer actuators. Actuators 2021; 10(4), 69.
[105] Pei Q, Rosenthal M, Stanford S, Prahlad H, Pelrine R. Multiple-degrees-of-freedom electroelastomer roll actuators. Smart Mater Struct 2004;13(5), N86.

[106] Wang H, Li L, Zhu Y, Yang W. Analysis and application of a rolled dielectric elastomer actuator with two degrees of freedom. Smart Mater Struct 2016;25(12), 125008.

[107] Li J, Liu L, Liu Y, Leng J. Dielectric elastomer spring-roll bending actuators: applications in soft robotics and design. Soft Robotics 2019;6(1):69–81.

[108] Kovacs G, Düring L, Michel S, Terrasi G. Stacked dielectric elastomer actuator for tensile force transmission. Sens Actuators A 2009;155(2):299–307. https://doi.org/10.1016/j.sna.2009.08.027.

[109] Maas J, Tepel D, Hoffstadt T. Actuator design and automated manufacturing process for DEAP-based multilayer stack-actuators. Meccanica 2015;50(11):2839–54.

[110] Kofod G, Wirges W, Paajanen M, Bauer S. Energy minimization for self-organized structure formation and actuation. Appl Phys Lett 2007;90(8), 081916.

[111] O'Brien B, McKay T, Calius E, Xie S, Anderson I. Finite element modelling of dielectric elastomer minimum energy structures. Appl Phys A 2008;94(3):507–14. https://doi.org/10.1007/s00339-008-4946-8.

[112] McGough K, Ahmed S, Frecker M, Ounaies Z. Finite element analysis and validation of dielectric elastomer actuators used for active origami. Smart Mater Struct 2014;23(9), 094002.

[113] Akbari S, Shea HR. An array of 100 μm × 100 μm dielectric elastomer actuators with 80% strain for tissue engineering applications. Sens Actuators A 2012;186:236–41.

[114] Chen Y, Zhao H, Mao J, Chirarattananon P, Helbling EF, Hyun N-SP, et al. Controlled flight of a microrobot powered by soft artificial muscles. Nature 2019;575(7782):324–9.

[115] Kim B, Jeong Y, Lim H-Y, Kim TS, Park J-O, Dario P, et al. Smart colonoscope system. In: IEEE/RSJ international conference on intelligent robots and systems. IEEE; 2002. p. 1367–72.

[116] Plante J-S. Dielectric elastomer actuators for binary robotics and mechatronics. Massachusetts Institute of Technology; 2006.

[117] Petralia MT, Wood RJ. Fabrication and analysis of dielectric-elastomer minimum-energy structures for highly-deformable soft robotic systems. In: 2010 IEEE/RSJ international conference on intelligent robots and systems. IEEE; 2010. p. 2357–63.

[118] Xing Z, Zhang J, McCoul D, Cui Y, Sun L, Zhao J. A super-lightweight and soft manipulator driven by dielectric elastomers. Soft Robotics 2020;7(4):512–20.

[119] Prechtl J, Kunze J, Bruch D, Seelecke S, Rizzello G. Bistable actuation in multi-DoF soft robotic modules driven by rolled dielectric elastomer actuators. In: 2021 IEEE 4th international conference on soft robotics (RoboSoft). IEEE; 2021. p. 82–9.

[120] Araromi OA, Gavrilovich I, Shintake J, Rosset S, Richard M, Gass V, et al. Rollable multi-segment dielectric elastomer minimum energy structures for a deployable microsatellite gripper. IEEE/ASME Trans Mechatron 2015;20(1):438–46.

[121] Zhao J, Zhang J, McCoul D, Cui Y, Lan X. A soft biomimetic module of elephant trunk driven by dielectric elastomers. In: 2018 IEEE international conference on robotics and biomimetics (ROBIO). IEEE; 2018. p. 349–54.

[122] Duduta M, Hajiesmaili E, Zhao H, Wood RJ, Clarke DR. Realizing the potential of dielectric elastomer artificial muscles. Proc Natl Acad Sci 2019;116(7):2476–81.

[123] Lau G-K, Heng K-R, Ahmed AS, Shrestha M. Dielectric elastomer fingers for versatile grasping and nimble pinching. Appl Phys Lett 2017;110(18):182906. 182906.

[124] Jung MY, Chuc NH, Kim JW, Koo IM, Jung KM, Lee YK, et al. Fabrication and characterization of linear motion dielectric elastomer actuators. In: Smart structures and materials 2006: electroactive polymer actuators and devices (EAPAD), 6168. SPIE; 2006. p. 616824.

[125] Kovacs G, Lochmatter P, Wissler M. An arm wrestling robot driven by dielectric elastomer actuators. Smart Mater Struct 2007;16(2), S306.

[126] Biddiss E, Chau T. Dielectric elastomers as actuators for upper limb prosthetics: challenges and opportunities. Med Eng Phys 2008;30(4):403–18.

[127] Kwak JW, Chi HJ, Jung KM, Koo JC, Jeon JW, Lee Y, et al. A face robot actuated with artificial muscle based on dielectric elastomer. J Mech Sci Technol 2005;19(2):578–88.

[128] Wang Y, Zhu J. Artificial muscles for jaw movements. Extreme Mech Lett 2016;6:88–95.

[129] Carpi F, De Rossi D. Bioinspired actuation of the eyeballs of an android robotic face: concept and preliminary investigations. Bioinspir Biomimet 2007;2(2), S500.

[130] Li L, Godaba H, Ren H, Zhu J. Bioinspired soft actuators for eyeball motions in humanoid robots. IEEE/ASME Trans Mechatron 2019;24(1):100–8.

[131] Goulbourne N, Frecker MI, Mockensturm EM, Snyder AJ. Modeling of a dielectric elastomer diaphragm for a prosthetic blood pum. In: Smart structures and materials 2003: electroactive polymer actuators and devices (EAPAD), 5051. SPIE; 2003. p. 319–31.

[132] Carpi F, Menon C, De Rossi D. Electroactive elastomeric actuator for all-polymer linear peristaltic pumps. IEEEASME Trans Mechatron 2010;15(3):460–70.

[133] Bowers AE, Rossiter JM, Walters PJ, Ieropoulos IA. Dielectric elastomer pump for artificial organisms. In: Electroactive polymer actuators and devices (EAPAD) 2011, 7976. SPIE; 2011. p. 797629.

[134] Pourazadi S, Ahmadi S, Menon C. Towards the development of active compression bandages using dielectric elastomer actuators. Smart Mater Struct 2014;23(6), 065007.

[135] Kelley CR, Kauffman JL. Exploring dielectric elastomers as actuators for hand tremor suppression. In: Electroactive polymer actuators and devices (EAPAD) 2017, 10163. SPIE; 2017. p. 1016322.

[136] Carpi F, Frediani G, De Rossi D. Opportunities of hydrostatically coupled dielectric elastomer actuators for haptic interfaces. In: Electroactive polymer actuators and devices (EAPAD) 2011, 7976. SPIE; 2011. p. 797618.

[137] Zhang R, Kunz A, Lochmatter P, Kovacs G. Dielectric elastomer spring roll actuators for a portable force feedback device. In: 2006 14th symposium on haptic interfaces for virtual environment and teleoperator systems. IEEE; 2006. p. 347 53.

[138] Choi HR, Lee SW, Jung KM, Koo JC, Lee SI, Choi HG, et al. Tactile display as a braille display for the visually disabled. In: 2004 IEEE/RSJ international conference on intelligent robots and systems (IROS), 2. IEEE; 2004. p. 1985–90.

[139] Boys H, Frediani G, Poslad S, Busfield J, Carpi F. A dielectric elastomer actuator-based tactile display for multiple fingertip interaction with virtual soft bodies. Electroactive polymer actuators and devices (EAPAD) 2017. 10163. SPIE; 2017, 101632D.

[140] Carpi F, Frediani G, De Rossi D. Electroactive elastomeric haptic displays of organ motility and tissue compliance for medical training and surgical force feedback. IEEE Transact Biomed 2009;56(9):2327–30.

[141] Carpi F, De Rossi D, Kornbluh R, Pelrine RE, Sommer-Larsen P. Dielectric elastomers as electromechanical transducers: Fundamentals, materials, devices, models and applications of an emerging electroactive polymer technology. Elsevier; 2011.

[142] Kornbluh RD, Pelrine R, Prahlad H, Wong-Foy A, McCoy B, Kim S, et al. From boots to buoys: promises and challenges of dielectric elastomer energy harvesting. In: Electroactivity in polymeric materials. Springer; 2011. p. 67–93.

Chapter 7

Fabrication of endoluminal medical devices

Mario Milazzo[a] and Cesare Stefanini[a,b]
[a]*The BioRobotics Institute, Scuola Superiore Sant'Anna, Pontedera, PI, Italy,* [b]*Healthcare Engineering Innovation Center, Department of Biomedical Engineering, Khalifa University of Science and Technology, Abu Dhabi, Unitted Arab Emirates*

7.1 Introduction

The application of technology in surgery and therapy is rapidly growing, leading to emerging products in different sectors. The need for application of novel technologies is, in the medical and surgical fields, a natural cause and consequence of a deep innovation process: In these sectors, surgeons, once they reach the limits of human capabilities, begin to extend their abilities by means of machine assistance. Man–machine synergies have allowed doctors to overcome the traditional surgical limitations, extending the surgeons' abilities to plan and carry out interventions with greater accuracy and less invasiveness, even allowing them to perform tasks that would be impossible with traditional methodologies.

This is the framework of minimally invasive surgery (MIS), a widely employed surgical approach consisting in using small tools as "extensions" of the doctors' hands, for diagnostic and treatment purposes, minimizing the distress to the patient and his external and internal tissues. Direct implications are an improved quality of the healthcare and a shorter hospitalization [1]. Aiming at meeting these challenging goals, which are in hard contrast with traditional surgery where surgeons deal with several issues due to the invasiveness of surgery (e.g., bleeding, pain, complications), recent technological advances have pushed scientists to provide novel less-invasive efficient tools for performing endoluminal operations [2,3]. These advanced devices have given the possibility to exploit both natural orifices and/or small-access holes, for inserting surgical devices, using also smart user interfaces (e.g., primary-secondary[a] configurations, haptic joysticks), thus improving the

[a] Due to terminology concerns, we propose this "primary-secondary" in place of "master-slave", a word choice largely employed in previous years.

Endorobotics. https://doi.org/10.1016/B978-0-12-821750-4.00007-4

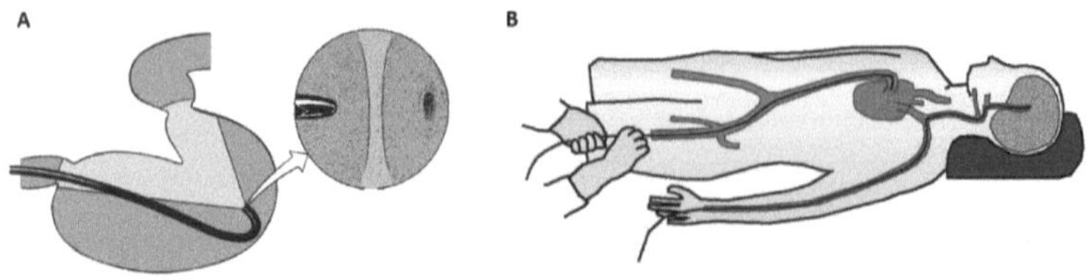

FIG. 7.1 Common use of endoluminal devices: visualizing and treating the gastrointestinal tract, hollow organs, and the cardiovascular system. *(Panel A—Reprinted with permission Cotton PB, Williams CB. Practical gastrointestinal endoscopy: the fundamentals. John Wiley & Sons; 2008. © 2013 John Wiley and Sons. Panel B—Reprinted with permission Haga Y, Muyari Y, Goto S, Matsunaga T, Esashi M. Development of minimally invasive medical tools using laser processing on cylindrical substrates. Electr Eng Jpn 2011;176(1):65–74. © 2011 John Wiley and Sons.)*

accuracy in the task. The most relevant applications concern the gastrointestinal tract, hollow organs, and cardiovascular apparatus (Fig. 7.1) [2–5].

Although MIS has been considered a valuable approach, there are four main challenges that every device in this field has to address: dimensions, materials properties, navigation, and sensing. Reduced dimensions are the first objective to meet since every surgical tool has to navigate and perform its tasks in lumens with dimensions that range from few millimeters down to fractions of millimeters, carrying all the required structural, actuation, and electronic components. Material properties of the external structures must abide by specific requirements in order to not be rejected, or damaged, or cause damages when in contact with sensitive tissues or body fluids (these latter with a pH that is often far from the neutral).

Therefore, materials have to be selected among the biocompatible machinable materials that may be tolerated and not contaminated by living tissues, and adequate to ensure structural stability throughout the navigation and task operativity. The last requirement for a MIS device is the ability to efficiently navigate in lumens and, possibly, to provide a sensory feedback. The main issue related to the locomotion consists in the adaptability to millimetric or submillimetric ducts in which steering can be difficult to perform (Fig. 7.2).

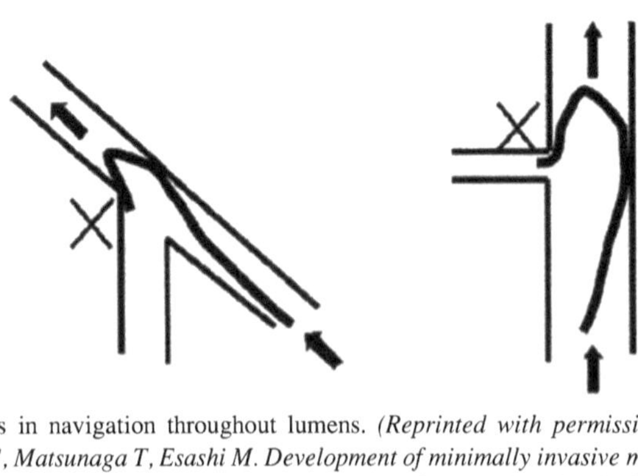

FIG. 7.2 Issues in navigation throughout lumens. *(Reprinted with permission from Haga Y, Muyari Y, Goto S, Matsunaga T, Esashi M. Development of minimally invasive medical tools using laser processing on cylindrical substrates. Electr Eng Jpn 2011;176(1):65–74. © 2011 John Wiley and Sons.)*

Solutions to overcome this problem have been the employment of magnetic fields to orient the tip of endoscopes [6,7], also supported by the parietal forces exerted when the device is in contact with the walls of the lumen [7], or the employment of small magnetic-driven capsules that can easily steer because of their small dimensions [8]. Another issue related to the navigation is the need to resist to the peristaltic activity of the gastrointestinal tract: In fact, diagnostic and treatment activities have to be performed independently of the involuntary muscular activity of the lumen that can disturb the surgical task. A potential solution to deal with this important requirement consists in endowing the devices with appendices in the forms of flaps or legs that, upon need, improve the navigation capabilities, and perform a stable anchoring on lumen walls without damaging the tissues [9,10].

From a design standpoint, most of the requirements for an efficient MIS endoluminal device converge to an efficient fabrication approach. Manufacturing has, indeed, important implications on the selection of the constitutive materials, and the following machinability for making specific design features (e.g., micro-/nanostructural components) that can be challenging to achieve because of the reduced dimensions. This chapter discusses the challenges and solutions to effectively develop and manufacture an endoluminal device, starting from the physical laws that govern the miniaturization of a device, to an overview of the different manufacturing approaches currently employed in the field. Lastly, after deploying a few relevant examples of fabrication for MIS endoluminal tools, we conclude with a summary of the main take-away messages and the future outlook in the field.

7.2 Developing endoluminal devices: From scaling laws to the fabrication

The design and fabrication of micromachines for MIS requires the application of the particular criteria of microengineering. It is not possible to simply scale-down a macromachine, because we have to consider that the scaling laws of different forces have different rates, thus changing their relative magnitudes in the microdomain. As a consequence, common physical intuition has to be substituted or supported by theoretical modeling and numerical simulations. Another topic that directly influences their development is their potential reusability in order to reduce waste and the indirect costs associated to single-use tools that, apparently, are cheaper than the multiuse ones. The manufacturability of microdevices is a complementary aspect in the microengineering scenario, since it determines the proper machinability of the constituent materials, thus ultimately affecting their structural and behavioral features (e.g., reusability).

7.2.1 Scale-related issues: Physical laws

In the design of traditional machines, the engineer can exploit the wide know-how which is available in technical literature, while skillful and experienced designers can quickly exploit their own expertise in pointing out the best technical solution to a given problem, basically relying on their own physical intuition which derives from the common experience. But when the size of the machines becomes smaller and smaller, the intuitive understanding of the most significant phenomena begins to fade. Actually, the physical laws which rule the microdomain are well known and are the same that applies to the macroworld. Roughly speaking, we may say that the "frontier of knowledge," i.e., the dimensional scale below which further experimental and theoretical investigations by physicians are required, can be located around the picoscale (10^{-12}m), where the realm of high energy physics begins. Coming back to the microengineering domain, Newton's mechanics and Maxwell's electromagnetism satisfy almost all the needs of the microengineer. Therefore, the difficulties of a microengineer are not only due to lack of knowledge but also to the change of the relative weight of the forces which act on the micromachines. This consideration, though simple in form, is rich in practical implications because the working of a machine depends essentially on the balance of all acting forces. For example, the speed of a car on a highway is the result of the balance of many types of frictional forces: If the magnitude of only one of these forces undergoes a variation, the performance of the car would change. Similar examples could be sketched of every kind of machine. If we scale-down the linear dimensions of a machine, we would observe rapid variations in the intensity of the acting force, very often leading to unexpectedly inefficient behaviors. A useful tool to investigate the scaling law of forces is represented by the scale exponent. Being "L" the typical or characteristic linear dimension of the machine, the scaling exponent of a force "F" is "n" if the force is proportional to L^n. Forces having the same scaling exponent do not change their relative magnitude in the scaling-down process. If a force has a positive exponent, it fades in small scales, while if the exponent is negative, the force becomes more and more important.

In Table 7.1, the scaling exponent of the most frequent forces in microengineering is reported. To assure that a machine keeps working through a scaling-down process, it is necessary to guarantee that the ratio between the relevant forces is kept constant. The ratios between the forces which are significant in different design tasks are expressed by adimensional numbers (e.g., in fluidic devices, the adimensional Reynolds number (*Re*) plays a central role since it gives the ratio between the inertial and viscous forces).

When a machine is scaled-down by a factor of ten, mass forces decrease their magnitude by a factor of 1000, while the pressure decrease is only 100. Consequently, the relative weight of material strength (ultimate stress × cross

TABLE 7.1 The scaling exponent of the most common forces.

Force	Scaling exponent	Examples
Inertia	3	• Coriolis force • D'Alembert's force
Mass	3	Weight
Pressure	2	• Cross sectional material strength • Fluidic actuators
Electrostatic	2	Ionic compounds
Viscosity	2	Viscous friction
Surface tension	1	Forces in liquid-air interfaces
Van der Waals	1	Contacts between polished surfaces
Intermolecular	−6	Covalent compounds

section area) with respect to mass force (e.g., the weight) is increased by a factor of ten. As a result, smaller structures can be very thin and they can easily withstand their own weight, while larger ones must be thicker and bulkier.

Miniaturizing a device means, first of all, resizing its main structure, which hosts and supports the different components and which can also exert actions on the external environment. The mechanical strength of this element has to face two kinds of loads: static and dynamic. In the microscale, however, the last effect becomes negligible or not dangerous, due to the low inertia of the moving parts that makes dynamic actions to be very small with respect to the strength of the structural material. In addition, loads can have an internal origin (i.e., actions caused by the supported components), or an external one (i.e., contact with other bodies). In any case, however, the ultimate cause for any arising force, transmitted inside the machine or not, is the interaction that it exchanges with the external world.

A necessary condition for a machine to withstand external physical actions is that the hardness of the structural parts has to be higher than that of the external bodies on which they act. That is the reason why, in nature, horses have hard hooves for walking on natural earth, and why those hooves have to be artificially hardened using iron if the ground is artificial. Concerning small systems, thinner cross sections are strong enough to withstand external actions because volume forces (e.g., weight) become less influent than contact forces with the downscaling.

Mechanisms in miniature machines are a very critical issue since the precise coupling between moving parts becomes more difficult than in the macroworld, where fabrication technologies are very accurate with respect to the dimensions of devices. This problem is enhanced by the behavior of friction in the microscale, which requires small diameter, "pivot-like" joints in order to preserve an acceptable efficiency. In fact, the physics of friction in the macroscale, for dimensions over 1 mm, is basically expressed by the Guillaume Amontons law, where the friction force is proportional to the normal contact force between two sliding surfaces by the coefficient of friction.

This model is not adequate in the microscale, where adhesive contact forces (electrostatic, Van der Waals, and capillary) become dominant and always present independently from the load conditions. That is the reason why the design of microsized bearing has to be "pivot-like," thus reducing the contact area and therefore the adhesive effects.

Possible solutions for smaller devices are the following: improving the fabrication accuracy reaching the molecular level, as in the case of nanotubes; exploiting slide-free contact between two moving parts as in piezoelectric wave actuators or wobble motors; substituting the continuous motion in mechanisms with the alternate one; and using flexible hinges as couplings.

Regarding the dynamic behavior of mechanisms, it is interesting to observe that power can be effectively transmitted in the microworld using elastic energy stored in a flexible organ. In fact, given the maximum stress and strain levels in the material, the stored energy is proportional to the volume of the organ, but the delivery time of this energy becomes smaller and smaller as the dimensions scale-down because resonance frequencies of structures scale with positive exponents. Therefore, more power per unit of volume can be transmitted elastically in the micro- than in the macrodomain. This is the reason why high-power density can be reached by piezoelectric resonant motors.

7.2.2 Reusability issues for endoluminal devices

Reusability of clinical devices has been a key topic of discussion throughout years. While, in the past, a repeated use of medical tools and consumables has been the origin of infections in patients due to an imperfect sterilization process, the advent of polymers has opened the possibility of cheap single-use devices to reduce complications. However, in view of a reduction of waste and its associated costs, a new trend for making clinical devices reusable has been observed in the last decade. New disinfection and sterilization techniques have been developed to overcome the limitations of the past, but the need to reach high temperatures (~120 °C), to remove all microorganisms and bacterial spores, has brought new issues to overcome. The main and most important concern is about the materials employed for clinical use: While metallic devices can sustain repeated cycles of heating/cooling, polymers can hardly bear more than one sterilization cycle since the deformations due to their material properties often make clinical tools inadequate for multiple uses. Moreover, as occurs in some cases like the endotracheal tubes, polymers are heated by the human body and are deformed even during their first use, thus making these tools unusable for even a second use independently of the required sterilization [11].

This issue also concerns endoluminal devices, and especially those used for the gastrointestinal tract where patient's stools can be the origin of surface contamination. Moreover, this issue is even more relevant in the COVID-19 era, as the respiratory apparatus can be vehicle of a deadly disease [12].

In addition to the traditional disinfection and sterilization methodologies, research studies have looked for solutions to give surfaces, in contact with potentially contaminated tissues, antimicrobial or antifouling properties. With the first term, scientists have identified all the features on a surface that are able to kill microorganisms or bacteria by using physical, chemical, or electrostatic approaches (e.g., shaping the microstructure, using antibiotics). Antifouling methodologies, in contrast, are those that prevent the attachment of foreign bodies, exploiting the physico-chemical properties of the surfaces and, thus, creating an invisible barrier. This avenue of research is particularly indicated for polymers, whose physico-chemical characteristics are easier to manipulate than those of metallic surfaces. Recent surveys have also shown that using devices with antimicrobial/antifouling surfaces has led to high rates of success in preventing infections, achieving an almost full reusability of the tools without a relevant increase of costs [13].

7.2.3 Fabrication technologies for endoluminal devices

Fabrication approaches for MIS endoluminal devices can be classified in five main clusters: micromachining, additive manufacturing technologies (i.e., 3D printing, lithography, and 4D printing), electro-discharge machining (EDM), injection molding, and laser machining. The following section will briefly discuss the opportunities and challenges for these approaches, also summarized in Table 7.2.

7.2.4 Micromachining

Micromachining is undoubtedly the most employed approach for making millimetric components almost independently of the constituent material, from metals to biological tissues (e.g., bone) [14–16]. It directly derives from traditional milling, and employs microtools to perform subtractive operations to achieve micrometric features on millimetric components. By using high-speed spindles, is it possible to achieve a high stability with an accurate and precise microstructural topography, a feature that becomes more important when downsizing the scale. The high quality of the manufactured workpieces directly implies a reduction of the fabrication time since parts do not often require an additional finishing process (e.g., polishing) [16].

However, micromachining possesses a number of issues that still require further investigations for a proper understanding and management. Among others, high chip-production rate, burr generation, temperature, and contact forces affect the cutting process and, indirectly, the finished products. Although these parameters may appear similar to those of traditional macroscale milling,

TABLE 7.2 Classification and assessment of fabrication strategies for endoluminal devices.

Fabrication strategies for endoluminal devices	Advantages	Disadvantages	Processable materials
Micromachining	• High precision • Highly diffused	• Contact forces that may create microdamages • Wear of the tools	• Polymers • Metals
Additive manufacturing technologies	• High flexibility in terms of shapes	• Low quality—postprocess required	• Process dependent
Electro-discharge machining (EDM)	• Absence of contact forces • High precision	• Narrow range of processable materials	• Electrically conductive
Injection molding	• High volumes • Cheap technology • High flexibility in terms of shapes	Mold required	• Polymers • Metals
Laser machining	• High precision	• Mainly 2D structures • Microstructurally damaged—postprocess required	• Polymers • Metals

they acquire a unique significance at *meso*-/microscale. For instance, cleanliness of the surfaces is important, because at small scales, it is easier to collect dirt particles or debris from the fabrication process that can obstruct holes or, more generically, induce interferences when assembling components. A complementary issue is given by the tribological features of the surfaces: Dimensions down to the micrometer can be critical when surfaces made of different materials are in contact since the characteristic dimensions of surface finishing may have the same order of magnitude as the overall dimensions, generating frictions potentially not compatible with the targeted application [17].

7.2.5 Additive manufacturing technologies: From 3D to 4D printing

Additive manufacturing technologies have revolutionized the design approach making it possible to fabricate components with complex shapes through a layer-by-layer approach that exploits the physico-chemical properties of solutions for their extrusion and/or hardening via laser or ultraviolet (UV) lights. This methodology, also called 3D printing, was presented by Langer and Vacanti in 1993 to create porous constructs (scaffolds) for tissue engineering purposes [18,19]. However, this approach has been extended to other fields to create parts with dimensions down to the millimeter scale in order to be employed as mechanical structures for mini-/microdevices [20], or to easily manufacture polymeric shells for endoluminal capsules [9]. The core of the fabrication relies on the material to print, generally called ink, which has to ensure specific rheological properties. Specifically, viscosity and density must be adequate for a correct extrusion from miniaturized nozzles without clogging, and, at the same time, the ink has to possess a proper shear tinning property to maintain the desired shape after printing. The most widely used inks for 3D printing are polymers, but novel research studies have shown the possibility to print metallic inks [21]. The main problems related to the fabrication with 3D printing pieces of equipment are the lack of precision, due to the intrinsic properties of the process (e.g., imperfect roundness of the holes), and the surface finishing that is often considered not adequate for the final application and requires an additional process. These issues have a relevant impact also in assembling 3D-printed components since the slight differences of the final dimensions from the ideal may prevent a correct pairing, especially when tolerances are based on backlash pairings. This scenario becomes particularly important when components require millimetric or submillimetric features, and this is the case for endoluminal devices, for which the 3D printing of structural components is not the first choice, usually.

A different additive manufacturing approach is given by stereolithography, in which the raw material is selectively hardened by a laser or UV light. This procedure has been largely exploited for metallic materials that can be difficult to process with other extrusion-based printing methods, making parts that have to be further processed in order to remove the waste and to achieve the required surface finishing [22].

Recently, a large number of studies have proposed an evolution of 3D printing, named 4D printing, in which the fourth dimension is meant to be the capability of the 3D-printed material to behave smartly in terms of the modification of the shape or other behavioral features. This particular phenomenon is triggered by the ability of the inks, often functionalized with nanoparticles [7], to react to external stimuli (e.g., heat, electro-magnetic fields) to better adapt to the targeted application [21]. This approach has been largely exploited also for endoluminal devices, which have exploited structural multistability or navigation driven by external magnetic fields [7,23,24].

7.2.6 Electro-discharge machining

EDM has been considered a real breakthrough in the manufacturing process. By exploiting the electrical conductivity of the workpieces, this process ensures the fabrication of components, even with micrometric features [25], through the erosion of the material induced by the difference of the polarization between the tool (i.e., a wire or a die) and the raw material. The key that differentiates this technology from the others is the complete absence of contact between the tool and the workpiece that are, instead, separated by a backlash in which sparks are generated and subtract material volumes [26]. EDM has a high potential to create miniaturized components [25], as for example those fabricated for capsules or other endoscopes [9], but its main limitation is the restricted number of processable materials that must be electrically conductive.

7.2.7 Injection molding

Injection molding is a fabrication process that is employed to fabricate parts from a granular particulate. The raw material is fed and melted in a machine that is able to inject the generated liquid into a mold with a designed topology. Based on the input material, the machine has to be able to raise the temperature up to a value that grants a full liquification of the granule. Therefore, the temperature reaches about 1500 °C when using metals, while 100 °C is sufficient to treat polymers [27,28]. The design of the molds is the key for a high-quality component since any defect is directly projected to the workpiece and can be the origin of microcracks, ultimately determining the failure of the part. Molds that are designed as the negative of the final product are usually fabricated via micromilling or EDM [27]. The latter is usually preferred for the high precision and the absence of waste that are possible to achieve, especially when dealing with micromolds. Applications for injection molding are, for instance, the parts composing cardiovascular stents [29].

7.2.8 Laser-based machining

Laser-based machining is an efficient procedure to create miniaturized 2D components that can be used for almost every material, after a proper tuning of the parameters and in particular the energy density. This specific feature directly affects the removal of material volumes creating patterns or features, also with the use of masks, and affects the microstructure of the areas close to the toolpath. This implies a different mechanical behavior of such parts, prone to damage upon dynamic loads, that require a further treatment process for a proper compensation [30].

7.3 Applications of endoluminal microdevices

In this section, we provide the most relevant applications of endoluminal devices using a classification based on their structure and that, somehow,

TABLE 7.3 Classification and assessment of endoluminal devices.

Endoluminal devices	Advantages	Disadvantages	Applications
Rigid systems	• Easy to fabricate • High-quality images • Low maintenance	• High stiffness • Not suitable for high-curvature lumens • No/limited steering	• Rhinoscopy • Cystoscopy • Laparoscopy • Stenosis
Articulated assemblies	• Possibility of primary-secondary system • Up to 6 degrees of freedom (DOFs)	• High stiffness • Not suitable for high-curvature lumens	General laparoscopic surgery and diagnostic
Flexible tethered system	• High adaptability • High steerability	• Contact forces throughout the lumen • High cost • High maintenance	Gastrointestinal, respiratory, and male urinary tracts
Wireless devices	• High steerability • No need of guiding thread • No contact forces throughout the lumen	• Need of an external source for guidance • High complexity • Mainly driven by peristalsis	Gastrointestinal tract

follows their evolution over time: from rigid systems to wireless devices. After presenting a general assessment in Table 7.3, we further discuss each class of endoluminal devices citing also a few remarkable studies focusing on the fabrication aspects.

7.3.1 Rigid systems

Rigid systems are the oldest devices employed in surgery. They have been mainly used as tools for monitoring the surface of organs and their vessels through creating an incision. Based on the specific targeted application, design criteria include viewing angle, magnification, and image quality. Usual employments are MIS devices for nose (rhinoscopy), bladder (cystoscopy), and abdomen (laparoscopy).

They are usually metallic tubular structures with fixed-angle shapes, endowed with lenses whose diameters range from 1 to 12mm. Micromilling and injection molding have been the most employed techniques to fabricate such devices. In contrast to the most recent advancements in the field, in which digital images are collected by flexible devices, the quality of the lens-generated images is still unsurpassed. However, despite their advantages, rigid systems have been abandoned in many applications for the gastrointestinal tract in favor of flexible devices able to reduce traumas and potential damages due to their intrinsic softer nature [31].

Another class of rigid systems largely employed in surgery is the one for treating stenosis. In this case, a metallic/polymeric stent is inserted with a deflated balloon in the stenosis site [32]. The operator inflates the balloon causing also the deformation of the stent that enlarges the cross section of the lumen, and keeps it in shape (Fig. 7.3) [33–35].

Self-expandable stents exploit the elasticity of the constituent materials, but their shape and stiffness are difficult to control, and previous studies have observed chronic recoil with a reduction of the stent area [36,37]. Liu et al. published the design of polymeric biodegradable stents made of poly(ε-caprolactone),

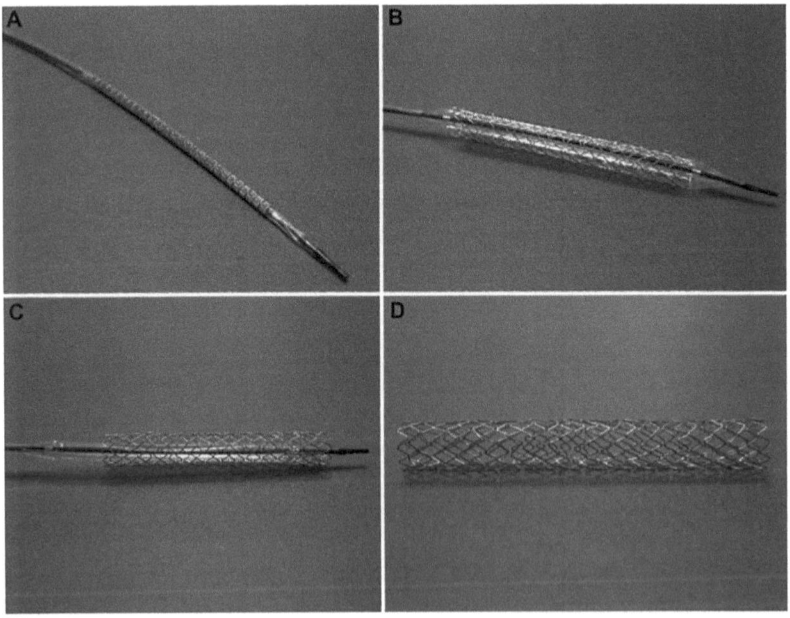

FIG. 7.3 Application of a stent. Panel A. The metallic stent is attached to the endoluminal balloon. Panel B. The balloon is inflated. Panel C. The balloon is discharged from air. Panel D. The stent is deployed. *(Reprinted with permission Simard T, Hibbert B, Ramirez FD, Froeschl M, Chen Y-X, O'Brien ER. The evolution of coronary stents: a brief review. Can J Cardiol 2014;30(1):35–45. © 2014 Elsevier.)*

fabricated with microinjection molding, with external diameters down to 2 mm. The shape used is not a simple cylinder but a special geometry specifically conceived for resisting external pressures from blood vessels. Tests carried out to assess the durability of the device showed no significant weight loss and collapse pressure reduction after 12 weeks [29].

7.3.2 Articulated assemblies

Articulated endoluminal devices are the direct evolution of rigid systems since they can be considered as assemblies of a number of rigid bodies connected by joints for improving the adaptability of the device to the surgical environment. They can be used as secondary systems controlled by doctors from a primary console that enables a full control, also with sensory feedback, to perform laparoscopic surgery and all the related operations (e.g., electrocautery task, grasping, wire loop management). In contrast to other solutions, articulated systems give operators the possibility to completely use almost all the dexterity required to perform surgical operations [38]. Using rigid tools, in contrast, any employment for visualizing and treating high-curvature tracts of the gastrointestinal tract is discouraged since the use of such systems would be unpleasant for the patient and, in parallel, potentially hurtful for the internal tissues composing the lumens [39].

The most famous example of articulated assembly is the da Vinci surgical system developed by Intuitive Surgical, Inc. (Sunnyvale, CA, USA), which has been installed worldwide in hospitals receiving good feedback from surgeons. Operators are comfortably seated at a primary console equipped with two haptic interfaces, pedals and a stereovision panel to provide a sensory feedback close to that of the in-person surgery. The cable-actuated platform (secondary side) operates while reducing any vibrational noise induced by the physiological tremor and ensuring more precise operations. Instruments on the secondary side may have up to 7 degrees of freedom (DOFs) with 90 degrees articulation, and based on the specific geometries, they are made via micromilling or EDM [40,41]. The main disadvantages of the da Vinci system are the high cost and the bulkiness that requires dedicated rooms for its installation. Being conceived for laparoscopic operations, the da Vinci equipment has now been allowed for other different clinical procedures compatible with its architecture [42,43].

Another remarkable example is given by the single-port laparoscopy bimanual robot platform developed by Petroni et al. in 2013. The authors developed a teleoperated bimanual robot for single-access interventions that consists of a 30-mm tube through which two 6 DOFs arms can operate from a bimanual haptic platform with stereoscopic camera (Fig. 7.4) [44,45]. The structural components of the robotic arms were manufactured via micromilling, while the gears actuating the joint were fabricated by highly specialized companies using wire EDM.

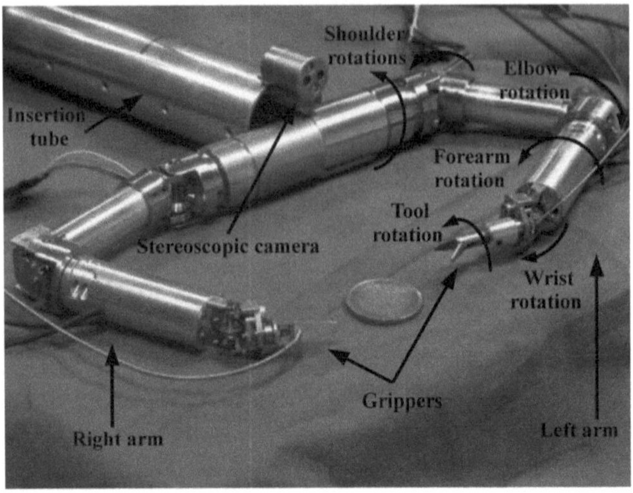

FIG. 7.4 Bimanual laparoscopic articulated device. *(Reprinted with permission Petroni G, Niccolini M, Menciassi A, Dario P, Cuschieri A. A novel intracorporeal assembling robotic system for single-port laparoscopic surgery. Surg Endosc 2013;27(2):665–70. © 2013 Springer.)*

7.3.3 Flexible tethered system

Flexible tethered systems have been the first real breakthrough in endoluminal applications. In contrast to rigid/articulated devices, flexible devices have revolutionized the diagnostic and treatment tasks because of their capability of adapting to the complex shapes of the lumens, especially in the intestinal tract characterized by several direction changes, thus limiting the damages of the tissues. The main disadvantages are the costs and the high maintenance required to ensure the correct performance of such systems. In fact, flexible devices are usually endowed with microtools and instrumentation for on-site tasks (e.g., lights, forceps, drains, needles). Images are usually collected by optical fibers or miniaturized cameras. General applications concern the gastrointestinal, respiratory, and male urinary tracts, which are difficult to access and are particularly delicate [31].

In 2011, Haga et al. described the design and the fabrication of a flexible 0.8-mm catheter actuated with a hydraulic suction mechanism. The manufacturing process involved a Ti-Ni superelastic alloy processed by laser machining: As reported in Fig. 7.5, the authors fabricated a number of different flexible structures for the intravascular surgery. Moreover, in order to reduce the friction during the insertion, silicone rubber was used to cover the device [2].

In 2017, Russo et al. presented the fabrication of a soft actuator and sensor device with distributed compliance able to be positioned on the tip of a flexible endoluminal device. They employed a mix of lithography and laser machining to create a layered composite actuated through a hydraulic source. This solution,

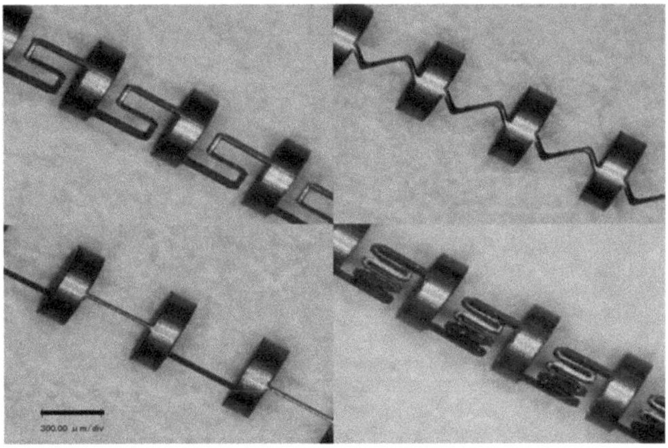

FIG. 7.5 Fabrication of a flexible catheter made of Ti-Ni superelastic alloy via laser machining: structural options. *(Reprinted with permission Haga Y, Muyari Y, Goto S, Matsunaga T, Esashi M. Development of minimally invasive medical tools using laser processing on cylindrical substrates. Electr Eng Jpn 2011;176(1):65–74. © 2011 John Wiley and Sons.)*

shown also in Fig. 7.6 both as a concept and in its use in *ex vivo* proof of concept experiment, increases the DOFs of the device enabling the possibility to perform surgical tasks exploiting a simple mechanism [46].

An intriguing study in this field was recently published by Kim et al. in 2019. The authors presented the fabrication of a submillimetric ferromagnetic soft continuum robot for exploring hard-to-reach lumens. This device can be manufactured either with 3D printing or with injection molding by using, as a raw material, a composite made of polydimethylsiloxane (PDMS) with uniformly dispersed NdFeB particles. In order to reduce the friction during the locomotion, a coating made of cross-linked hydrophilic polymers (polydimethylacrylamide—PDMAA) was applied with a pregel solution and eventually cured with UV light. Locomotion is provided manually by the operator who exploits the contact forces with the walls of the lumen, while steerability is enabled by an external magnetic field. Although this device is very promising, all tests were carried out *in vitro* on dedicated transparent benches, and therefore, it has not been possible to test the accuracy and precision when physiological tissues are interposed and there is no visual feedback [7].

A final example is given by a tethered device fabricated via injection molding able to perform manipulation tasks. This device is composed of a main body made of PDMS in which two electrodes of graphene polylactic acid (GPLA) are embedded as a phase transition material. Finally, a permanent magnet is positioned on the tip of the system to perform steering upon a magnetic external source. The final assembly has a diameter equal to 6mm but, although promising, the locomotion and steering capability have not been tested *in vivo* to date [47].

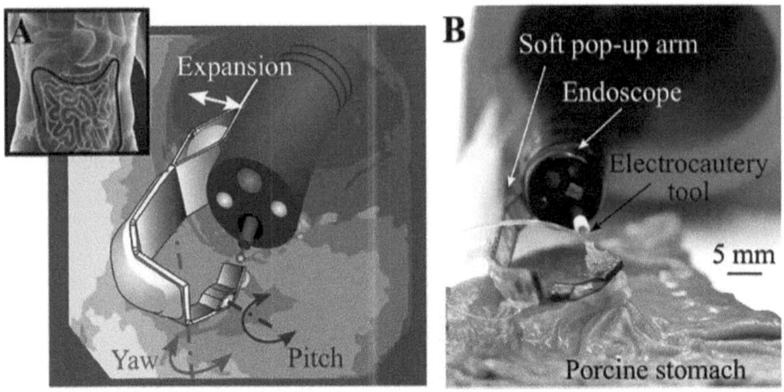

FIG. 7.6 Developing an improved tethered system through the addition of a multifunctional structure. Panel A. Concept of the flexible layered structure for performing endoluminal treatments. Panel B. *Ex vivo* test of the fabricated device in a porcine animal model. *(Reprinted with permission Russo S, Ranzani T, Walsh CJ, Wood RJ. An additive millimeter-scale fabrication method for soft biocompatible actuators and sensors. Adv Mater Technol 2017;2(10):1700135. © 2017 John Wiley and Sons.)*

7.3.4 Wireless devices

Wireless devices are traditionally capsules that are inserted transorally or transanally in a patient in order to visualize, take pictures, and sometimes operate in the gastrointestinal tract before being expelled by the colon. Throughout their journey, capsules are able to examine all the parts of the gastrointestinal tract, from the esophagus to the final tract of the intestine, being controlled by external sources (mainly electromagnetic fields) and/or driven by peristalsis, achieving a high precision on their controllability without requiring the use of a cable (Fig. 7.7) [8].

This feature makes the wireless devices more highly preferred than other solutions since they are practically and psychologically accepted by patients [48–50]. Capsules may have dimensions up to the order of 25 mm × 10 mm, be endowed with LEDs, which provide the required lighting, surgical tools (if any), miniaturized cameras, a cell battery, and the related electronics. In some cases, retractable appendices similar to legs are used to anchor the capsule during the task and prevent a migration from the work site induced by the peristaltic forces [9].

In order to provide an active locomotion for capsules, Quirini et al. designed a legged capsule able to navigate lumens through actuated superelastic legs. While the main structure was fabricated with micromilling, the four 15mm retractable legs, helpful also to resist the peristalsis when the capsule is at the treatment site, were manufactured via wire EDM. This procedure ensured the making of millimetric features such as elastic flextures with high precision. The same fabrication process was used for generating the profiles of the miniaturized gears required for the leg actuation mechanism [9].

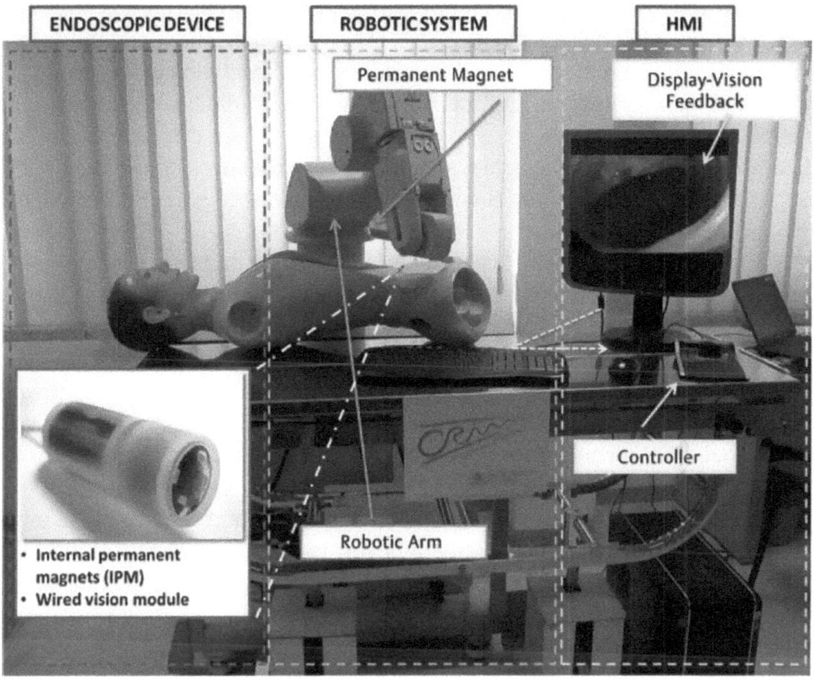

FIG. 7.7 System architecture to control a capsule through an external magnetic field. Test on a bench. *(Reprinted with permission Ciuti G, et al. Robotic versus manual control in magnetic steering of an endoscopic capsule. Endoscopy 2010;42(2):148. © 2010 Georg Thieme Verlag KG.)*

From the idea of ingestible capsules, Nagy et al. developed an endoluminal device composed of a number of 11 mm × 26 mm capsules that, once in the stomach, are able to self-assemble to perform the clinical task. Later, the structure is disassembled in order to be expelled through the intestine. The module structures, which included permanent magnets that served as joints, were fabricated through stereolithography. The actuator modules for locomotion and task performance were designed with a combination of memory-shape alloys and traditional miniaturized components (e.g., springs) that were either commercially available or manufactured in house with micromilling and 3D printing [23].

7.4 Conclusions and future outlook

In 2000, Dr. Majid Samii, one of the world's top neurosurgeons, suggested that three technological improvements were expected to be developed and improved: teleoperation, robotics, and miniaturized instruments to perform tasks in areas with corners whose dimensions are smaller than 1 or 2 mm. In 20 years, robotics has moved toward this direction and has had a strong impact in a large number of fields, from industrial to biomedical applications, improving the capabilities of doctors to noninvasively perform the tasks with instruments that are able to ensure high precision and accuracy [40,41,51,52]. This chapter details opportunities and challenges for developing MIS endoluminal devices from the fabrication standpoint, highlighting how most of the requirements (e.g., dimension scaling, biocompatibility, reusability) are a consequence of the selected manufacturing processes. These affect the achievability of reduced dimensions and submillimetric features, and the machinability of materials that have impact on the (bio)stability of the MIS tools.

From micromilling to the latest advances in 3D/4D printing, scientists have looked for the best strategies to fabricate high-quality parts for easy-to-assemble endoluminal systems that efficiently perform clinical tasks without being invasive or causing damage in the treated lumens. These objectives have a concrete repercussion on the global healthcare system, from the patient who becomes more confident in the novel surgical strategies to the doctors who can better focus on specific aspects of the operation without concentrating on tremors or other inconveniences.

Although the employment of endoluminal devices has been one of the most successful approaches in clinics, there is still lack of a universal solution for every possible application: Articulated systems can easily operate in localized areas for laparoscopic surgery, especially in hollow organs, using the several available DOFs enabled by the mechanical joints, but cannot achieve the flexibility in use of soft tethered endoscopes that have had a significant role in observing and treating pathologies of the gastrointestinal tract. However, similar to engineered wireless capsules, flexible endoscopes cannot achieve the full dexterity required for specific treatments which, instead, is almost completely enabled by articulated devices. In our opinion, the diversity and nongeneralizability of the tools is not, necessarily, an issue: We envision new surgical approaches that can exploit the synergistic use of different types of endoluminal devices, simultaneously accessing lumens to visualize and operate on internal areas. Moreover, the advancements in data sharing and communication could pave the way to surgical tasks performed by teams geographically dispersed in order to fully exploit the complementary competences of doctors through the primary-secondary configurations of the equipment.

Besides the achievements and the difficulties faced to date, we see as a greater challenge the possibility to autonomously navigate throughout complex lumens in which the combination of reduced diameters and small curvatures may be difficult to manage. Two examples of such a scenario are given by the navigation inside the cochlea or the brain vasculature. In the first case, we have a snail-shaped 3D structure with a lumen possessing a variable millimetric diameter, in which cochlear implants have to be placed. The brain, in contrast, is a structure with lumens possessing locally variable curvatures, which could be traveled by flexible devices. In both cases, the common approach concerns the employment of endoluminal devices that are able to penetrate the lumens exploiting the forces from the lumen walls. This is the reason why surgical treatments in such areas are particularly delicate, and would require an additional improvement in terms of controllability for navigating inside the structures without touching the parietal walls. This specification could require either additional driving sources (e.g., magnetic fields, ultrasounds), sensitive integrated sensory feedbacks, or miniaturized features able to locally adapt to the lumens without stressing them. A first preliminary solution was given by Kang et al., who developed a continuum robot able to avoid obstacles and be self-sustaining [53]. However, this study cannot be easily downscaled because of the limitations given by fabrication issues related to the piezoelectric actuators that, if miniaturized, could not support the structural loads. Moreover, this solution would require additional features to enable the stability on the targeted site without relying on the intrinsic structural stability of the tool. Other potential solutions would require further investigations in terms of electro-mechanical interactions and compatibility when downscaling a device but, more importantly, an adequate analysis of the fabrication challenges in order to ensure and elevate the capabilities of MIS endoluminal devices for the specific targeted application.

A final thought is related to the costs for an endoluminal device. The final cost of a clinical tool is not only the simple sum of the costs of the material, the components, and the manufacturing procedures. Other cost items should be considered and may have a significant role in the long term. Among others, the reusability of a device has a particular importance since this aspect defines the selection of the constituent materials and the fabrication process. A single-use tool is definitely a cheaper solution in the short term, usually fabricated via injection molding or additive manufacturing, but can be more expensive in the long term since the accumulation of used devices (waste) and their disposal have a significant cost. Conceiving MIS tools that can be reusable has a great impact not only on the healthcare costs but also on the environment. We live in the era that aims at the goal of zero waste thus limiting the human footprint on the natural resources. Single-use polymeric devices move toward the opposite direction. In fact, synthetic polymers are the results of oil manufacturing, a precious resource that is limited, over-exploited, and has side products that are harmful for nature. Therefore, this scenario has to stimulate scientists and companies to develop new low-/zero-carbon materials able to exploit their properties to efficiently fulfill the expected tasks without indirectly being detrimental for the human-natural system.

Acknowledgments

The authors acknowledge the European Commission and the Italian Government. EU FP7 “Integ-μ” Project (CP-IP 214013-2), EU FP5-LIFE QUALITY “MINOSC” Project (QLG5-CT-2001-02150), EU ICT-2007.3.6—Micro/nanosystem “ARAKNES” Project (224565), and FIRB bando Idee Progettuali 2007—Sistemi avanzati di manifattura per la lavorazione di prodotti miniaturizzati con processi combinati (Multitasking).

References

[1] Mack MJ. Minimally invasive and robotic surgery. JAMA 2001;285(5):568–72.

[2] Haga Y, Muyari Y, Goto S, Matsunaga T, Esashi M. Development of minimally invasive medical tools using laser processing on cylindrical substrates. Electr Eng Jpn 2011;176(1):65–74.

[3] Cotton PB, Williams CB. Practical gastrointestinal endoscopy: the fundamentals. John Wiley & Sons; 2008.

[4] Romanelli JR, Desilets DJ, Earle DB. NOTES and endoluminal surgery. Springer; 2017.

[5] Thompson CC, Ryou M, Soper NJ, Hungess ES, Rothstein RI, Swanstrom LL. Evaluation of a manually driven, multitasking platform for complex endoluminal and natural orifice transluminal endoscopic surgery applications (with video). Gastrointest Endosc 2009;70(1):121–5.

[6] Steiner JA, Hussain OA, Pham LN, Abbott JJ, Leang KK. Toward magneto-electroactive endoluminal soft (MEESo) robots. In: ASME 2019 Dynamic Systems and Control Conference October 8–11, 2019, Park City, Utah, USA; 2019.

[7] Kim Y, Parada GA, Liu S, Zhao X. Ferromagnetic soft continuum robots. Sci Robot 2019;4(33).

[8] Ciuti G, et al. Robotic versus manual control in magnetic steering of an endoscopic capsule. Endoscopy 2010;42(2):148.

[9] Quirini M, Menciassi A, Scapellato S, Stefanini C, Dario P. Design and fabrication of a motor legged capsule for the active exploration of the gastrointestinal tract. IEEE/ASME Trans Mechatron 2008;13(2):169–79.

[10] Simi M, Valdastri P, Quaglia C, Menciassi A, Dario P. Design, fabrication, and testing of a capsule with hybrid locomotion for gastrointestinal tract exploration. IEEE/ASME Trans Mechatron 2010;15(2):170–80.

[11] Jayabalan M. Sterilization and reprocessing of materials and medical devices—reusability. J Biomater Appl 1995;10(1):97–112.

[12] Rai P. Disinfection of endoscopy and reusability of accessories. J Dig Endosc 2020;11(1):61.

[13] Milazzo M, et al. Biodegradable polymeric micro/Nano-structures with intrinsic antifouling/antimicrobial properties: relevance in damaged skin and other biomedical applications. J Funct Biomater 2020;11(3). https://doi.org/10.3390/jfb11030060.

[14] Milazzo M, Danti S, Inglese F, Berrettini S, Stefanini C. Micro-machining of bovine bone for otologic applications. In: European society for precision engineering and nanotechnology, conference proceedings—18th international conference and exhibition, EUSPEN 2018; 2018. p. 375–6.

[15] Milazzo M, et al. Ossicular replacement prostheses from banked bone with ergonomic and functional geometry. J Biomed Mater Res B Appl Biomater 2017;105(8):2495–506.

[16] De Cristofaro S, et al. High-speed micro-milling: novel coatings for tool wear reduction. Int J Mach Tools Manuf 2012;63:16–20.

[17] Maboudian R, Ashurst WR, Carraro C. Tribological challenges in micromechanical systems. Tribol Lett 2002;12(2):95–100.

[18] Milazzo M, et al. Additive manufacturing approaches for hydroxyapatite-reinforced composites. Adv Funct Mater 2019;29(35). https://doi.org/10.1002/adfm.201903055, 1903055.

[19] Mota C, et al. Multiscale fabrication of biomimetic scaffolds for tympanic membrane tissue engineering. Biofabrication 2015;7(2). https://doi.org/10.1088/1758-5090/7/2/025005, 025005.

[20] Pourghodrat A, Nelson CA. Disposable fluidic actuators for miniature in-vivo surgical robotics. J Med Devices 2017;11(1):0110031–8.

[21] Morouço P, et al. Four-dimensional (bio-) printing: a review on stimuli-responsive mechanisms and their biomedical suitability. Appl Sci 2020;10(24):9143.

[22] Levinson HJ. Principles of lithography. vol. 146. SPIE Press; 2005.
[23] Nagy Z, et al. Assembling reconfigurable endoluminal surgical systems: opportunities and challenges. Int J Biomechatronics Biomed Robot 2009;1(1):3–16.
[24] Zarek M, Mansour N, Shapira S, Cohn D. 4D printing of shape memory-based personalized endoluminal medical devices. Macromol Rapid Commun 2017;38(2), 1600628.
[25] Masaki T, Kawata K, Masuzawa T. Micro electro-discharge machining and its applications. In: IEEE proceedings on micro electro mechanical systems, an investigation of micro structures, sensors, actuators, machines and robots; 1990. p. 21–6.
[26] Rajurkar KP, Sundaram MM, Malshe AP. Review of electrochemical and electrodischarge machining. Procedia CIRP 2013;6:13–26.
[27] Rosato DV, Rosato MG. Injection molding handbook. Springer Science & Business Media; 2012.
[28] Piotter V, Bauer W, Benzler T, Emde A. Injection molding of components for microsystems. Microsyst Technol 2001;7(3):99–102.
[29] Liu S-J, Chiang F-J, Hsiao C-Y, Kau Y-C, Liu K-S. Fabrication of balloon-expandable self-lock drug-eluting polycaprolactone stents using micro-injection molding and spray coating techniques. Ann Biomed Eng 2010;38(10):3185–94.
[30] Chryssolouris G. Laser machining: theory and practice. Springer Science & Business Media; 2013.
[31] Schneider A, Feussner H. Biomedical engineering in gastrointestinal surgery. Academic Press; 2017.
[32] Canonge J, et al. Comprehensive review of physician modified aortic stent grafts: technical and clinical outcomes. Eur J Vasc Endovasc Surg 2021;61:560–9.
[33] Kutryk MJB, Serruys PW, editors. Stents currently available. In: Coronary stenting current perspectives. London: Martin Dunitz; 1999. p. 17–65.
[34] Palmaz JC, Sibbitt RR, Reuter SR, Tio FO, Rice WJ. Expandable intraluminal graft: a preliminary study. Work in progress. Radiology 1985;156(1):73–7.
[35] Simard T, Hibbert B, Ramirez FD, Froeschl M, Chen Y-X, O'Brien ER. The evolution of coronary stents: a brief review. Can J Cardiol 2014;30(1):35–45.
[36] Serruys PW, et al. A bioabsorbable everolimus eluting coronary stent system (ABSORB): 2-year outcomes and results from multiple imaging methods. Lancet 2009;373(9667):897–910.
[37] Ormiston JA, Webster MWI, Armstrong G. First-in-human implantation of a fully bioabsorbable drug-eluting stent: the BVS poly-L-lactic acid everolimus-eluting coronary stent. Catheter Cardiovasc Interv 2007;69(1):128–31.
[38] der Putten EP, Goossens RHM, Jakimowicz JJ, Dankelman J. Haptics in minimally invasive surgery—a review. Minim Invasive Ther Allied Technol 2008;17(1):3–16.
[39] Vitiello V, Lee S-L, Cundy TP, Yang G-Z. Emerging robotic platforms for minimally invasive surgery. IEEE Rev Biomed Eng 2012;6:111–26.
[40] Mucksavage P, Kerbl DC, Pick DL, Lee JY, McDougall EM, Louie MK. Differences in grip forces among various robotic instruments and da Vinci surgical platforms. J Endourol 2011;25(3):523–8.
[41] Guthart GS, Salisbury JK. The intuitive/sup TM/telesurgery system: overview and application. In: Proceedings 2000 ICRA. Millennium conference. IEEE international conference on robotics and automation. Symposia proceedings (Cat. No. 00CH37065), vol. 1; 2000. p. 618–21.
[42] Bodner J, Wykypiel H, Wetscher G, Schmid T. First experiences with the da Vinci™ operating robot in thoracic surgery. Eur J Cardiothorac Surg 2004;25(5):844–51.
[43] Tewari A, Srivasatava A, Menon M, V. I. P. team. A prospective comparison of radical retropubic and robot-assisted prostatectomy: experience in one institution. BJU Int 2003;92(3):205–10.

[44] Petroni G, et al. A novel robotic system for single-port laparoscopic surgery: preliminary experience. Surg Endosc 2013;27(6):1932–7.

[45] Petroni G, Niccolini M, Menciassi A, Dario P, Cuschieri A. A novel intracorporeal assembling robotic system for single-port laparoscopic surgery. Surg Endosc 2013;27(2):665–70.

[46] Russo S, Ranzani T, Walsh CJ, Wood RJ. An additive millimeter-scale fabrication method for soft biocompatible actuators and sensors. Adv Mater Technol 2017;2(10):1700135.

[47] Park J, Lee H, Kee H, Park S. Magnetically steerable manipulator with variable stiffness using graphene polylactic acid for minimally invasive surgery. Sensors Actuators A Phys 2020;309:112032.

[48] Swain P, Iddan GJ, Meron G, Glukhovsky A. Wireless capsule endoscopy of the small bowel: development, testing, and first human trials. In: Biomonitoring and endoscopy technologies, vol. 4158; 2001. p. 19–23.

[49] Meron GD. The development of the swallowable video capsule (M2A). Gastrointest Endosc 2000;52(6):817–9.

[50] Iddan G, Meron G, Glukhovsky A, et al. Wireless capsule endoscopy. Nature 2000;405:25.

[51] Tannous M, et al. Haptic-based touch detection for collaborative robots in welding applications. Robot Comput Integr Manuf 2020;64:101952.

[52] Ramacciotti M, Milazzo M, Leoni F, Roccella S, Stefanini C. A novel shared control algorithm for industrial robots. Int J Adv Robot Syst 2016;13(6), 1729881416682701.

[53] Kang B, Kojcev R, Sinibaldi E. The first interlaced continuum robot, devised to intrinsically follow the leader. PLoS One 2016. https://doi.org/10.1371/journal.pone.0150278.

Chapter 8

Modeling and control strategies for flexible devices☆

M. Taha Chikhaoui[a] and Benoît Rosa[b]
[a]Univ. Grenoble Alpes, CNRS, UMR 5525, VetAgro Sup, Grenoble INP, TIMC, Grenoble, France,
[b]CNRS, ICube Laboratory, UMR7357 CNRS-Strasbourg University, Illkirch-Graffenstaden, France

8.1 Introduction

Continuum robots are slender, flexible objects that describe smooth three-dimensional curves and can undergo high strains. With embedded actuators (for instance, shape-memory alloys or pneumatic bellows) or mechanical transmission elements (for instance, cables or tubes) inside their structures, one can control part of the continuum robot shape. In the context of endorobotic systems, continuum robots allow the robotization of interventional tools, in order to perform complex surgical tasks, while keeping motors outside of the patient. Examples of such systems are shown in Fig. 8.1. In both systems, flexible robotized arms with surgical tools at the tip are inserted in the working channel of a main endorobot. A camera allows observing the surgical scene, as well as getting feedback for a better control of the flexible arms. Finally, the fetal surgery system in Fig. 8.1a shows an example of an electromagnetic sensor embedded at the tip of the robotized arm for direct positional feedback.

Contrarily to classical robots, which are composed of rigid links connected by joints, continuum robots offer an infinite number of degrees of freedom (DOFs) ([3]). As a consequence, modeling and control approaches may differ radically from typical robotic approaches. This chapter first discusses the main modeling approaches developed over the last decades, from simple geometric models to advanced mechanics-based and data-driven approaches. In a second part, recent continuum robot control strategies are discussed, spanning from conventional end-effector controllers to more advanced theories regarding full-body/shape control and hybrid position/force controllers.

☆This document is the results of the research project funded by grants ANR-11-LABX-0004-01 and ANR-18-CE19-0012.

Endorobotics. https://doi.org/10.1016/B978-0-12-821750-4.00008-6

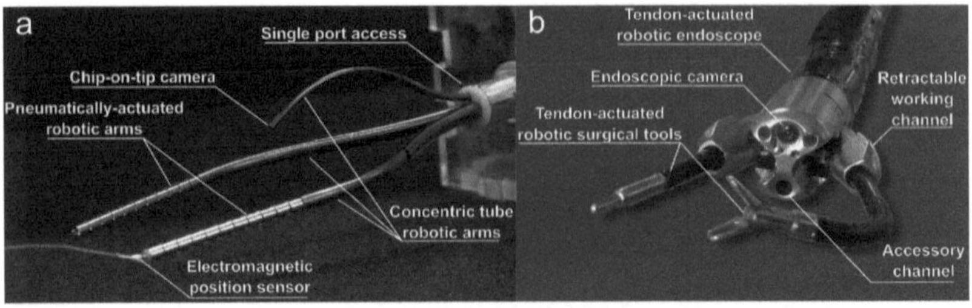

FIG. 8.1 Endoscopy systems with continuum robots for controlling the surgical tools. (a) Fetal surgery system from Vandebroek et al. [1]. (b) The STRAS robotic endoscopy system from Zorn et al. [2].

The reader should note that a few notable review papers have been published on various aspects related to this chapter. Of note are a review from Webster III and Jones [4] on constant curvature models, a general overview from Burgner-Kahrs et al. [3] of medical continuum robots, and a review on control approaches for continuum robots by Chikhaoui and Burgner-Kahrs [5]. This chapter will not redevelop all details touched upon in those reviews. Rather, advances that have been made since the time those reviews were written are pinpointed.

8.2 Modeling of continuum robots

As described in the general introduction of this chapter, continuum robots typically take a tubular shape, but can embed various types of actuators and mechanical transmission elements, such as smart materials (e.g., shape-memory alloys), fluidic actuators, tendons, or precurved elastic elements. Over the time, various models have been developed for such robots [3].

The goal of a robot model is to relate the actuation vector q (for instance, cable tensions and/or displacements for a tendon-actuated continuum robot, or TACR) and its derivatives with respect to time, to geometric/kinematic parameters of the robot (e.g., its shape, static equilibrium, and dynamic behavior), i.e., its state. Forward models describe the state achieved from a given actuation vector q, whereas inverse models compute the actuation vector q required to achieve a given state (note that time derivatives of the state and/or the actuation vector may be used for differential kinematics or dynamic models).

Most of the times, the state of a continuum robot includes its spatial configuration, i.e., the minimal set of parameters that allow describing the 3D pose of any point along the robot body. In some cases, the robot control Jacobian or compliance matrices can be accessed directly through modeling, without resorting to perturbation methods. For standard rigid-link robots, the configuration can be described by a finite set of parameters, whose cardinality is the number of DOFs of the robot. Continuum robots, however, can take continuously curving shapes and therefore have, formally, an infinite number of DOFs. In order to describe their geometric shape in space, it is then needed to use an arc-length parameterized function, often noted $g(s)$. If $F(0)$ is the reference frame at the base of the robot, its z vector pointing toward the tangent direction of the robot's centerline, then $g(s) \in SE(3)$ describes the transformation from $F(0)$ to a virtual body frame $F(s)$ sliding along the robot's centerline (i.e., the neutral line, which does not experience any extension under pure bending). The arc-length parameter s ranges from 0 at the base of the robot to $s = L$ at its tip. An example of arc-length parameterization is shown in Fig. 8.2a.

As $F(s)$ slides along the robot's centerline, it experiences an angular rate of change per unit of arc length $u = [u_x\ u_y\ u_z]^T$. u is the curvature vector, u_x and u_y the local components of curvature due to bending of the robot, and u_z the torsional component. For instance, in Fig. 8.2a, one can observe a rotation around the local z vector (in green) between $F(s)$ and $F(L)$, which denotes a torsional twist, i.e., $u_z \neq 0$. Similarly, if the robot body is extensible or compressible, the components of the vector $v = [v_x\ v_y\ v_z]^T$ describe, respectively, the two lateral shear deformations and the extension/compression of the robot's centerline along its local tangent. Recognizing that u and v act, with respect to s, as the rotational and positional velocity vectors act with respect to t, allows the reader to apply common robotic approaches pertaining to Lie algebra [6,7].

The following of this chapter details the main methods that have been used in the literature in order to model continuum robots (see Fig. 8.2).

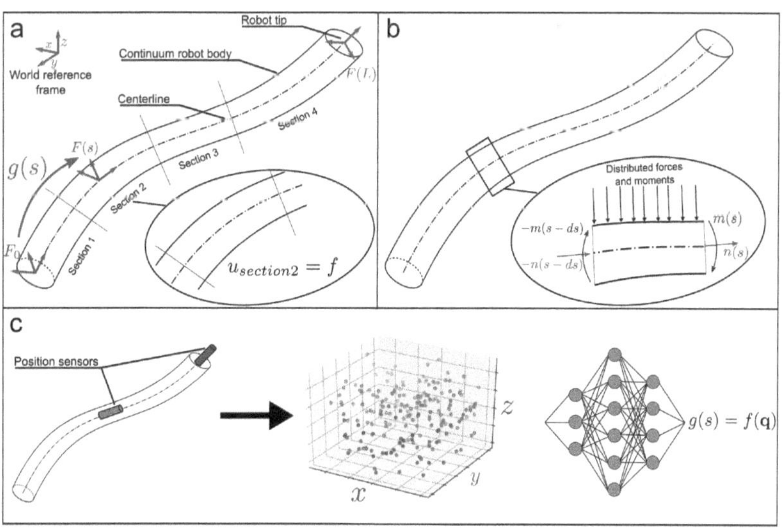

FIG. 8.2 Three main modeling approaches for continuum robots. (a) Geometry-based modeling, where each section of the robot has a function f approximating the curvature vector; (b) mechanics-based kinematics, which typically use classical elasticity theories; and (c) data-based modeling with two position sensors in the middle and distal tip of the continuum robot. (a) Also shows an example of arc-length parameterization.

8.2.1 Geometry-based modeling

Geometry-based approaches, also referred to as kinematic framework by Burgner-Kahrs et al. [3], are often used as a way of simplifying continuum robot models. The core idea is to make the assumption that the continuum robot shape can be described by a specific geometric function.

The simplest of the geometry-based models is the *constant curvature* model. This model makes the assumption that the curvature vector u is constant with respect to s, often without any torsion (i.e., $u_z = 0$). The constant curvature modeling approach has been first described by Hannan and Walker [8] and has been widely used in the early developments of continuum robots due to its simplicity. In fact, one can show that the constant curvature model reduces to three parameters: the norm of the constant curvature vector κ, the angle ϕ of the bending plane with respect to a constant reference, and the length l of the bending section. This modeling approach is simple and computationally efficient, and the inverse kinematic problem can be solved analytically.

Continuum robots, however, often have a complex shape that cannot be described as a single constant curvature section. For this reason, Jones and Walker [9] have extended the model to piecewise constant curvature. This extension makes the assumption that the variation of curvature along the robot length is small, so that the robot shape can be described as a sequence of constant curvature sections (see also Fig. 8.2a). As shown by Webster III and Jones [4], the inverse kinematic problem using a piecewise constant curvature model can be solved efficiently in closed form.

While many mathematical frameworks have been proposed to derive and solve constant curvature models, Webster III and Jones [4] proposed a unified formulation. The authors derive the model as the composition of a robot-specific and a robot-independent mapping. The robot-specific mapping makes the link between the joint variables q and the configuration variables (κ, ϕ, l), while the robot-independent mapping computes $g(s)$ from the configuration variables. It should, however, be noted that the set of configuration variables used introduces a *virtual* singularity in the kinematic model at the very center of the robot's workspace, when the robot is straight. Recently, Della Santina et al. [10] proposed an alternative parameterization for the configuration space, improving the stability of kinematic and dynamic computations when using a piecewise constant curvature model.

Whereas constant curvature modeling seems to be an entirely geometric approach, it is not devoid of mechanical sense. In fact, simple rod mechanics such as Euler-Bernoulli beams show that a constant moment applied along a uniform beam creates a constant curvature deflected shape. This moment can be the result, for instance, of an inextensible tendon pulling along the side of the beam without any friction. For instance, Devreker et al. [11] have shown that, under a constant curvature modeling assumption, one can model and optimize a continuum robot actuated by fluidic McKibben muscles. In a similar context, Runge et al. [12] recently showed that a constant curvature kinematic model can efficiently approximate the output of a finite element simulation.

Overall, the accuracy of constant curvature models is limited by many underlying assumptions. We have seen that mechanical models produce constant curvature outputs in limited cases, and in the absence of friction, external forces applied on the robot, or mechanical coupling effects between sections/actuators. Such effects could be modeled using the piecewise constant curvature approach, at the price of an increased number of sections [13]. Alternatively, more complex geometric functions such as Euler [14] or Pythagorean-Hodograph [15] functions have been proposed. Such approaches allow describing more complex shapes, while keeping an underlying function formulation, which simplifies computations. Such approaches are, however, very recent, and for the moment, the tuning process to obtain the model parameters has been robot-specific, sometimes requiring machine learning algorithms.

8.2.2 Mechanics-based modeling

In opposition to geometry-based approaches, mechanics-based models make use of the general principle of mechanics. Different formalisms have been studied to that end. Lumped-parameter models, also called pseudo-rigid body models, consist in describing the body of a robot as a series of small rigid elements connected by springs and dampers. Such models can be combined with energy methods in order to predict the shape of catheters [11,16], as well as the force and contacts acting on them using external sensors [17].

A more common approach is to use results from beam mechanics in order to model continuum robots. Indeed, continuum robots can be considered as an assembly of rods of different properties (stiffness, Poisson's ratio, etc.). One can then use classical elasticity theories in order to model continuum robots. The more common is the Euler-Bernoulli classical beam theory, which has, for instance, been used by Camarillo et al. [18] to model TACRs. Since a few years, however, the special Cosserat rod model has been employed with success for various continuum robot models. The reader is referred to the review paper by Burgner-Kahrs et al. [3] for a general historical presentation of Cosserat rod models. For a general curved rod experiencing a distributing loading torque τ and force f per unit length, and neglecting the inertial terms, the body-frame equations write:

$$\frac{dm}{ds} = \tau - [u]m - [v]n \tag{8.1}$$

$$\frac{dn}{ds} = f - [u]n, \tag{8.2}$$

where $[u]$ is the skew-symmetric matrix associated with the vector u, and m and n are, respectively, the internal moment and force in the rod at the arc length s.

One needs to combine the Cosserat rod equations with a material constitutive model in order to connect displacements in the rod to internal forces and moments. Linear elasticity models linking m to u and n to v (i.e., Hooke's law) are usually employed. Finally, the equations obtained describe a set of ordinary differential equations, which can be integrated on the rod arc length following a set of boundary conditions depending on the inputs (the actuator displacements and/or external forces, for instance). As a result, the Cosserat rod formalism is applicable to a large variety of continuum robots and was therefore used with success for concentric tube robots [6,19], TACRs [20,21], or parallel continuum robotic structures [22,23]. Cosserat rod models have also been coupled with energy considerations in order to optimize the stability of tube pairs in concentric tube robots by Ha et al. [24]. In similar applications, Peyron et al. [25] showed that energy minimization and high-order continuation methods can yield stability metrics for magnetically steered continuum robots.

A special note should be made about tendon-driven robots. Until recently, existing mechanics-based models for tendon-driven continuum robots required tendon forces as an input [6,26]. Such models are accurate, but obtaining tendon forces on a real tendon-driven robot requires using specific sensors, which are difficult to integrate in medical robots. Oliver-Butler et al. [27] showed that considering tendons as extensible Cosserat rods could improve the accuracy of the model, as well as relate tendon forces and their extensibility. This last property allows deriving the robot's shape and stiffness under prescribed tendon displacements.

It should be noted that Cosserat rod models are generally more precise than Euler-Bernoulli beam models, since they inherently handle large, nonlinear deformations. Solving such models, however, requires solving sets of differential equations, almost always with split boundary conditions. Such boundary value problems (BVPs) require iteratively solving the initial value problem while making guesses about unknowns in the boundary condition in order to converge. As a result, Cosserat rod models are often slower to solve. This trade-off between accuracy and computational speed was studied in detail for tendon-driven multisection robots by Chikhaoui et al. [28]. Moreover, Cosserat rod models can also help the user gain knowledge of more information about the robot's state than just its shape. Rucker and Webster [29] as well as Smoljkic et al. [30] show that complementing the initial value problem solved within the BVP allows computing arc-length parameterized Jacobian and compliance matrices, which can prove very useful for robot control.

All the abovementioned models make the assumption of a perfect system. In practice, many systems exhibit different kinds of nonlinearities, such as internal friction (for instance, between the tendon and its routing sheath/disks, or between tubes for a concentric tube robot) or material nonlinearities (nonlinear stress–strain relation). Such nonlinearities are most of the time very difficult to incorporate in the models. For instance, concentric tube robot models make the assumption that NiTi tubes stay in the linear elastic area of the stress–strain curve [19,31]. Considering nonlinear effects in the robot model is still an ongoing challenge, even in the case of offline finite element analysis [32,33]. In the context of soft robots, hyperelastic material behavior has been integrated in piecewise constant curvature models by Shiva et al. [34].

Internal friction in the system is also an important contributor of mechanical uncertainties. This is especially true in TACRs where friction between the tendon and its sheath can be important and reduce drastically the model accuracy. Integrating friction forces in mechanics-based models proved, however, a very challenging task. In fact, an underlying assumption to mechanic frameworks is that forces and moments are conservative, which is not the case of friction forces. For this reason, until recently, friction had been integrated only in pseudo-rigid body [11] or piecewise constant curvature models [35–37]. Recent advances by Gao et al. [38] show that friction can be incorporated in the Cosserat rod model of a TACR when considering cable routing through regularly spaced thin disks at the side of the robot, under the assumption that friction occurs at single points (tendon entry and exit in the disk). While a definitive step forward, this approach is not applicable for endorobots in which tendons interact with their routing sheath over their whole length (such as robotized flexible endoscopes). For concentric tube robots, Ha et al. [39] introduced friction effects in a Cosserat rod formulation by relaxing the usually accepted zero-clearance model between tubes. Using this approach, friction effects cannot be fully predicted, but can be bounded.

Finally, a small number of research groups have developed dynamic models for continuum robots. Inertial and gravitational effects are indeed often neglected because of the reduced size and limited mass of continuum robots, especially in surgical and endoscopic contexts, where such systems are miniaturized. Godage et al. [40] show that dynamic models can be efficiently computed for continuum arms. The approach uses energy considerations to derive spatial dynamic models for variable length multisection continuum robots. The formulation requires, however, a number of simplifying assumptions (constant curvature, kinematically independent sections). More recently, Renda et al. [41] and Till et al. [42] have shown that Cosserat rod models can be used efficiently for real-time dynamics of continuum robots, paving the way to interesting research in the future. Liu et al. [43] also showed that frictional effects in TACRs can be incorporated into dynamical models under certain conditions. Similarly, real-time finite element simulations allow modeling continuum robot's kinematic and dynamic behavior [33]. Finally, notable recent approaches have departed from the BVP resolution method typically using shooting methods by employing the Lagrangian formulation in order to derive efficient dynamic models for continuum robots [44,45], paving the way to interesting research in the future.

8.2.3 Data-based models

The general idea behind a data-based model is to learn a kinematic model (forward, inverse, and/or differential) directly from data. Such data-based models can be obtained using various techniques, from functional approximations to machine learning algorithms.

The first applications of such approaches did not use data coming from external sensors. Indeed, the challenge was then to reduce the computational time of complex, computationally intensive mechanics-based models, for real-time usage. Using a wavelet-based decomposition, Gravagne and Walker [46] showed that modal analysis could be used to reduce the dimensionality of the inverse kinematics problem for a redundant continuum robot. A similar approach was used by Godage et al. [47] in order to efficiently compute kinematics and dynamics of extensible constant curvature continuum robots, based on the output of a finite element model. Dupont et al. [19], in the context of concentric tube robots, showed that the output of the mechanics-based model could be efficiently approximated by a product of truncated Fourier series. This approach allowed a significant reduction of the computational time, and the formulation of the inverse kinematics problem as a simple root finding problem.

With the increase in computational power, functional approximation techniques have become less useful. Data-based models, however, have another main interest: by using sensory data as an input, one can hope to learn complete models, including the effect of nonlinearities and physical phenomena, which are not modeled by state-of-the-art mechanics-based models. In this approach, one will first gather a database of measured/estimated robot states (tip position or robot shapes, for instance) while the robot workspace is spanned. Using recursive or machine learning algorithms, one can then learn various features of the robot's state. Data-based models can be categorized into two types: offline and adaptive algorithms.

Fully offline algorithms will try and learn the robot model from previously collected data. Two main challenges need to be overcome. Firstly, one needs to collect sufficiently diverse robot data to allow the algorithm to properly learn the robot behavior. In practice, uniform sampling of the input space would be required. In the case of learning forward kinematics, for instance, a simple uniform sampling of the joint space can be used [48]. However, as shown by Ha et al. [49], time-dependent behavior (such as friction effects) may bias the results. Learning the inverse kinematic mapping may prove more difficult, because it is not trivial to efficiently and uniformly span the task space. Strategies such as rejection sampling or goal-directed exploration [50] have been proposed to that end. Secondly, once an appropriate dataset has been acquired, one must learn the robot model from it. Several learning algorithms have been studied, from probabilistic models such as Gaussian mixture models [51] to neural network approaches [51–54]. Interestingly, Reinhart et al. [55] propose not to learn the whole kinematic model, but only the error between the analytical model and the data. A similar idea is used by Kuntz et al. [56] with deep neural networks. Indeed, the model first learns from a large dataset of model-based simulated shapes. In a second step, the network learns the difference between the model learned in the first step and real-world data. This transfer learning approach allows learning from a significantly smaller real-world dataset than what would be needed if learning the whole robot shape from scratch. Abovementioned approaches, however, do not take time dependency into account. While recent studies (such as [57], in the context of continuum robot compliance computation) have shown that recursive neural network approaches can help capturing time-dependent effects, applying such algorithms for learning the full kinematic model of a continuum robot, as well as collecting suitable data to this end, is still a challenge to date.

A second approach in data-based modeling is to learn from observations of the robot as they are collected, in real time. Indeed, as described before, nonlinear effects such as friction create short-term time-dependent effects. Long-term time-dependent effects can also occur, for instance, if uncertainties in the robot calibration procedure modify the robot behavior at each new usage. An example of such long-term behavior is evidenced by Poignonec et al. [58] in the context of endorobots. The authors show that the hysteresis loops of the bending segment of a TACR vary from day to day due to modifications of the overall robot shape during initialization. In order to cope with such time varying effects, so-called adaptive learning algorithms have been developed. The overall idea is that sensory data acquired by embedded sensors, cameras, or medical imaging devices, such as X-ray, are fed to the algorithm in order to adapt the model in real time. In the context of concentric tube robots, for instance, Kim et al. [59] have shown that tip positional data can be used in order to refine the Fourier parameters of the functional approximation from Dupont et al. [19], improving the precision of the model. Jang et al. [60] have devised a similar approach by locally refining the tube parameters in order to predict the tip position. Fagogenis et al. [48] improved that idea by using a locally weighted probabilistic regression algorithm, which allows learning local correction maps and weighing the importance of data points in time. Recently, Yu et al. [61] proposed a similar approach using Gaussian mixture models in order to create an adaptive model for catheter control. Again, such models are at the moment only updated based on topologic data (i.e., joint-space and task-space position/orientation, possibly arc-length parameterized) and do not explicitly take time-dependent behavior into account.

Finally, it should be noted that the two approaches presented before are not mutually exclusive. Offline methods have the advantage of being perhaps more complete, allowing to capture more complex behaviors since the training time is not limited. On the other hand, adaptive methods allow correcting the model in real time if/when sensory data are available. Following the same approach as Fagogenis et al. [48], one could envision a hybrid approach where a full, complex model is learned offline, and a lightweight adaptive model runs in real time in order to learn a corrective function to the offline model, reverting to the offline model output when sensory data are not available.

8.2.4 Challenges and future directions

From geometry-based models to mechanics-based and data-based models, a large quantity of research studies have shown progress for building robust models of continuum robots. However, several challenges remain, and new exciting directions of research have been started recently as well.

Constant curvature models and geometry-based models are generally well-understood and allow simple and efficient computation of continuum robot models, including complex effects such as dynamic behavior. The accuracy of such models is, however, inherently limited by the fact that, like pseudo-rigid-body models, the discretization step is fixed beforehand. Indeed, the number of constant curvature sections must be chosen beforehand, limiting the accuracy range of such models. Conversely, mechanics-based models are independent from the discretization step and therefore can be computed at any required accuracy depending on the user's need. Such models are, however, more computationally intensive, even though the increase of computational power decreases the gap as time advances. Moreover, incorporating complex nonlinear effects in those models has proven challenging and is still an ongoing line of research. Finally, data-based models have the advantage to be partially or completely nonparametric, allowing to model complex behavior learned from robot data. The challenge then moves from *modeling* the robot behavior to *collecting* datasets for learning meaningful robot behaviors. Learning time-dependent effects, in the short term and long term, is also an ongoing challenge.

Finally, one should note that although continuum robots are commonly used in endorobotics contexts, an increasing number of interesting mechanisms is also developed in the soft robotics community as described in a recent review paper by Runciman et al. [62]. Although the modeling approaches presented in this chapter are directly applicable for soft robots, new modeling approaches will need to be developed in the future.

8.3 Control strategies

Continuum robots' behavior is governed by complex, internal physical phenomena and interactions with the environment. While such phenomena are inherently safe for continuum robots, the challenge lies in reaching a satisfactory degree of control over the robot's state. The purpose of robot control is to modify a robot from a given state to a desired one. When only sensors are used, the current state is called a measured one; when sensors are combined to model knowledge, the current state is said to be observed; and when a user is directly inferring the robot's state, it is said to be perceived. Generally, a robot's state is localized in a given point on its body. This is mostly defined by the end-effector position and/or orientation, generated force/stiffness, and/or kinematic performances (e.g., manipulability). However, some applications may require including multiple discrete points and/or a continuous description. In classical rigid-link robots, these discrete points generally include joints. This led to the development of full-body control approaches (i.e., controlling multiple/all robot joints) in humanoid robots and hyper-redundant robots.

When it comes to flexible devices and continuum robots, the discretization in multiple joints does not necessarily hold and is altered by the absence of defined joints. It is yet possible to define characterizing discrete points when such continuum robots consist of multiple sections. More generally, the aforementioned state definition applies to flexible devices and is completed by the state at the robot's shape level (i.e., the smooth curves they describe). Contrarily to classical serial robots, the motion of the distal joints/sections usually affects that of the proximal ones for continuum robots. These coupling effects, added to model nonlinearities, increase the complexity of full-body/shape control.

Adding to the previous challenges, concerns of sensor integration in continuum robots to describe their state are raised in the community. Characteristics such as small dimensions, absence of a free lumen, and additional stiffness introduced by sensors prevent taking advantage of the full potential of current sensor technologies, as described by Shi et al. [63].

This section highlights recent control strategies for continuum robots, as end-effector control. Further approaches for multiobjective control are detailed, with a focus on shape control and position/force control. Finally, future trends in continuum robot control are introduced, including data-based approaches.

8.3.1 End-effector control

End-effector position control is widely used, and relatively, rich literature exists. Representing the first step toward mastering the behavior of continuum robots, such controllers are inspired from—or straightforward applications of—classical rigid-link robot control. The goal is to move continuum robots' end-effector to a desired position or pose, should it be three-dimensional or planar. Therefore, similar rigid robot control categorization applies to continuum robots, covering open-loop controllers and closed-loop ones.

As is the case for other robotics applications, open-loop control of continuum robots relies essentially on model inversion. Therefore, the performances of the controller strictly depend on model accuracy and inversion stability (cf. Section 8.2). Indeed, model inversion is not always straightforward especially for mechanics-based approaches, mainly due to the absence of closed-form solutions and to the complexity of numerical inversion, added to possible inherent redundancies and non-linearities. Open-loop controllers represent the first type of controllers applied in continuum robotics. Nevertheless, safety requirements in medical robotics limited their application. Even if open-loop control is not widely used in continuum robotics, it is beneficial for planning purposes and represents a plausible solution when sensors cannot be integrated. Further, path planners provide a global awareness on the robot's workspace and a priori knowledge on feasibility and performances of the tasks. Fig. 8.3a depicts a standard open-loop control scheme. A planner, similar to those above-mentioned, defines the sequence of desired state variables. When passed through the robot model, it is possible to verify motion feasibility of such states, their compatibility with the environment requirements, and fulfillment of tasks. An open-loop controller will then invert such inputs in order to infer the suitable actuators variables to be fed to the continuum robot (should it be a real setup or in simulation environment). Possible disturbances will interfere with the robot and deviate its behavior, which is identified punctually or continuously as a current state.

In this scope, path planning using nearest-neighbor graphs for concentric tube continuum robots is applied by Niyaz et al. [64] to follow surgical trajectories based on Cosserat rod modeling. While robust path planners applied to the robot's shape are still investigated, planning can also be restricted to the robot's end-effector with the same goal. This is possible if a given continuum robot is intrinsically able to perform follow-the-leader motion. An illustrative example by Henselmans et al. [65] includes a continuum robot mechanically

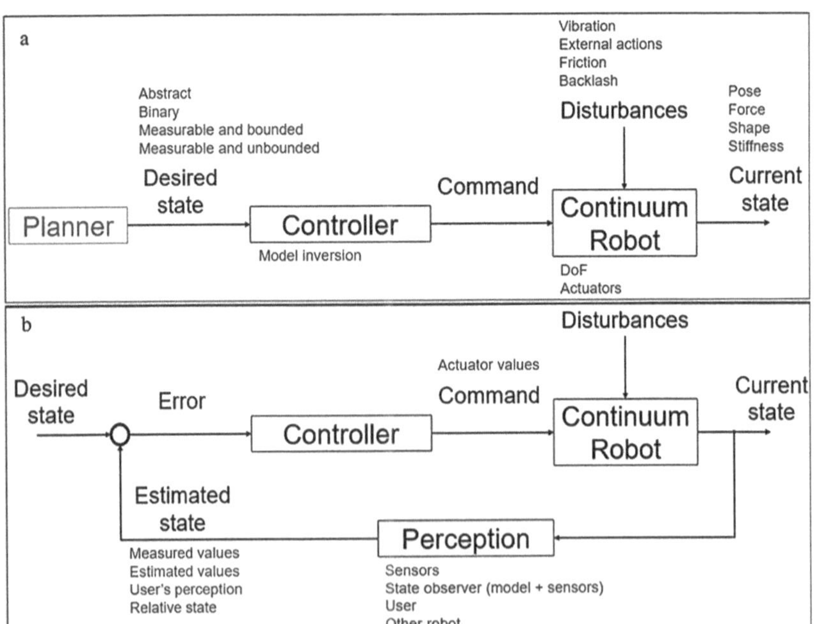

FIG. 8.3 Generic control block diagrams for (a) open-loop and (b) closed-loop.

designed to follow-the-leader and is applied to follow C-curves and S-curves, with and without insertion guide.

Girerd and Morimoto [66] recently proved an open-loop scheme controller effective, by using Fourier-based inverse kinematics (cf. Section 8.2.3) allowing a fast convergence of Newton-Raphson algorithm. This is applied on a hand-held, three-concentric-tube continuum robot for position control, and investigated for percutaneous abscess drainage. Nevertheless, this approach was compared to teleoperation control, where the latter highlighted superior performances.

When sensor integration is possible and effective (cf. [63]), closed-loop control or feedback control is generally more reliable. Fig. 8.3b highlights the main components of a closed-loop control scheme, when the controller block includes model knowledge. At each control loop, the estimated state is updated, as well as the robot configuration. An error with respect to the desired state is then computed and corrected for in the control block. Therefore, combining the global awareness of motion planners with feedback is a powerful tool. Khadem et al. [67] apply such paradigm by online motion planning using ultra-sound image processing data with kinematic model knowledge based on finite element modeling (FEM). This is validated for needle steering in prostate brachytherapy with obstacle avoidance capabilities. Thus, planning can be also part of closed-loop controllers; nevertheless, as there may be diversity in its application (offline or online), its representation is omitted in Fig. 8.3b for convenience.

Other various control examples emerged recently as the use of robust control using a computationally efficient dynamic model (generalized-a method) by Alqumsan et al. [68]. The aim is to control the end-effector in a plane using a TACR with straight and helical tendon routing. The control is based on differential inverse kinematics together with RGB-D camera feedback. It is worth mentioning that controlling the dynamics is a growing research area that recently led to closed-loop control of a cable-driven soft robot. Thieffry et al. [69] design a robust reduced-order Luenberger observer based on FEM and model reduction. The model is linear, and thus, the method is valid in the neighborhood of an equilibrium point in order to control the robot's end-effector. A comparable approach, developed by Ataka et al. [70], suggests pose control on a soft eversion (growing) robot using model-based control (differential inverse kinematics) for planar motion. With a similar visual feedback and model knowledge, yet using Jacobian transpose, Boehler et al. [71] perform position control of a TACR in automated mode. The framework is augmented with a user interface to highlight detected/tracked features. Further, Charreyron et al. [72] use a comparable visual feedback in a shared autonomy scenario based on differential inverse kinematics. The operator is able to control the insertion of a magnetically steerable micro-cannula via teleoperation, while bending is performed automatically to perform subretinal injections. Linearized controllers are also investigated by Campisano et al. [73] using Cosserat rod kinematic model in a quasi-static closed-loop scheme. They use combined feedback from tip pose for path following and from actuators to compensate for nonlinearities of the actuation system to limit model deviations. Comparison between open-loop and closed-loop is performed on the "Hydrojet" soft continuum robot, actuated by water jet propulsion investigated for upper gastrointestinal cancer screening.

Building a reliable model in the scope of control can be challenging; therefore, Bruder et al. [74] use Koopman-based system identification for model predictive control. The application regards planar trajectory tracking of a two-segment soft robot. Khadem et al. [75] build a model predictive controller for concentric tube continuum robots and highlight better performances in tip trajectory tracking with respect to a standard Jacobian-based controller. The controller is autonomous and allows for a more general awareness (in a given time horizon), in order to avoid non-recoverable configurations such as joint limits, instabilities, and singularities.

Other approaches suggest to compensate, at least estimate, modeling errors, especially when the latter is approximated. Wang et al. [76] use a Denavit-Hartenberg-based kinematic model, in which inaccuracy is actively estimated via an unscented Kalman filter. The goal is to control a robotized ureteroscope's end-effector (its tip angle) using a PD controller. In a comparable manner, yet with a piecewise constant curvature dynamic model, Della Santina et al. [10] simulate model-based control of a soft robot's 3D tip position. A dedicated state parameterization allows to adapt such sophisticated model. Model estimation is further investigated by Mo et al. [77], as they develop a second order sliding mode control, using an online Jacobian recursive estimation algorithm with image feedback. This approach is reported to perform better than model-less prior approaches, in terms of 2D trajectory tracking. The method is applied for laser beam steering using a tendon-actuated continuum manipulator inserted within the working channel of an endoscope. Chitalia et al. [78] investigate compensation of model inaccuracies, using a force controller together with a hysteresis compensator component for closed-loop control based on force feedback. They further include a disturbance observer to grasp the high friction in the distal joint and tendon slack induced due to the design of the system. Performances are evaluated on a 2 DOFs, notched, tendon-actuated continuum robot for pediatric neurosurgery.

Using alike model approximations, Franco and Garriga-Casanovas [79] derive an equivalent rigid-link model and a port-Hamiltonian formulation, coupled with a position sensor for disturbance estimation. This allows to control the tip angle of a single-segment, pneumatically actuated soft robot with external planar disturbance compensation. Modeling errors are considered as disturbances and are compensated for. The main idea is based on designing an energy-shaping controller, highlighting better performances than standard PID controllers.

Conversely to path planners, Jacobian-based control provides local awareness but real-time performances. This opens the scope for teleoperation schemes, extensively used in medical robotics. Sarli et al. [80] introduce the use of image feedback for eye-to-hand teleoperation using differential inverse kinematics. The controller is augmented with a redundancy resolution algorithm, applied to a three-segment, push-pull rod-based continuum robot. The variability of the possible 7 DOFs is linked to the deployment of the robot outside the resectoscope sheath and limits the motion of given segments.

8.3.2 Multiobjective control

Controlling a continuum robot includes its end-effector position/pose, as well as the generated forces. While the latter control shares many similarities with classical rigid-link robotics, the specificity of continuum robots (i.e., high compliance, complex shape, and small dimensions) raises further challenges in terms of mastering the effects of generated forces that influence the robot's shape. The latter characteristic together with the related distributed stiffness enriches the state description of continuum robots. This section introduces multiobjective control such as multitasking and multipoint/shape control.

Shape monitoring and control can be performed in different ways. The stricter approach for minimal volume coverage is undeniably follow-the-leader motion, where the robot's body is constrained to follow the path described by the end-effector (e.g., Garriga-Casanovas and Rodriguez y Baena [81] provide related kinematic analysis in 3D). Such approach is recently validated by Girerd et al. [82] for concentric tube continuum robots using simultaneous localization and mapping (SLAM) to perform planar motion and obstacle avoidance.

More tolerant shape controllers suggest to control closely related features, instead of every point on the robot. For instance, Sabetian et al. [83] introduce self-collision detection and avoidance for dual-arm concentric tube continuum robots. Such requirements are mapped into joint (hence, actuator)-level constraints, and a redundancy resolution scheme is adopted. Validation is performed in simulations where accurate shape measurements are assumed feasible.

Similarly to the previous approach, obstacle avoidance while reaching a given target is performed for needle steering. As the model is based on constant curvatures, Lapouge et al. [84] introduce a state observer combining ultrasound image processing with multirate unscented Kalman filtering, in order to estimate the needle's pose and curvatures (hence the shape). The needle is steered through an appropriate path planner to perform both tasks.

A less explicit shape controller is developed by Zhang et al. [85] for a push-pull rod-based continuum robot. Indeed, the structure is capable of conforming to preset configurations. A transition control scheme is developed to handle the particularity of the predefined configurations through teleoperation. Similarly, a constant curvature assumption is adopted in this work.

While shape control regards exclusively robot points, multitasking is referred to as controlling state descriptors belonging to different spaces. An example of such scheme includes task-space control combined with joint-space monitoring. Berthet-Rayne et al. [86] compare inverse kinematics approaches for a so-called quasi-continuum or hyper-redundant robot via teleoperation evaluation. Multitasking includes controlling the robot's end-effector while avoiding joint limits using Jacobian pseudo-inverse, linear programming, and sparse computation (via pseudo-L0 norm and iterative computation). Zhang et al. [87] is another illustrative example of visual, multi-task control that combines eye-in-hand and eye-to-hand strategies. The objectives are to track a trajectory and keep contact with a surface using an endorobot, while controlling the orientation, depth, and endoscopic 2D image position, together with including visibility and actuator limits. The control objectives are formulated as a quadratic programming problem applied to an optical coherence tomography probe inserted within the working channel of an endorobot used to assess the boundaries of cancerous lesions in colorectal surgery.

Hybrid position/force control represents a different multiobjective control scheme. Contrarily to multipoint control, the challenge is related to controlling two entities that belong to different, dual spaces. This is illustrated with controlling a concentric tube robot through trajectory tracking with end-effector while improving robot's manipulability in order to avoid actuation limits and tube snapping. An additional objective, set by Khadem et al. [88], is to apply forces at the end-effector level by pulling a spring. The controller is based on differential inverse kinematics and null-space optimization, while feedback is performed via stereo-vision. A similar approach of hybrid position/force control of the end-effector of a magnetic flexible endoscope (a magnetic capsule actuated by an external permanent magnet) in a realistic colon phantom is introduced by Scaglioni et al. [89]. Model predictive control is implemented with an online optimization scheme for preplanned trajectories.

Hybrid position/force control provides solid advantages in several applications requiring contact with the environment. Not only such paradigm enables to ensure reference forces to be applied on a given object, but also allows to interpret the same forces in cutting-edge applications for navigation. Fagogenis et al. [90] use a thigmotactic algorithm—so-called haptic vision—by tracking low-force contact to navigate on the heart walls. Such approach is performed in autonomous control of an intracardiac catheter, in the form of a concentric tube continuum robot. The controller is based on null-space formulation, where continuous contact tracking is a higher priority task and lateral motion generation has a lower priority, being by automatic robot vision.

A further application of force control regards a haptic input device developed by Frazelle et al. [91]. Based on combining measurement cables with actuation cables in an extensible TACR-like device (master), the approach's purpose is to control a pneumatically actuated soft robot (slave), having three segments each. The control is based on a force-based model to provide feedback for the user on the presence of planar obstacles. Such work tackles the challenge of teleoperating a continuum structure, where natural landmarks are hardly identifiable (smoothness of the shape) and intuitiveness is drastically decreased (complex 3D motion), with respect to classical rigid-link robots. It should be noted that teleoperation is at the interface of open-loop and closed-loop control, as the user's perception itself is closing the loop and synchronizing the motion. However, in a rigorous automatic control definition, no systematic feedback interpretation and error measurement are applied; therefore, it is considered open-loop control in this scope.

With the evolution of continuum robots in unstructured environments, Lin et al. [92] address the challenge of collision detection. A feedback force control of a variable stiffness pneumatically actuated soft robot is applied. Detection of external forces is performed through feedback of inertial measurement units (IMU) combined with the knowledge of a constant curvature model. Compensation is achieved through reaction-based control where the robot is moved in opposite direction of the contact while reducing its stiffness, until a tolerated contact force is estimated.

Further, capabilities of continuum robots for force sensing are utilized for hybrid force position control using redundancy resolution. Yasin and Simaan [93] develop an admittance/position control loop with force control based on force feedback by joint-level force sensing. Comparable performances to using commercial force sensors are highlighted. This approach is applied on a multi-backbone continuum robot for force-controlled ablation and knot tightening.

Impedance control that showed undeniable advantages in classical rigid-link robotics is investigated for continuum robots. He et al. [94] developed simulations of impedance control of a planar single-segment TACR based on a constant curvature model, derived dynamically through a pseudo-rigid body model. The objective is to control the end-effector position, velocity, and contact force. The presented work highlights the benefits of developing a variable impedance controller, while pointing limitations of using pure position feedback and the need of a robust state observer and/or a more accurate model.

An analogous approach to shape control can be elaborated hereby with stiffness control. Even though controlling a distributed stiffness along a continuum robot is rigorously not equivalent to controlling distributed forces, the former is implicitly considered a generalization of force control. Koehler et al. [95] highlight stiffness control for a planar continuum robot based on FEM. To do so, an appropriate mapping between end-effector and actuator compliance is determined, and the stiffness projection is applied in a low-frequency loop, while the faster loop controls the motors to render the commanded stiffness.

8.3.3 Recent trends

Considering the challenges identified in Section 8.2, model knowledge is yet to completely cover all the aspects of continuum robot's behavior. Therefore, data-based approaches are developed to feed the control loop. In this scope, Kalidindi et al. [96] introduce simulations of offline approximate inverse kinematics obtained by actuator babbling and an online adaptive controller based on neural network is developed. The adaption is applied to possible physical changes (i.e., one module failure) of a soft robot in order to maintain trajectory tracking. The authors highlight better tracking performances with respect to conventional open-loop and closed-loop inverse kinematics.

Conversely to a pure data-based approach, learning robot's model and control can be performed in a hybrid fashion and rely partly on model knowledge. Lee et al. [97] combine online learning of the inverse model by initiation with FEM analysis data. Such approach is performed in a feedback control scheme for a pneumatically actuated soft robot, tracked by electromagnetic (EM) sensors. Wang et al. [98] develop a closed-loop controller that combines trained inverse kinematics (offline) and adaptive PID controller (online), under uncertainties of system dynamics and external disturbances. The PID parameters are adjusted based on a back propagation neural network and the controller is combined with a state observer that discriminates between loaded and free-space robot states. This approach is validated on a tendon-actuated continuum manipulator investigated for transoral surgery.

In a comparable hybrid approach, Ma et al. [99] apply feedback position control of a tendon-driven flexible endoscope using EM sensors with a priori knowledge of a constant curvature model. The data-based part regards online estimation of the Jacobian matrix. The controller, based on shared autonomy, combines teleoperation (direct control) and automatic control in different free-space and loaded scenarios. Ma et al. [100] illustrated a possible use-case via visual servoing on the da Vinci Research Kit equipped with a flexible endoscope. The robot Jacobian is calculated based on constant curvature modeling. Closely related approaches are introduced by Wu et al. [101], where model-free control is performed through Jacobian estimation. The specificity in this case is the redundancy resolution scheme performed through visual feedback for a concentric tube continuum robot.

Other technical challenges include the sensory strategy. Due to the inherent limited dimensions of continuum robots, integrating external sensors represent a major limitation in perceiving the robot state. In this scope, proprioception is a promising approach to assist controlling continuum robots. Alatorre et al. [102] investigate this approach on an ionic liquid-based two-segment tendon-actuated continuum robot for closed-loop position control, using resistive proprioception concept via bio-compatible stretch sensors. The authors prove this method capable of drastically improving the position control accuracy and repeatability over open-loop controllers.

The capabilities of soft and continuum robots are yet to be fully considered. Not only is flexibility a major feature in terms of interaction safety but also in the scope of motion generation. Obstacles have usually been considered as a limitation in the robots' environment, and the focus was to avoid, circumvent, and bypass such barriers. Recently, this paradigm has drastically evolved, leading to consider such obstacles as an advantage the robot can take to enrich the generated motion. In other words, obstacles become an advantage, a virtual rigid extension of continuum robot kinematics (an additional workspace is covered when considering interaction with obstacles). In this scope, Zhang et al. [103] use obstacles to control a planar cable-driven catheter through differential inverse kinematics, based on FEM considering contacts. Such control is performed in closed-loop via an electromagnetic tracking system (forces being estimated by FEM), by solving a quadratic programming problem combined with a linear complementary problem. This method assumes both the robot's and the contact objects' models are known and accurate.

A different illustration of this paradigm is introduced by Greer et al. [104] in terms of path planning. When in contact with such planar soft growing robots, obstacles help steering the robot's motion and improve its stability. The planner differentiates between tip contact and body collision and considers their effects in the following extension steps by assuming a pivot motion. The path is updated subsequently using a differential kinematic model. A comparable approach is applied by Selvaggio et al. [105] to a planar, pneumatically actuated, extensible soft robot. Based on a constant curvature kinematic model, the path planner defines a new segment (should it be straight or constantly curved) at each contact point with an obstacle. The appropriate input values are found through inverse kinematics (free-space segments) and a Cosserat rod model (constrained segments), augmented with a lookup table. Both methods also assume perfect a priori knowledge of the environment and an accurate robot model.

8.3.4 Discussions

Continuum robots are distinguished with a relatively more complex state than that of classical rigid-link robots. Their state parameterization, added to inherent nonlinearities and coupling effects, as well as sensor integration, represents an ever-increasing challenge in the scope of robot control. Continuum robot control first focused on straightforward end-effector control, should it concern position, pose, and/or forces. Regular closed-loop control, based on differential inverse kinematic control, was mainly adopted, using Jacobian matrices and eventually compliance matrices. Following that, and with the assessment of modeling performances, path planners and model error estimators completed the initial approaches. Later on, continuum robot control included multiple objectives, allowing to grasp some of the capabilities of such structures. Should it concern kinematic redundancies or stiffness, control approaches span from regular feedback control where only limited model knowledge is required; to anticipative actions and planning using model predictive control based on higher confidence on model accuracy (although backed-up with state observers). Such approaches highlighted capabilities of constrained navigation and versatile operation of continuum robots. Major advances were performed in the scope of shape control, as well as collision detection and handling interaction with the environment in general, including obstacle avoidance and impedance control. Abovementioned challenges regarding model reliability impacted continuum robot control as well. Data-based approaches show increasing efficiency in this scope and reduce the burden of accurate model computation with various model-free control schemes including Jacobian estimation and FEM initialization.

Ultimately, continuum robotics community would reach a more complete state control and benefit from the advantage of flexible, dexterous devices. Such tremendous advances include steps toward 3D shape control and the use of obstacles as an advantage to further navigate with continuum robots. Further, recent results on force and impedance control pave the way for generalized stiffness control, which would grasp the full potential of continuum robots.

Nevertheless, major research is yet expected in order to handle the specificity of each continuum robot technology and finally achieve a generic control framework, as is the case for classical rigid-link robots. The growing design and development community is continuously introducing innovative structures, and modeling and control research need to progressively converge to standard, reliable frameworks.

8.4 Conclusions

This chapter highlights recent advances in modeling and control of continuum robots. Behind the amazing progress made on both aspects in recent years lies a technical reality: continuum robots can take a large spectrum of shapes and aspects. Moreover, design approaches and paradigms evolve quickly, with a development cycle much faster than that of the modeling and control paradigms. As a result, and as evidenced in this chapter, there is not, to date, a standard modeling and control approach (as could be, for instance, Denavit-Hartenberg parameterization and Jacobian-based control for classical rigid-link robots). Moreover, both modeling and control, especially with recent data-based approaches, are intrinsically linked to advances on sensor integration and estimation algorithms (cf. [106]).

Established, common frameworks for modeling and control of continuum robots are, however, slowly coming to light. A first step in this direction takes the form of modeling and control frameworks with public, open-source code [21,42,107–109]. Similar efforts start to take place in the hardware community, with the notable example of the Soft Robotics Toolkit [110], which aims at creating a repository of common resources for building soft robotics platforms of various types.

While the road ahead is still long, such software and hardware approaches will likely contribute to strengthening the community with shared software resources and platforms in order to advance and capitalize on research on continuum robot design, control, and modeling. Coupling these efforts with application-oriented communities and linking to clinical applications and training—similarly to the work done on laparoscopic robotic surgery with the da Vinci Research Kit community [111]—will ultimately lead to integrated efforts and common frameworks for precise, reliable modeling and control frameworks.

References

[1] Vandebroek T, Ourak M, Gruijthuijsen C, Javaux A, Legrand J, Vercauteren T, Ourselin S, Deprest J, Vander Poorten E. Macro-micro multi-arm robot for single-port access surgery. In: 2019 IEEE/RSJ international conference on intelligent robots and systems (IROS). IEEE; 2019. p. 425–32.

[2] Zorn L, Nageotte F, Zanne P, Legner A, Dallemagne B, Marescaux J, de Mathelin M. A novel telemanipulated robotic assistant for surgical endoscopy: preclinical application to ESD. IEEE Trans Biomed Eng 2017;65:797–808.

[3] Burgner-Kahrs J, Rucker DC, Choset H. Continuum robots for medical applications: a survey. IEEE Trans Robot 2015;31:1261–80. https://doi.org/10.1109/TRO.2015.2489500.

[4] Webster III RJ, Jones BA. Design and kinematic modeling of constant curvature continuum robots: a review. Int J Robot Res 2010;29:1661–83. https://doi.org/10.1177/0278364910368147.

[5] Chikhaoui MT, Burgner-Kahrs J. Control of continuum robots for medical applications: state of the art. In: International conference on new actuators; 2018. p. 154–64.

[6] Rucker DC. The mechanics of continuum robots: model-based sensing and control. Ph.D. thesis, Vanderbilt University; 2011.

[7] Lynch KM, Park FC. Modern robotics. Cambridge University Press; 2017.

[8] Hannan M, Walker I. Novel kinematics for continuum robots. In: Advances in robot kinematics. Springer; 2000. p. 227–38.

[9] Jones BA, Walker ID. Kinematics for multisection continuum robots. IEEE Trans Robot 2006;22:43–55.

[10] Della Santina C, Bicchi A, Rus D. On an improved state parametrization for soft robots with piecewise constant curvature and its use in model based control. IEEE Robot Automat Lett 2020;5:1001–8. https://doi.org/10.1109/LRA.2020.2967269.

[11] Devreker A, Rosa B, Desjardins A, Alles EJ, Garcia-Peraza LC, Maneas E, Stoyanov D, David AL, Vercauteren T, Deprest J, et al. Fluidic actuation for intra-operative in situ imaging. In: IEEE/RSJ international conference on intelligent robots and systems; 2015. p. 1415–21.

[12] Runge G, Wiese M, Günther L, Raatz A. A framework for the kinematic modeling of soft material robots combining finite element analysis and piecewise constant curvature kinematics. In: IEEE international conference on control, automation and robotics; 2017. p. 7–14.

[13] Mahl T, Mayer AE, Hildebrandt A, Sawodny O. A variable curvature modeling approach for kinematic control of continuum manipulators. In: American control conference. IEEE; 2013. p. 4945–50.

[14] Gonthina PS, Kapadia AD, Godage IS, Walker ID. Modeling variable curvature parallel continuum robots using Euler curves. In: IEEE international conference on robotics and automation; 2019. p. 1679–85.

[15] Singh I, Amara Y, Melingui A, Mani Pathak P, Merzouki R. Modeling of continuum manipulators using Pythagorean hodograph curves. Soft Robot 2018;5:425–42.

[16] Greigarn T, Çavuşoğlu MC. Pseudo-rigid-body model and kinematic analysis of MRI-actuated catheters. In: IEEE International Conference on Robotics and Automation; 2015. p. 2236–43.

[17] Venkiteswaran VK, Sikorski J, Misra S. Shape and contact force estimation of continuum manipulators using pseudo rigid body models. Mech Mach Theory 2019;139:34–45.

[18] Camarillo DB, Milne CF, Carlson CR, Zinn MR, Salisbury JK. Mechanics modeling of tendon-driven continuum manipulators. IEEE Trans Robot 2008;24:1262–73.

[19] Dupont PE, Lock J, Itkowitz B, Butler E. Design and control of concentric-tube robots. IEEE Trans Robot 2010;26:209–25.

[20] Neumann M, Burgner-Kahrs J. Considerations for follow-the-leader motion of extensible tendon-driven continuum robots. In: IEEE international conference on robotics and automation; 2016. p. 917–23.

[21] Rao P, Peyron Q, Lilge S, Burgner-Kahrs J. How to model tendon-driven continuum robots and benchmark modelling performance. Front Robot AI 2021;7:223.

[22] Mahoney AW, Anderson PL, Swaney PJ, Maldonado F, Webster RJ. Reconfigurable parallel continuum robots for incision less surgery. In: IEEE/RSJ international conference on intelligent robots and systems; 2016. p. 4330–6.

[23] Orekhov AL, Aloi VA, Rucker DC. Modeling parallel continuum robots with general intermediate constraints. In: IEEE international conference on robotics and automation; 2017. p. 6142–9.

[24] Ha J, Park FC, Dupont PE. Optimizing tube precurvature to enhance the elastic stability of concentric tube robots. IEEE Trans Robot 2016;33:22–37.

[25] Peyron Q, Boehler Q, Rabenorosoa K, Nelson BJ, Renaud P, Andreff N. Kinematic analysis of magnetic continuum robots using continuation method and bifurcation analysis. IEEE Robot Automat Lett 2018;3:3646–53.

[26] Dalvand MM, Nahavandi S, Howe RD. An analytical loading model for n-tendon continuum robots. IEEE Trans Robot 2018;34:1215–25.

[27] Oliver-Butler K, Till J, Rucker C. Continuum robot stiffness under external loads and prescribed tendon displacements. IEEE Trans Robot 2019;35:403–19.

[28] Chikhaoui MT, Lilge S, Kleinschmidt S, Burgner-Kahrs J. Comparison of modeling approaches for a tendon actuated continuum robot with three extensible segments. IEEE Robot Automat Lett 2019;4:989–96.

[29] Rucker DC, Webster RJ. Computing jacobians and compliance matrices for externally loaded continuum robots. In: IEEE international conference on robotics and automation; 2011. p. 945–50.

[30] Smoljkic G, Reynaerts D, Vander Sloten J, Vander Poorten E. Compliance computation for continuum types of robots. In: IEEE/RSJ international conference on intelligent robots and systems; 2014. p. 1066–73.

[31] Rucker DC, Jones BA, Webster III RJ. A geometrically exact model for externally loaded concentric-tube continuum robots. IEEE Trans Robot 2010;26:769–80.

[32] Baek C, Yoon K, Kim DN. Finite element modeling of concentric-tube continuum robots. Struct Eng Mech 2016;57:809–21.

[33] Bieze TM, Largilliere F, Kruszewski A, Zhang Z, Merzouki R, Duriez C. Finite element method-based kinematics and closed-loop control of soft, continuum manipulators. Soft Robot 2018;5:348–64.

[34] Shiva A, Sadati SH, Noh Y, Fraś J, Ataka A, Würdemann H, Hauser H, Walker ID, Nanayakkara T, Althoefer K. Elasticity versus hyperelasticity considerations in quasi-static modeling of a soft finger-like robotic appendage for real-time position and force estimation. Soft Robot 2019;6:228–49.

[35] Kato T, Okumura I, Kose H, Takagi K, Hata N. Tendon-driven continuum robot for neuroendoscopy: validation of extended kinematic mapping for hysteresis operation. Int J Comput Assit Radiol Surg 2016;11:589–602.

[36] Subramani G, Zinn MR. Tackling friction-an analytical modeling approach to understanding friction in single tendon driven continuum manipulators. In: IEEE international conference on robotics and automation; 2015. p. 610–7.
[37] Yuan H, Zhou L, Xu W. A comprehensive static model of cable-driven multi-section continuum robots considering friction effect. Mech Mach Theory 2019;135:130–49.
[38] Gao A, Zou Y, Wang Z, Liu H. A general friction model of discrete interactions for tendon actuated dexterous manipulators. J Mech Robot 2017;9.
[39] Ha J, Fagogenis G, Dupont PE. Modeling tube clearance and bounding the effect of friction in concentric tube robot kinematics. IEEE Trans Robot 2018;35:353–70.
[40] Godage IS, Medrano-Cerda GA, Branson DT, Guglielmino E, Caldwell DG. Dynamics for variable length multisection continuum arms. Int J Robot Res 2016;35:695–722.
[41] Renda F, Boyer F, Dias J, Seneviratne L. Discrete Cosserat approach for multisection soft manipulator dynamics. IEEE Trans Robot 2018;34:1518–33.
[42] Till J, Aloi V, Rucker C. Real-time dynamics of soft and continuum robots based on Cosserat rod models. Int J Robot Res 2019;38:723–46.
[43] Liu Z, Zhang X, Cai Z, Peng H, Wu Z. Real-time dynamics of cable-driven continuum robots considering the cable constraint and friction effect. IEEE Robot Autom Lett 2021;6 (4):6235–42.
[44] Boyer F, Lebastard V, Candelier F, Renda F. Dynamics of continuum and soft robots: a strain parameterization based approach. IEEE Trans Robot 2020;37:847–63.
[45] Tariverdi A, Venkiteswaran VK, Martinsen ØG, Elle OJ, Tørresen J, Misra S. Dynamic modeling of soft continuum manipulators using lie group variational integration. PLOS One 2020;15, e0236121.
[46] Gravagne I, Walker ID. Kinematics for constrained continuum robots using wavelet decomposition. In: Robotics 2000; 2000. p. 292–8.
[47] Godage IS, Branson DT, Guglielmino E, Medrano-Cerda GA, Caldwell DG. Shape function-based kinematics and dynamics for variable length continuum robotic arms. In: IEEE international conference on robotics and automation; 2011. p. 452–7.
[48] Fagogenis G, Bergeles C, Dupont PE. Adaptive nonparametric kinematic modeling of concentric tube robots. In: IEEE/RSJ international conference on intelligent robots and systems; 2016. p. 4324–9.
[49] Ha J, Fagogenis G, Dupont P. Effect of path history on concentric tube robot model calibration. In: Hamlyn symposium on medical robotics; 2017. p. 77–8.
[50] Rolf M, Steil JJ. Efficient exploratory learning of inverse kinematics on a bionic elephant trunk. IEEE Trans Neural Netw Learn Syst 2013;25:1147–60.
[51] Xu W, Chen J, Lau HY, Ren H. Data-driven methods towards learning the highly nonlinear inverse kinematics of tendon-driven surgical manipulators. Int J Med Robot Comput Assist Surg 2017;13, e1774.
[52] Grassmann R, Modes V, Burgner-Kahrs J. Learning the forward and inverse kinematics of a 6-DOF concentric tube continuum robot in SE(3). In: IEEE/RSJ international conference on intelligent robots and systems; 2018. p. 5125–32.
[53] Porto RA, Nageotte F, Zanne P, de Mathelin M. Position control of medical cable- driven flexible instruments by combining machine learning and kinematic analysis. In: IEEE international conference on robotics and automation; 2019. p. 7913–9.
[54] Thuruthel TG, Falotico E, Cianchetti M, Laschi C. Learning global inverse kinematics solutions for a continuum robot. In: Symposium on robot design, dynamics and control. Springer; 2016. p. 47–54.
[55] Reinhart RF, Shareef Z, Steil JJ. Hybrid analytical and data-driven modeling for feed- forward robot control. Sensors 2017;17:311.

[56] Kuntz A, Sethi A, Webster RJ, Alterovitz R. Learning the complete shape of concentric tube robots. IEEE Trans Med Robot Bionics 2020.
[57] Jakes D, Ge Z, Wu L. Model-less active compliance for continuum robots using recurrent neural networks. In: IEEE/RSJ international conference on intelligent robots and systems; 2016.
[58] Poignonec, T., Zanne, P., Rosa, B., Nageotte, F., 2020. Towards in situ backlash estimation of continuum robots using an endoscopic camera. IEEE robotics and automation letters.
[59] Kim C, Ryu SC, Dupont PE. Real-time adaptive kinematic model estimation of concentric tube robots. In: IEEE/RSJ international conference on intelligent robots and systems; 2015. p. 3214–9.
[60] Jang C, Ha J, Dupont PE, Park FC. Toward on-line parameter estimation of concentric tube robots using a mechanics-based kinematic model. In: IEEE/RSJ international conference on intelligent robots and systems; 2016. p. 2400–5.
[61] Yu B, Fernández JDG, Tan T. Probabilistic kinematic model of a robotic catheter for 3D position control. Soft Robot 2019;6:184–94.
[62] Runciman M, Darzi A, Mylonas GP. Soft robotics in minimally invasive surgery. Soft Robot 2019;6:423–43.
[63] Shi C, Luo X, Qi P, Li T, Song S, Najdovski Z, Ren H, Fukuda T. Shape sensing techniques for continuum robots in minimally invasive surgery: a survey. IEEE Trans Biomed Eng 2016;64:1665–78. https://doi.org/10.1109/TBME.2016.2622361.
[64] Niyaz S, Kuntz A, Salzman O, Alterovitz R, Srinivasa S. Following surgical trajectories with concentric tube robots via nearest-neighbor graphs. In: Xiao J, Kröger T, Khatib O, editors. International symposium on experimental robotics. Cham: Springer International Publishing; 2020. p. 3–13. https://doi.org/10.1007/978-3-030-33950-0_1.
[65] Henselmans PW, Smit G, Breedveld P. Mechanical follow-the-leader motion of a hyper-redundant surgical instrument: proof-of-concept prototype and first tests. Proc Inst Mech Eng 2019;233:1141–50. https://doi.org/10.1177/0954411919876466. 31526098.
[66] Girerd C, Morimoto TK. Design and control of a hand-held concentric tube robot for minimally invasive surgery. IEEE Trans Robot 2021;37:1022–38. https://doi.org/10.1109/TRO.2020.3043668.
[67] Khadem M, Rossa C, Usmani N, Sloboda RS, Tavakoli M. Robotic-assisted needle steering around anatomical obstacles using notched steerable needles. IEEE J Biomed Health Inform 2018;22:1917–28. https://doi.org/10.1109/JBHI.2017.2780192.
[68] Alqumsan AA, Khoo S, Norton M. Robust control of continuum robots using Cosserat rod theory. Mech Mach Theory 2019;131:48–61. https://doi.org/10.1016/j.mechmachtheory.2018.09.011.
[69] Thieffry M, Kruszewski A, Duriez C, Guerra T. Control design for soft robots based on reduced-order model. IEEE Robot Automat Lett 2019;4:25–32. https://doi.org/10.1109/LRA.2018.2876734.
[70] Ataka A, Abrar T, Putzu F, Godaba H, Althoefer K. Model-based pose control of inflatable eversion robot with variable stiffness. IEEE Robot Automat Lett 2020;5:3398–405. https://doi.org/10.1109/LRA.2020.2976326.
[71] Boehler Q, Gage DS, Hofmann P, Gehring A, Chautems C, Spahn DR, Biro P, Nelson BJ. REALITI: a robotic endoscope automated via laryngeal imaging for tracheal intubation. IEEE Trans Med Robot Bionics 2020;1. https://doi.org/10.1109/TMRB.2020.2969291.
[72] Charreyron SL, Boehler Q, Danun A, Mesot A, Becker M, Nelson BJ. A magnetically navigated microcannula for subretinal injections. IEEE Trans Biomed Eng 2020;1. https://doi.org/10.1109/TBME.2020.2996013.

[73] Campisano F, Caló S, Remirez AA, Chandler JH, Obstein KL, Robert J.Webster I, et al. Closed-loop control of soft continuum manipulators under tip follower actuation. Int J Robot Res 2021;40:923–38. https://doi.org/10.1177/0278364921997167.

[74] Bruder D, Gillespie B, Remy CD, Vasudevan R. Modeling and control of soft robots using the Koopman operator and model predictive control. arXiv e-prints 2019. arXiv:1902.02827.

[75] Khadem M, O'Neill J, Mitros Z, da Cruz L, Bergeles C. Autonomous steering of concentric tube robots via nonlinear model predictive control. IEEE Trans Robot 2020;1–8. https://doi.org/10.1109/TRO.2020.2991651.

[76] Wang X, Bie D, Han J, Fang Y. Active modeling and compensation for the hysteresis of a robotic flexible ureteroscopy. IEEE Access 2020;1. https://doi.org/10.1109/ACCESS.2020.2984424.

[77] Mo H, Wei R, Ouyang B, Xing L, Shan Y, Liu Y, et al. Control of a flexible continuum manipulator for laser beam steering. IEEE Robot Autom Lett 2021;6:1074–81. https://doi.org/10.1109/LRA.2021.3056335.

[78] Chitalia Y, Jeong S, Yamamoto KK, Chern JJ, Desai JP. Modeling and control of a 2-dof meso-scale continuum robotic tool for pediatric neurosurgery. IEEE Trans Robot 2021;37:520–31. https://doi.org/10.1109/TRO.2020.3031270.

[79] Franco E, Garriga-Casanovas A. Energy-shaping control of soft continuum ma- nipulators with in-plane disturbances. Int J Robot Res 2020. https://doi.org/10.1177/0278364920907679. 0278364920907679.

[80] Sarli N, Del Giudice G, De S, Dietrich M, Herrell SD, Simaan N. Turbot: a system for robot-assisted transurethral bladder tumor resection. IEEE/ASME Trans Mech 2019;1. https://doi.org/10.1109/TMECH.2019.2918137.

[81] Garriga-Casanovas A, Rodriguez y Baena F. Complete follow-the-leader kinematics using concentric tube robots. Int J Robot Res 2018;37:197–222. https://doi.org/10.1177/0278364917746222.

[82] Girerd C, Kudryavtsev AV, Rougeot P, Renaud P, Rabenorosoa K, Tamadazte B. Slam-based follow-the-leader deployment of concentric tube robots. IEEE Robot Automat Lett 2020;5:548–55. https://doi.org/10.1109/LRA.2019.2963821.

[83] Sabetian S, Looi T, Diller ED, Drake J. Self-collision detection and avoidance for dual-arm concentric tube robots. IEEE Robot Automat Lett 2019;1. https://doi.org/10.1109/LRA.2019.2933194.

[84] Lapouge G, Troccaz J, Poignet P. Multi-rate unscented Kalman filtering for pose and curvature estimation in 3D ultrasound-guided needle steering. Control Eng Pract 2018;80:116 24. https://doi.org/10.1016/j.conengprac.2018.08.014.

[85] Zhang S, Li Q, Yang H, Zhao J, Xu K. Configuration transition control of a continuum surgical manipulator for improved kinematic performance. IEEE Robot Automat Lett 2019;4:3750–7. https://doi.org/10.1109/LRA.2019.2928762.

[86] Berthet-Rayne P, Leibrandt K, Gras G, Fraisse P, Crosnier A, Yang GZ. Inverse kinematics control methods for redundant snake-like robot teleoperation during minimally invasive surgery. IEEE Robot Automat Lett 2018;3:2501–8. https://doi.org/10.1109/LRA.2018.2812907.

[87] Zhang Z, Rosa B, Caravaca-Mora O, Zanne P, Gora MJ, Nageotte F. Image-guided control of an endoscopic robot for oct path scanning. IEEE Robot Automat Lett 2021;6:5881–8. https://doi.org/10.1109/LRA.2021.3087085.

[88] Khadem M, O'Neill J, Mitros Z, da Cruz L, Bergeles C. Autonomous steering of concentric tube robots for enhanced force/velocity manipulability. In: IEEE/RSJ International Conference on Intelligent Robots and Systems; 2019. p. 2197–204. https://doi.org/10.1109/IROS40897.2019.8967593.

[89] Scaglioni B, Previtera L, Martin JW, Norton J, Obstein K, Valdastri P. Explicit model predictive control of a magnetic flexible endoscope. IEEE Robot Automat Lett 2019;1. https://doi.org/10.1109/LRA.2019.2893418.

[90] Fagogenis G, Mencattelli M, Machaidze Z, Rosa B, Price K, Wu F, Weixler V, Saeed M, Mayer JE, Dupont PE. Autonomous robotic intracardiac catheter navigation using haptic vision. Sci Robot 2019;4. https://doi.org/10.1126/scirobotics.aaw1977.

[91] Frazelle CG, Kapadia AD, Walker ID. A haptic continuum interface for the teleoperation of extensible continuum manipulators. IEEE Robot Automat Lett 2020;5:1875–82. https://doi.org/10.1109/LRA.2020.2970642.

[92] Lin N, Wu P, Wang M, Wei J, Yang F, Xu S, Ye Z, Chen X. IMU-based active safe control of a variable stiffness soft actuator. IEEE Robot Automat Lett 2019;4:1247–54. https://doi.org/10.1109/LRA.2019.2894856.

[93] Yasin R, Simaan N. Joint-level force sensing for indirect hybrid force/position control of continuum robots with friction. Int J Robot Res 2021;40:764–81. https://doi.org/10.1177/0278364920979721.

[94] He G, Fan Y, Su T, Zhao L, Zhao Q. Variable impedance control of cable actuated continuum manipulators. Int J Control, Automat Syst 2020. https://doi.org/10.1007/s12555-019-0449-y.

[95] Koehler M, Okamura AM, Duriez C. Stiffness control of deformable robots using finite element modeling. IEEE Robot Automat Lett 2019;4:469–76. https://doi.org/10.1109/LRA.2019.2890897.

[96] Kalidindi HT, Thuruthel TG, Laschi C, Falotico E. Cerebellum-inspired approach for adaptive kinematic control of soft robots. In: IEEE international conference on soft robotics; 2019. p. 684–9. https://doi.org/10.1109/ROBOSOFT.2019.8722735.

[97] Lee KH, Fu DK, Leong MC, Chow M, Fu HC, Althoefer K, Sze KY, Yeung CK, Kwok KW. Nonparametric online learning control for soft continuum robot: an enabling technique for effective endoscopic navigation. Soft Robot 2017;4:324–37. https://doi.org/10.1089/soro.2016.0065.

[98] Wang Z, Wang T, Zhao B, He Y, Hu Y, Li B, et al. Hybrid adaptive control strategy for continuum surgical robot under external load. IEEE Robot Autom Lett 2021;6:1407–14. https://doi.org/10.1109/LRA.2021.3057558.

[99] Ma X, Wang P, Ye M, Chiu PWY, Li Z. Shared autonomy of a flexible manipulator in constrained endoluminal surgical tasks. IEEE Robot Automat Lett 2019;4:3106–12. https://doi.org/10.1109/LRA.2019.2924851.

[100] Ma X, Song C, Chiu PW, Li Z. Visual servo of a 6-DOF robotic stereo flexible endoscope based on da vincix research kit (dVRK) system. IEEE Robot Automat Lett 2020;5:820–7. https://doi.org/10.1109/LRA.2020.2965863.

[101] Wu K, Zhu G, Wu L, Gao W, Song S, Lim CM, Ren H. Safety-enhanced model- free visual servoing for continuum tubular robots through singularity avoidance in confined environments. IEEE Access 2019;1. https://doi.org/10.1109/ACCESS.2019.2891952.

[102] Alatorre D, Axinte D, Rabani A. Continuum robot proprioception: the ionic liquid approach. IEEE Trans Robot 2021;1–10. https://doi.org/10.1109/TRO.2021.3082020.

[103] Zhang Z, Dequidt J, Back J, Liu H, Duriez C. Motion control of cable-driven continuum catheter robot through contacts. IEEE Robot Automat Lett 2019;4:1852–9. https://doi.org/10.1109/LRA.2019.2898047.

[104] Greer JD, Blumenschein LH, Okamura AM, Hawkes EW. Obstacle-aided navigation of a soft growing robot. In: IEEE international conference on robotics and automation; 2018. p. 4165–72. https://doi.org/10.1109/ICRA.2018.8460777.

[105] Selvaggio M, Ramirez LA, Naclerio ND, Siciliano B, Hawkes EW. An obstacle- interaction planning method for navigation of actuated vine robots. In: IEEE international conference on robotics and automation; 2020.
[106] Mahoney AW, Bruns TL, Swaney PJ, Webster RJ. On the inseparable nature of sensor selection, sensor placement, and state estimation for continuum robots or "where to put your sensors and how to use them". In: IEEE international conference on robotics and automation; 2016. p. 4472–8.
[107] Duriez C, Coevoet E, Largilliere F, Morales-Bieze T, Zhang Z, Sanz-Lopez M, Carrez B, Marchal D, Goury O, Dequidt J. Framework for online simulation of soft robots with optimization-based inverse model. In: IEEE international conference on simulation, modeling, and programming for autonomous robots. IEEE; 2016. p. 111–8.
[108] Sadati SH, Naghibi SE, Shiva A, Michael B, Renson L, Howard M, Rucker CD, Althoefer K, Nanayakkara T, Zschaler S, Bergeles C, Hauser H, Walker I. Tmtdyn: a matlab package for modeling and control of hybrid rigid–continuum robots based on discretized lumped systems and reduced-order models. Int J Robot Res 2019; [in press].
[109] Janabi-Sharifi F, Jalali A, Walker ID. Cosserat rod-based dynamic modeling of tendon-driven continuum robots: a tutorial. IEEE Access 2021;9:68703–19.
[110] Holland DP, Berndt S, Herman M, Walsh CJ. Growing the soft robotics community through knowledge-sharing initiatives. Soft Robot 2018;5:119–21.
[111] Kazanzides P, Chen Z, Deguet A, Fischer GS, Taylor RH, DiMaio SP. An open- source research kit for the da vinci surgical system. In: IEEE international conference on robotics and automation; 2014. p. 6434–9.

Chapter 9

Ultrasound technology for capsule endoscopy

Alexandru C. Moldovan[a], Mihnea V. Turcanu[a], Srinjoy Mitra[b], and Sandy Cochran[a]
[a]*University of Glasgow, Glasgow, United Kingdom,* [b]*University of Edinburgh, Edinburgh, United Kingdom*

9.1 Introduction

Engineered by Iddan et al. [1] and realized commercially by Given Imaging [2], video capsule endoscopy (VCE) was approved for human use in 2001, with approximate physical device dimensions, diameter $\O \approx 10\,\text{mm}$, and length $L \approx 30\,\text{mm}$. The key innovation over conventional endoscopy was to alter the means of positioning from manual control from outside the body to natural passage through the gastrointestinal (GI) tract, transmitting optical data en route and allowing information to be obtained from the full length of the gut. The natural passage requires administration orally and motion in general toward the anus and excretion. Other VCE devices have subsequently become available, specific models have been optimized for different parts of the GI tract [3], and clinical uses have been developed [4].

Previously, optical endoscopy was developed over a lengthy period and endoscopic ultrasound (EUS) subsequently emerged as a significant advance, benefitting from the same minimally invasive access to the body but with the additional ability to image tissue in the wall of and beyond the GI tract. Suiting it to this purpose, ultrasound (US) is safe, widely recognized as inexpensive, and adds diagnostic value through its ability to image deeper inside the tissue. US devices can be miniaturized and operated in real time [5]. Other intraluminal US systems also exist, though with many detailed differences, e.g., intravascular ultrasound (IVUS), transesophageal echocardiography (TEE), and intracardiac echocardiography (ICE).

The combination of easier administration through the mouth and the ability to image beyond the surface of the GI tract make a compelling case for US capsule endoscopy (USCE) that is strengthened by the potential for intracorporeal robotic control [6]. However, major technological barriers exist and this has

Endorobotics. https://doi.org/10.1016/B978-0-12-821750-4.00009-8

kept USCE within the research domain rather than being available through commercial implementation for clinical use. The remainder of this chapter considers the present status of USCE, including imaging, therapy, and ancillary technologies.

Section 9.2 begins with a brief, general overview of US technology. Relevant details of US transducers are presented, along with an outline of how they are incorporated into systems and related technologies that are needed for therapeutic purposes. Section 9.3 outlines the demands placed on robotics as a technology complementing USCE. US imaging for diagnostic USCE is covered in some detail in Section 9.3, and Section 9.4 describes potential implementations of therapeutic US. Future developments are considered in Section 9.5, and conclusions are drawn in Section 9.6.

9.2 Medical ultrasound technology

A medical US system [7] includes a geometrically defined array of transducers, to which a user interface, control and transmission (TX) electronics are connected. US is generated by the array either to insonate a region to be imaged for diagnostic purposes or to deliver energy for therapy. If the system is for diagnosis, the same transducers or others are connected to reception (RX) electronics to amplify, filter, and digitize US signals received by the transducer array. The digital signals are subsequently manipulated mathematically to form the image viewed by the clinician and/or interpreted with artificial intelligence.

There are three challenges to realize a diagnostic US system for USCE: construction of the transducer array and sufficiently small electronics; provision of energy to the intracorporeal system; and wireless bidirectional TX/RX data flow. For a therapeutic system, provision of energy is potentially a greater problem than for diagnosis, as therapeutic US typically needs higher intensity insonation [8], and the wireless data flow is reduced. These challenges are mediated by partitioning of the electronics between the intracorporeal (intracapsular) and extracorporeal functions and components.

9.2.1 Ultrasonic transducers

A key part of any US system is the active material or device, which is typically but not always piezoelectric. Piezoelectric material transduces the mechanical oscillations of an incident US field into an electrical signal through the direct piezoelectric effect, and *vice versa* through the indirect piezoelectric effect [9]. Another technology in US transduction, based on thin-plate capacitors, was created in the mid-1990s, now realized as capacitive micromachined ultrasonic transducers (CMUTs) [10]. These devices convert mechanical to electrical energy and *vice versa* through modulation of the electric field between their electrodes. CMUTs are an attractive alternative to traditional piezoelectric materials due to improved bandwidth and sensitivity, and they provide enhanced potential for integration with electronics because their manufacturing process uses microelectromechanical systems (MEMS) techniques, allowing them to be postprocessed on silicon integrated circuits [11].

The resonant frequency of an US transducer directly influences the resolution of the image that it is used to create. Higher frequencies allow smaller features of the imaged tissue to be differentiated. Lower frequencies, in the range of hundreds of kHz to a few MHz, are generally used for targeted drug delivery (TDD) through sonoporation, whereas frequencies in the 5–20 MHz range are employed to image organs beyond the GI tract with sufficient penetration. Generally, frequencies above the conventional limit of 20–30 MHz are useful for transmural gut imaging due to resolution better than 120 μm, but the penetration depth is reduced because of higher attenuation [12]. Piezoelectric devices capable of achieving such frequencies are termed microultrasound (μUS) transducers.

The other constituent parts of a medical US device are electrodes applied to the active material, electrical interconnects between the transducer and the control electronics, acoustic matching and backing layers, an acoustic lens to focus the US field, and a waterproof casing. The matching layer is added to improve energy transfer between the active material and the tissue, whereas the backing layer is used to damp reverberations and increase the axial resolution of the US scans. Fig. 9.1 illustrates the main components of a medical US imaging array.

The simplest transducer contains one active element, of rectangular or circular geometry, with a flat surface. This design has a fixed focal distance and does not allow the acoustic field to insonate a particular region, commonly called beam steering. It limits the imaging range of the device to only one axis, which corresponds to depth, with the received signal known as an A-scan. Unfocused transducers have poor spatial resolution in all three dimensions because their natural focus is wide. This can be mitigated with the use of a spherically shaped active element (Fig. 9.2A) or by attaching an acoustic lens to a flat transducer (Fig. 9.2B). Beam steering with single-element transducers can be obtained by mechanical translation and rotation of the probe, but the focal

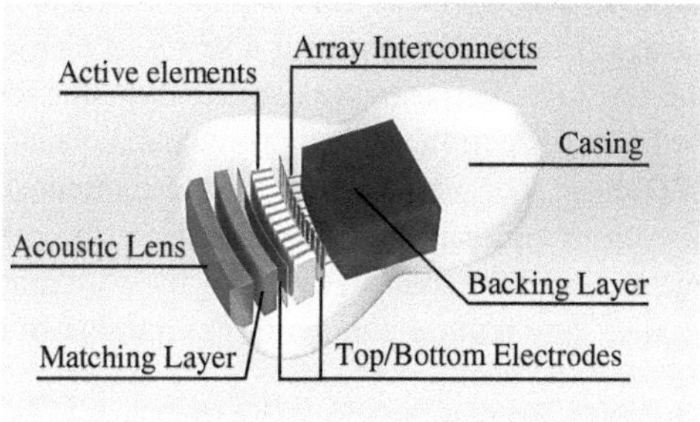

FIG. 9.1 Principal components of a medical ultrasonic imaging probe.

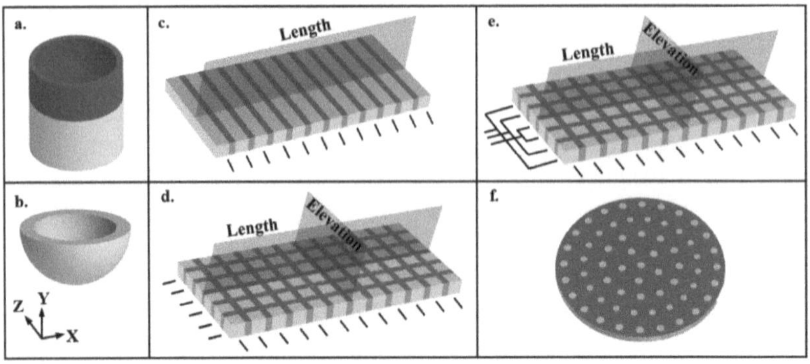

FIG. 9.2 Common transducer geometries used in medical ultrasound (*light colors* represent piezoelectric materials, and *dark colors* represent passive materials): (A) bowl (spherical) single-element transducer; (B) circular element with acoustic lens; (C) 1D Linear array; (D) 2D array; (E) 1.5D array; (F) circular array.

distance cannot be modified without altering the geometry of the device. Probe rotation allows the transducer to capture the entire cross-section of the GI tract, but this setup is prone to motion artifacts [13].

In contrast, US arrays provide dynamic beam control through their ability to electronically focus and steer the field at various angles and depths in the tissue without requiring physical displacement of the probe. These devices are therefore able to perform 2D, planar scans in a process known as B-mode imaging or 3D volumetric scans of tissue and are a valuable tool in modern echography. Arrays comprise a series of piezoelectric elements or CMUTs, which can be addressed individually or in groups, depending on the scanning method that is employed. TX focusing is based on constructive interference of the fields produced by the individual elements and is adjusted by modifying the phase delays applied to the array elements [14]. Focusing, i.e., beam forming, on reception is performed on the data recorded from all elements and is based on the position of the target.

The imaging ability of an US array depends on the number of elements and their layout. 1D arrays (Fig. 9.2C) comprise a series of elements spanning the lateral length of the aperture. These arrays achieve dynamic focusing and beam steering only in the lateral direction and can produce 2D planar images of tissue. 2D arrays (Fig. 9.2D) have individually addressable elements in both the lateral and elevational directions and can thus steer the beam in both directions, providing a 3D volume image of the tissue. However, they are much more complex than 1D arrays because they require a much larger number of individual piezoelectric elements.

Generally, 1D arrays have between 32 and 512 elements, whereas 2D arrays can have as many as 10,000 elements. Intermediate array designs, such as 1.5D arrays (Fig. 9.2E), were developed to reduce electronic complexity and the number of individual interconnects by grouping several elements in the elevational direction. However, their steering capability is reduced in the elevational direction. Another array geometry, commonly used for forward facing probes in IVUS catheters and with potential for embedding into USCE devices, is the circular array (Fig. 9.2F), in which the active elements are placed radially [15].

9.2.2 Ultrasound electronics

Conventional diagnostic medical US electronics design has been simplified in the past 20 years by the development of US chipsets by companies such as Analog Devices (Norwood, MA, United States) and Texas Instruments (Dallas, TX, United States). These incorporate the range of functions for US TX, RX, and image formation shown in Fig. 9.3, allowing innovative US systems to be designed and built by small engineering teams such as systems now available that communicate with a display/user-interface device either wirelessly or *via* a standard connection, such as USB.

While such chipsets are extremely effective in many different ultrasound systems, they are not applicable in USCE for several reasons. A key issue is that the chipset is designed for flexibility rather than being optimized for USCE, leading to dimensions too large to fit within a capsule endoscopy device; moreover, energy consumption is too high.

A further constraint is that USCE requires wireless communication. While this is not ruled out by the use of a chipset, the system partitioning that is an inevitable corollary would require the communication channel to sit either between the digitizer and the RX beam former, with a very large bandwidth, or between the processor and the display, requiring too many system components within the capsule itself.

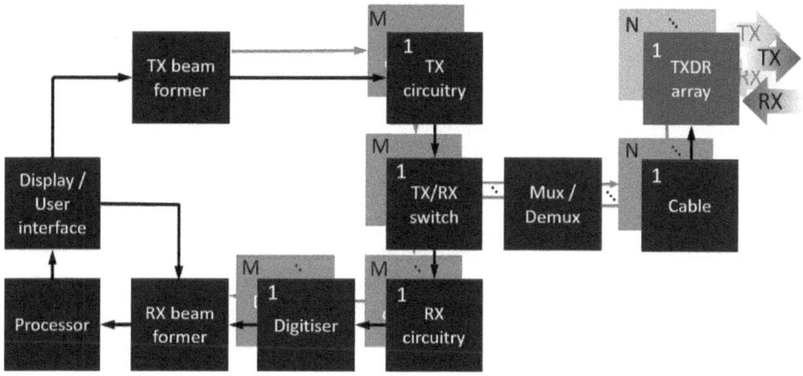

FIG. 9.3 Generic block diagram of US array electronics for linear array operation. Key: *Mux/Demux*, multiplexer/demultiplexer; *TXDR*, transducer array; *M*, number of channels in electronics; *N*, number of transducers in array.

It is instructive to consider the quantity of data that may be required for USCE. Assuming an US frequency, $f = 20\,\text{MHz}$, and the wavelength, $\lambda \approx 80\,\mu\text{m}$. An array with transducers positioned with pitch λ would comprise $N_E \approx 384$ elements to fully populate the circumference of a capsule with $\text{Ø} \approx 10$ mm for radial imaging. With conventional $4\times$ oversampling, digitizing depth of 12 bit/sample, and a physical imaging depth of 10 mm, the data to generate a single image frame would total $N_\Sigma \approx 600\,\text{kByte}$.

For comparison, a 256-level (8 bit) grayscale image with a pixel dimension of $\sim\lambda^2$ and annular geometry, outer diameter $\text{Ø}_{OD} \approx 30\,\text{mm}$, and inner diameter $\text{Ø} \approx 10\,\text{mm}$, representing a radial image, would total $N_I \approx 100\,\text{kByte}$. Further assuming a maximum USCE device speed of $1\,\text{mm}\,\text{s}^{-1}$ through the gut and lateral resolution, i.e., along the length of the GI tract, of 2λ, this highly simplified analysis suggests data rates of $R_{D\Sigma} \approx 3.75\,\text{MByte}\,\text{s}^{-1}$ for unprocessed data and $R_{DI} \approx 0.625\,\text{MByte}\,\text{s}^{-1}$ for image data. In all cases, the potential benefits of data compression and contemporary techniques such as compressive sensing have been neglected.

9.2.3 Diagnostic USCE energy consumption and self-heating

Energy consumption and heat dissipation are key problems in USCE. Such devices have a limited energy source which must allow sufficient operating lifetime, while avoiding harm to the patient due to overheating. Research results remain to be established in this field.

Low-energy systems similar to those found in VCE are presently impossible because of the mechanical nature of US imaging, which requires larger amounts of energy during transmission than the light source in a VCE device. It is therefore essential to limit the energy dissipated in USCE devices to avoid heating of the exterior of the capsule to a dangerous level [16]. Additionally, US propagation itself is associated with heating *via* attenuation and its energy intensity, measured in $\text{W}\,\text{cm}^{-2}$, is limited by healthcare regulations [17].

In relation to operating lifetime, energy can be supplied for USCE either *via* an internal battery or through external power delivery. The use of batteries is well understood, although it must be noted that very few battery types are certified for intracorporeal use. Wireless energy delivery from external sources is also well-researched, with various new techniques in development, including US-powered implants. However, the generally continuous motion of USCE devices through the gut adds difficulty in supplying enough power to such a small volume within the human body without crossing the specific absorption rate (SAR) limits imposed by FCC.

Another difficulty with US transducers is that they require higher operating voltages than common optical devices, specifically for biasing of CMUTs, as mentioned in the previous section, and for US TX with piezoelectric devices. This issue is significant because there is no ideal method to provide the necessary voltage. The use of multiple battery cells in series occupies space, and the source voltage amplitude cannot be controlled without designated circuits. Voltage upconversion from a low voltage source is another alternative, but it also requires designated electronics such as a charge pump. The efficiency of such devices is limited and leads to reduced overall efficiency and increased heating. Better efficiency is obtainable with the inclusion of extra components such as transformers, inductors, and capacitors, but the cost in use of space within the capsule is high.

9.2.4 Therapeutic technologies

Unlike diagnostic US, therapeutic ultrasound [18] may not raise concerns about data rates since a therapeutic capsule may require only simplified localization information. However, other concerns arise in the therapeutic domain, including the therapeutic mechanism. High-intensity focused US (HIFU) is commonly used for focused US surgery (FUS), but this technique is likely to be unsuitable since there is a high risk of necrosis through the thickness of the GI tract and subsequent perforation. US-mediated TDD is the favored alternative, with microbubbles (MBs) as a mediating technology. Clinically approved MBs are gas-filled bubbles with diameter in the range 0.5–10 μm [19]. Their shells can consist of lipids, polymers, lipopolymers, and proteins singly or combined. The gas is usually perfluorocarbon or sulfur hexafluoride [20].

MBs can be used in both diagnosis and therapy. In diagnosis, they are used as contrast agents: once injected intravenously, they can be detected because their low density and dynamic behavior make them reflect US waves more effectively than surrounding tissue. As their size and rheology are similar to those of red blood cells, they can be used to investigate blood flow in organs and tumors [21]. They normally have a short half-life once injected and are therefore administered only prior to US intervention.

For therapeutic purposes, MBs are used as cavitation nuclei. When exposed to US, the dimensions of the MBs oscillate according to the pressure changes in the US wave. Oscillations are most vigorous at the resonant frequency of the MBs, and this can lead to cavitation, which can affect the surrounding tissue, including transient modulation of the permeability of the GI epithelium [22]. MBs can also be conjugated to drugs, such as chemotherapeutic agents [23], and USCE has the potential to deliver them to target locations. Once the conjugated MBs reach the relevant site, they are insonated, the bubbles cavitate, and the drug is released. Variations on this approach are also possible, such as the dynamic combination of therapeutic molecules with molecular packaging, e.g., cyclodextrins, and targeted release with the application of mild hyperthermia and MB-enhanced cavitation [24].

9.2.5 Robotic capsules

In VCE, it is generally assumed that the passage of the device through the GI tract depends upon natural processes, such as peristaltic motion, from mouth to anus. However, after exploration of many different technologies [25], the use of magnetism for positioning and localization has emerged as particularly promising [26]. Coupled to an external robot, this allows control of position and orientation, both highly attractive for USCE [27].

Unlike VCE, USCE will often require precise positioning and localization to generate useful information. An exception to this may be in the use of a circumferential transducer array, such as the outline configuration from which calculations were made in Section 9.2.2, but this still requires practical demonstration. However, the configuration of most initial promise for USCE is likely to combine an array with fewer elements, to reduce the US system complexity and quantity of data to be transferred, with some form of active, touchless robotic positioning. This has already been demonstrated in principle in an oversize, tethered device [27].

Robotic positioning also requires precise localization. Previously, this has been explored using electromagnetic, magnetic, and even optical means but none of these is sufficiently precise for touchless robotic control. An alternative is to integrate localization with the robotic technique in use, as is possible with magnetic positioning [28].

Two operating modes can be envisaged for USCE in combination with robotics. One is to control the motion of the device for scanning, in resemblance to conventional external US imaging [29]. The other is to use robotically controlled USCE for more detailed imaging of a region of interest. In both cases, and with a transducer array of limited extent, robotic control will require positioning of the device within the GI tract and orientation so that the US array has a continuous acoustic path [5] to and from the gut wall.

9.3 USCE for diagnostic ultrasound imaging

As outlined previously, VCE is a well-established procedure in which a capsule, sufficiently small to ingest and capable of optical imaging, collects as many as 50–60,000 images [30] during its passage, transmitting them wirelessly to an external receiver. USCE has the potential to overcome the difficulty that VCE can image only the surface of the lumen and does not allow visualization of deeper lesions [31]. As all forms of capsule endoscopy aim to reduce invasiveness and discomfort to the patient and allow full passage through the gut, the maximum device dimensions should be $L < 33\,\text{mm}$ and $\text{Ø} < 12\,\text{mm}$ [12]. Furthermore, to be completely autonomous, the device must include components to supply or harvest energy, wireless communication with an antenna and related electronics and, potentially, positioning, orientation, and localization components. As well as these modules, the very small volume inside the capsule must also house the US array, the electronics, and the power source (Fig. 9.4) [32].

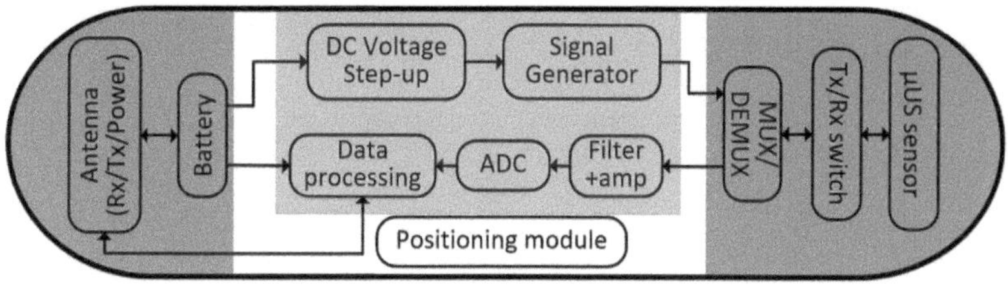

FIG. 9.4 Block diagram depicting the constituent parts of an ultrasonic imaging endoscopic capsule [32].

9.3.1 Imaging geometries

The simplest conventional clinical EUS devices contain a single-element US transducer which rotates 360 degrees around the long axis of the endoscope to capture a radial cross-section of the GI tract. Fabrication is relatively simple compared to array-based devices because of the reduced complexity, but the frame rate is limited by the mechanical rotation of the system and motion artifacts are introduced [33]. Nevertheless, given its simplicity, this approach has been successfully embedded and demonstrated in USCE prototypes [12,31,34].

Recent EUS devices include either an axially oriented curvilinear array [35], capable of a maximum scanning angle of 180 degrees along the curvature direction [36], or a radial array (Fig. 9.5A, [37]) to cover the 360 degrees circumference of the gut in a single scan. Because of the larger scanning angle, radial EUS imaging generally requires less time than the axial curvilinear setup and is better suited to the GI tract [38]. Manufacturing of both types of arrays for EUS is a complex procedure with high failure rates caused by the bending of extremely sensitive and thin piezoelectric elements [39], and implementation of a curved array into a capsule adds further challenges to the already complex tasks in miniaturization. Indeed, because USCE is an emerging technology, there have been no reports yet of successful integration and manufacturing of USCE devices including curvilinear or radial arrays.

A simpler approach to array integration into USCE devices is the use of axial transducers (Fig. 9.5B) rather than circumferential ones (Fig. 9.5C). This reduces the burden of specialized manufacturing and allows greater focus on the other essential goal in USCE, to develop suitable microelectronics. The disadvantage of the axial array approach is that it can provide images only of axial sections of the gut. It thus requires mechanical rotation of the USCE device to image a specific region or capture an entire cross-section of tissue, indicating the fundamental importance of robotic control.

9.3.2 Single-element transducers

9.3.2.1 *Fixed transducers*

Fixed transducer USCE devices comprise one or a number of independent piezoelectric elements to record A-scans inside the GI tract when kept stationary

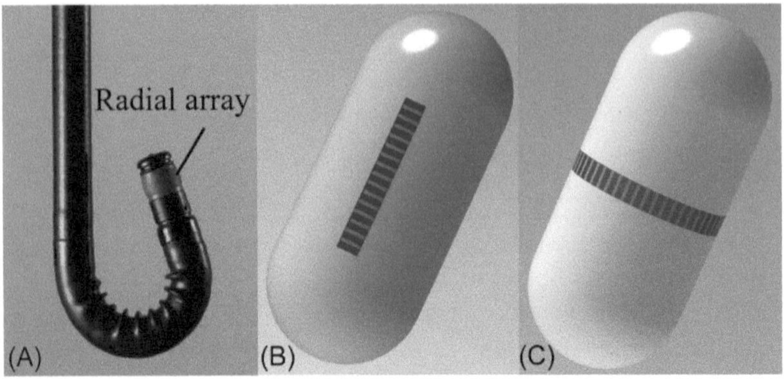

FIG. 9.5 Ultrasonic array geometries employed in endoscopes and capsules: (A) commercial radial array echoendoscope; (B) capsule concept incorporating an axial array and (C) a radial array. *((A) is from Uchida, D., Kato, H., Matsumoto, K. et al. Single-session esophagogastroduodenoscopy and endoscopic ultrasound using a forward-viewing radial scan ultrasonic endoscope. BMC Gastroenterol 19, 220 (2019).)*

and which can be mechanically translated in order to image 2D sections. A fixed-focus capsule (Fig. 9.6A) was developed and fabricated to evaluate the capabilities and feasibility of high frequency (≥30 MHz) μUS in USCE [16]. The device included three single-element μUS transducers distributed axially and a fourth transducer placed in diametric opposition to the other three. The four piezoelectric elements were designed to operate at 30 MHz and had a fixed focal distance of 6 mm in a capsule with $\varnothing = 10$ mm and $L = 30$ mm. For the purposes of the study, the capsule was connected to a tether, $\varnothing_{OD} = 3$ mm, which housed a microcoaxial cable for each transducer. A printed circuit board (PCB) was embedded in the capsule to amplify the received signals but electrical excitation, further amplification and filtering of the received signals, data logging, and analysis were performed with extracorporeal hardware.

The fixed-focus capsule was evaluated *in vivo* on porcine models due to their similarity in histology and physiology with the human GI tract. The capsule was tested *in vivo* in both the esophagus and the small bowel of multiple pigs. For esophageal tests, the capsule was inserted orally, whereas small bowel access was achieved through a stoma. The capsule was then manually retracted, in "pull-back" mode, while the transducers captured A-scans of the GI tract. 2D images were constructed by combining individual line scans and interpolating between them.

The study found that the esophageal wall pressed against the capsule surface more than anticipated, which meant that the esophageal tissue was situated in the near field of the focused transducers, lowering the quality of the image due to near field effects (Fig. 9.6B). However, imaging the small bowel was easier because of the wider lumen which allowed the focal zone of the transducers to be aligned with the small bowel tissue. Therefore, the USCE device was able to

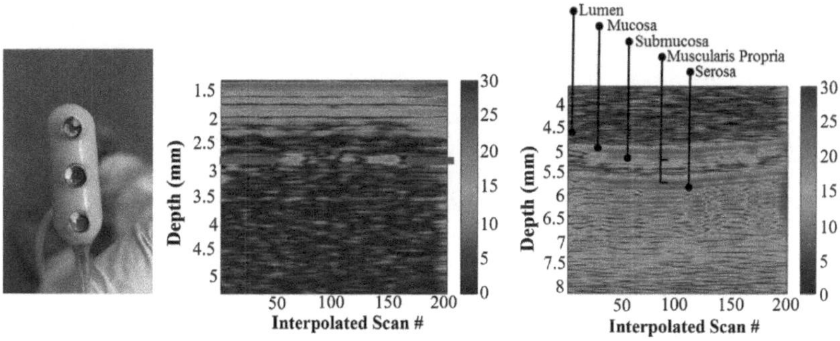

FIG. 9.6 (A) Image of the Sonocap μUS capsule depicting the three single-element transducers positioned axially; (B) 2D ultrasound image of porcine esophagus, with the top of the image closer to the μUS transducer. The near-surface tissue appears defocused due to its position in the near field of the probe, whereas clear subsurface features are visible at depths between 2.5 and 3.5 mm (indicated by *red arrows*); (C) 2D ultrasound image of porcine small bowel, with the top of the image closer to the μUS transducer, obtained for a capsule translation of 12 cm. Constitutive small bowel tissues can be clearly differentiated in the 2D image. *(From H. S. Lay et al., "In-Vivo Evaluation of Microultrasound and Thermometric Capsule Endoscopes," in IEEE Transactions on Biomedical Engineering, vol. 66, no. 3, pp. 632–639, March 2019, https://doi.org/10.1109/TBME.2018.2852715.)*

produce clear 2D images of the target (Fig. 9.6C), which allowed differentiation of the various tissue layers that form the small bowel. The results indicate that the focal distance of future USCE devices should be reduced for esophageal applications or the US transducers should be recessed inside the capsule to allow several millimeters of separation within the lumen. Additionally, different transducer focal zones are required depending on the imaged section of the GI tract. A further finding was that a saline drip was essential for acoustic coupling, suggesting that human ingestion should be accompanied by water. Finally, a power ceiling of 100 mW was determined to prevent a temperature increase of the capsule above the 43°C safe-use threshold for medical US devices [40].

9.3.2.2 Rotating transducers

Several research groups (Wang et al. [31], Qiu et al. [12], Lee et al. [41]) have developed capsules embedding mechanically rotated, single-element, focused transducers on phantoms, as well as on *ex vivo* and *in vivo* tissues. All prototypes achieved 360 degrees radial rotation, were tethered, and were controlled with external equipment.

The first endoscopic capsule to house a rotating single-element transducer for radial B-mode imaging of the GI tract (Fig. 9.7A) was developed and tested *ex vivo* by Wang et al. in 2017 [31]. The device, Ø = 15 mm and L = 25 mm, included a curved 39 MHz US probe with a fixed focal distance of 7 mm, rotated by a geared, rotary solenoid coil motor. The rotation was oscillatory rather than continuous and allowed data acquisition at 0.5 fps. The diameter of the capsule

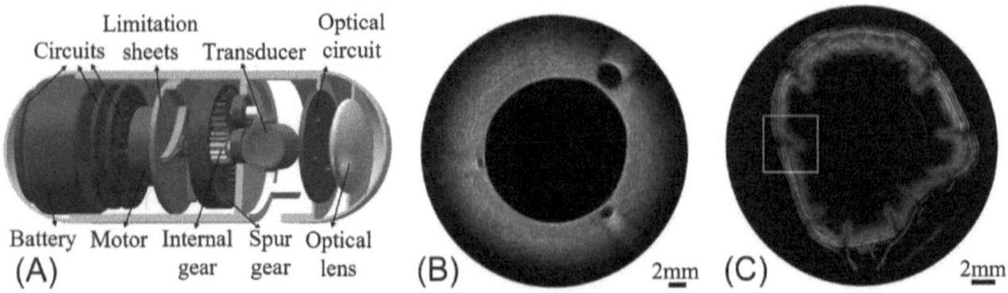

FIG. 9.7 (A) The component parts of the capsule developed by Wang et al.; (B) B-mode scan of the agar phantom depicting the three features contained in it; (C) *ex vivo* image of porcine small bowel showing the different layers of the gut. *(From X. Wang et al., "Development of a Mechanical Scanning Device With High-Frequency Ultrasound Transducer for Ultrasonic Capsule Endoscopy," in IEEE Transactions on Medical Imaging, vol. 36, no. 9, pp. 1922–1929, Sept. 2017, https://doi.org/10.1109/TMI.2017.2699973.)*

was constrained by the 3D printed gear system, making it too large for human applications but it was used for animal trials.

Bespoke imaging and motor control electronics were developed on external PCBs because the study focused on the mechanical stage and US transducer rather than on the electronics. The imaging resolution of the capsule was evaluated with a phantom comprising twelve wires of 15 μm diameter distributed circumferentially and another phantom with five wires distributed linearly. The linear wire distribution indicated that the transducer had axial resolution, $R_{AX}=57.8\,\mu m$, and lateral resolution, $R_{LAT}=181.6\,\mu m$. Another phantom was made of agar and contained three anechoic features with diameters of 0.9, 1.4, and 3.0 mm placed at separations of 120 degrees. The penetration depth in the agar, $D_{PEN}=6\,mm$, and the US probe showed clear differentiation of all three features (Fig. 9.7B). Finally, the capsule was used to perform cross-sectional imaging of *ex vivo* porcine small bowel tissue immersed in a water tank. The resulting B-mode scans resolved several tissue layers found in the gut (Fig. 9.7C).

A capsule comprising a single-element, mechanically rotated $f=30\,MHz$ μUS transducer with a hemispherical front surface that focused the acoustic field at 8 mm was reported by Qiu et al. (Fig. 9.8A and B) [12]. The $\text{Ø}=10\,mm$ and $L=30\,mm$ device was tethered to $\text{Ø}_{OD}=1.7\,mm$ Teflon tube that housed the electrical cables and allowed controlled motion of the capsule axially in the GI tract. The transducer was rotated at an incremental angle every 0.2 s and thus acquired A-scans at a rate of 5 Hz. An external image processing module was developed for interfacing with the tethered capsule, and motor control was achieved with two analog switch channels. The transmission system used single pulses at an amplitude of ±60 V. The device was tested *in vivo* in porcine esophagus *via* oral insertion and in the small bowel through a stoma. Prior to the experiments, it was characterized using a phantom comprising five 25 μm diameter tungsten wires demonstrating $R_{AX}=69\,\mu m$ and $R_{LAT}=262.5\,\mu m$.

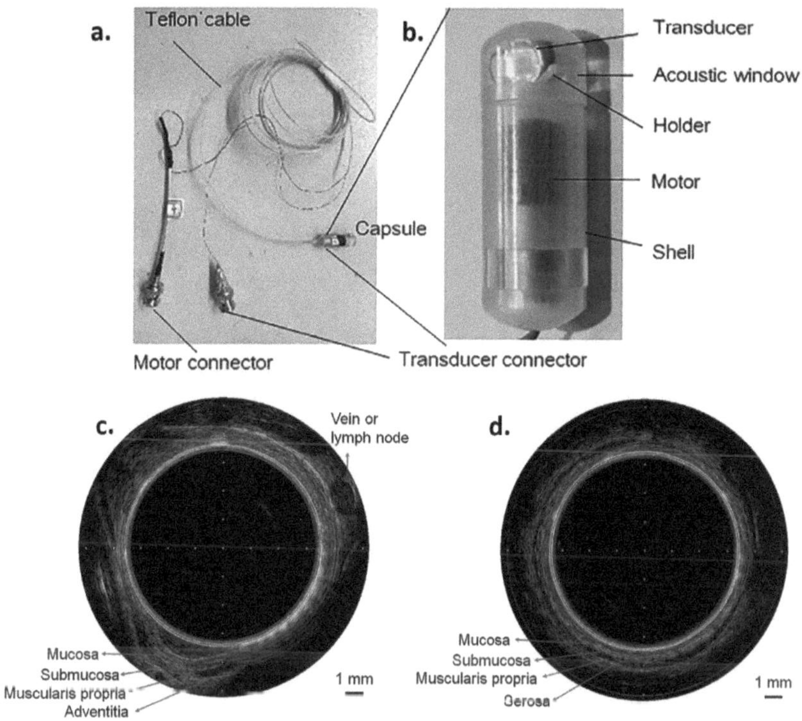

FIG. 9.8 (A and B) Rotating single-element transducer capsule for radial scanning of the GI tract developed by Qiu et al.; (C) radial B-mode scan of the esophagus depicting tissue layer differentiation in the bottom left corner, reduced image clarity in the top right corner generated by increased bowel pressure on the capsule and imaging artifacts caused by a probable air gap or reduced coupling in the bottom right corner; (D) radial B-mode scan of the small bowel showing clear differentiation of the tissue layers. *(From Qiu Y, Huang Y et al., J. Ultrasound capsule endoscopy with a mechanically scanning micro-ultrasound: a porcine study. Ultrasound Med Biol 2020:796–804.)*

Single B-mode radial images of the intestine were constructed by combining ~800 single A-scans. These highlighted the different layers of both the esophagus (Fig. 9.8C) and the small bowel (Fig. 9.8D). However, it was noted that the natural folds of the esophagus were compressed around the capsule during translation, causing loss of clarity for the tissue proximal to the transducer. Further artifacts were attributed to gas in the GI tract and reduced acoustic coupling with the tissue.

Lee et al. [41,42] developed a $Ø = 10\,mm$ and $L = 40\,mm$ capsule (Fig. 9.9A) comprising a $f = 10\,MHz$ unfocused disk transducer connected to a geared stepper motor to acquire radial B-mode images of the surrounding tissue. The aim of the project was to investigate acoustic coupling with the GI tract as a function of peristalsis and to identify issues associated with the procedure. The driving and

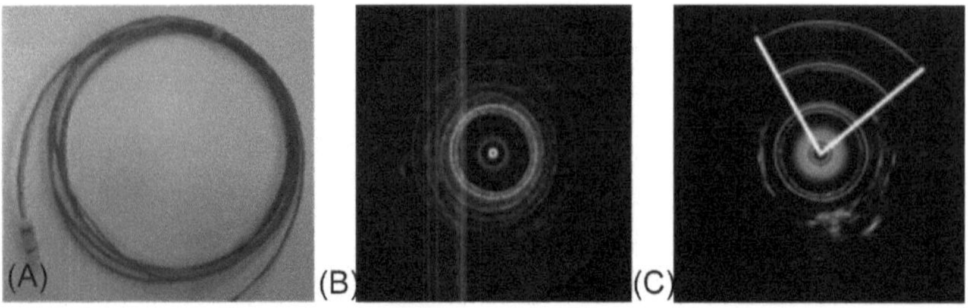

FIG. 9.9 (A) Capsule prototype comprising a mechanically rotated single-element unfocused transducer developed by Lee et al.; (B) mode images of a section of *ex vivo* (C) and *in vivo* small bowel tissue depicting several constitutive layers. *(From J. H. Lee et al., "Towards wireless capsule endoscopic ultrasound (WCEU)," 2014 IEEE International Ultrasonics Symposium, 2014, pp. 734–737, https://doi.org/10.1109/ULTSYM.2014.0181.)*

receiving electronics were developed on an external PCB, connected to the capsule through a 4m long, $Ø_{OD}=3.3$ mm tether and were interfaced with a PC through a commercial field-programmable gate array (FPGA) board. Initial tests were performed on a phantom comprising 15 monofilament fishing lines, diameter 300 μm, and demonstrated the device $R_{AX}<300$ μm and $R_{LAT}\approx380$ μm.

These figures were sufficient to distinguish layers of porcine esophagus and small bowel tissue both *ex vivo* (Fig. 9.9B) and *in vivo* (Fig. 9.9C). *Ex vivo* trials also showed the capsule walls produced echoes visible in the radial B-mode images. Nonetheless, near circumferential visualization of the tissue was possible with adequate probe-tissue separation. Furthermore, the *in vivo* trials showed good coupling with the walls of the GI tract during peristaltic contraction; however, the peristaltic relaxation and the presence of gas in the GI tract hampered imaging. The study was insufficient to confirm if images produced by an autonomous capsule with no position or pressure control on the lumen could provide satisfactory images of the gut.

9.3.3 Transducer arrays

Single-element devices, whether stationary or mechanically scanned, provided crucial evidence to support further development of USCE devices. Specifically, they showed that acoustic coupling can be achieved between the transducer and the wall of the GI tract in different regions with minimal or no irrigation and that viable ultrasonic frequencies allowed layers of tissue to be distinguished, thus providing information on pathology. USCE development relating to transducer arrays is at an earlier stage, with work just beginning.

To guide the design of circumferential arrays for USCE, a virtual prototyping system was developed by Lay et al. [43] for assessment of electronic and acoustic capabilities. Part of the study proposed a device comprising an

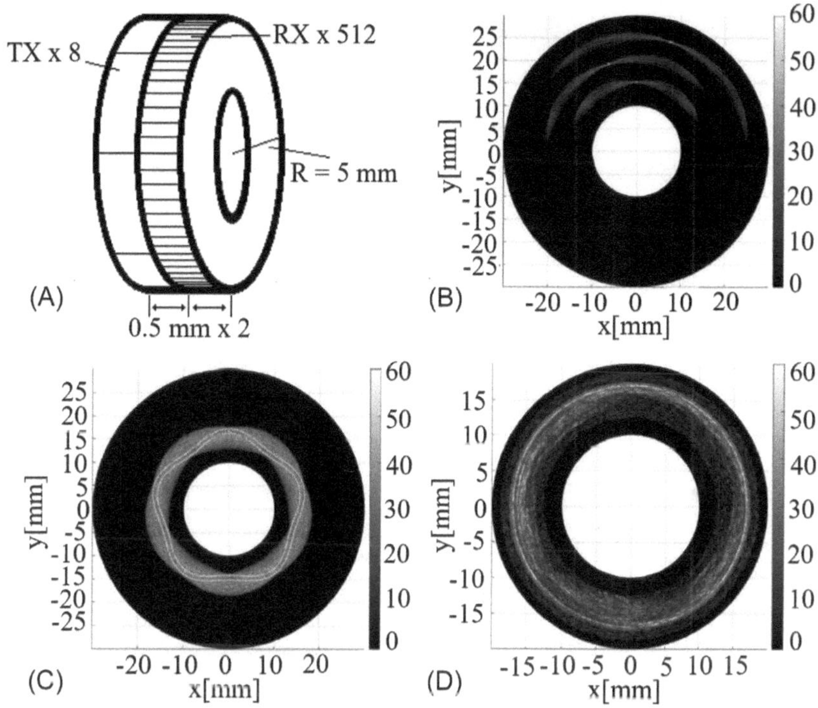

FIG. 9.10 (A) Schematic of the proposed radial transducer for echoendoscopy showing the 8-segment unfocused transmission ring transducer and the 512-element imaging array. (B) Simulated B-scan of the three wire targets; (C) simulated B-scan of the two-layered tissue phantom showing the sinusoidal structure of the modeled tissue; (D) simulated B-scan of the small bowel model adapted from a real acoustic scan. *(From Lay HS, Cox BF, Seetohul V, Demore CEM, Cochran S. Design and simulation of a ring-shaped linear array for microultrasound capsule endoscopy. IEEE Trans Ultrason Ferroelectr Freq Control 2018:589–99.)*

8-segment, unfocused TX ring transducer with $f=25$ MHz and a 512-element circumferential RX array (Fig. 9.10A). As well as modeling the piezoelectric transducers, the simulation was also able to include the influence of the analog front-end (AFE) electronics on the imaging performance (Fig. 9.3). The simulated device provided unfocused TX and dynamic RX beam. Simulated B-mode images were produced for a three-wire-target phantom (Fig. 9.10B) and a two-layered phantom with a sinusoidal variation in its structure (Fig. 9.10C). An in silico gut phantom (Fig. 9.10D) was also constructed, using data recorded from *ex vivo* porcine small bowel imaged with a 45 MHz probe for better simulation accuracy.

In the experimental domain, a $f=5$ MHz, $N=128$ element CMUT radial array was developed and characterized by Memon et al. [44] and Wang et al. [45] (Fig. 9.11) with the aim of being incorporated into an autonomous USCE device. A key aspect of this array is that it is flexible, allowing the curvature that

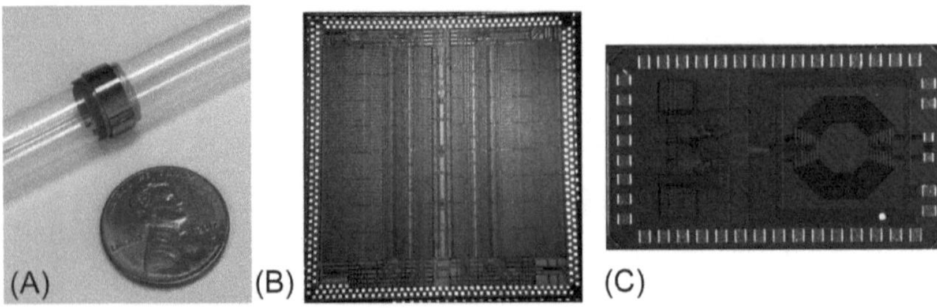

FIG. 9.11 (A) Flexible CMUT array developed by Memon et al., wrapped around a 1 cm diameter rod; (B) image of the 128 channel transmit/receive chip; (C) image of the wireless antenna. *(From F. Memon et al., "Capsule ultrasound device: Further developments," 2016 IEEE International Ultrasonics Symposium (IUS), 2016, pp. 1–4, https://doi.org/10.1109/ULTSYM.2016.7728851.)*

will be needed for USCE devices. The researchers also developed a $6 \times 6\,\text{mm}^2$ ASIC (application-specific integrated circuit) with 128 TX/RX channels and $1 \times 1.76\,\text{mm}^2$ wireless TX electronics with a data rate of 10 Mbps. The imaging chip used a 14-stage charge pump to boost the DC voltage from the battery level to the ~17 V needed for the CMUT device. Individual array elements were characterized and shown to produce satisfactory surface acoustic pressures at a DC bias voltage, $V_{BIAS} = -20\,\text{V}$, and excitation voltage, $V_{EXCITE} = 18\,\text{V}$. However, these devices have not yet been integrated into a capsule and no images have been reported.

9.4 USCE for ultrasound therapy

As we argued in Section 9.1, the addition of EUS to optical endoscopy as weapons in the clinician's armamentarium suggests a similar addition of USCE to VCE to increase patient comfort and allow diagnosis along the full length of the GI tract. There is no such similarity in terms of ultrasound therapy. Instead, this potential capability has emerged from research into US-mediated TDD (USmTDD), motivated by the need to increase the gamut of drugs that can be delivered orally and ingested through conventional intestinal biological mechanisms. A parallel can thus be drawn between Fig. 9.4, representing a possible diagnostic USCE device, and Fig. 9.12, representing a corresponding therapeutic device [32].

9.4.1 Technologies

Transducers used for imaging generally have wide bandwidth to achieve high spatial resolution; operate at high frequencies, in the range 4–12 MHz for conventional US and above 20 MHz for μUS; and produce very low time-averaged US intensity to ensure safety. Their high bandwidth leads to significant

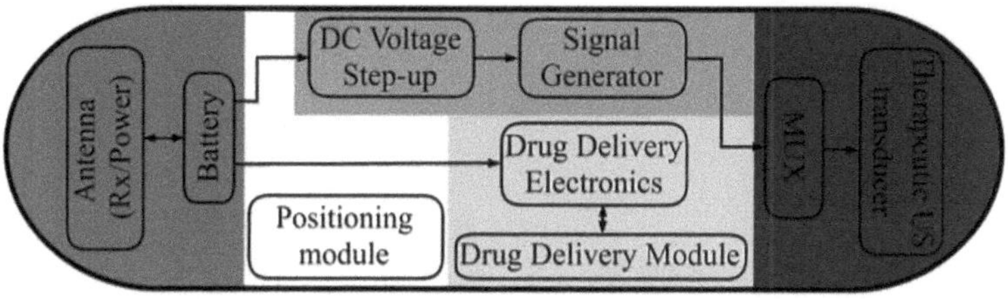

FIG. 9.12 Block diagram of a conceptual capsule for autonomous therapy [32].

inefficiency in many designs. In contrast, therapeutic transducers have a narrow bandwidth, operate at a lower frequency, in the range 0.2–4 MHz, and produce higher time-averaged US intensity to achieve the therapeutic effect. Normally, therapeutic transducers are highly efficient, minimizing self-heating and thermal bioeffects.

As for diagnostic USCE, prototype capsules have been produced with single-element transducers [46,47] as discussed in Section 9.4.2. US array designs for therapeutic intracorporeal sonoporation have also been developed [8] based on miniature 1–3 connectivity 1D phased arrays. In both cases, simulations were consistent with characterization results, showing the capability to produce sufficient US intensity for sonoporation but below HIFU intensity levels.

9.4.2 Single-element transducers

Investigation into the use of single-element transducers for therapeutic USCE began with *in vitro* experiments. A miniature focused US transducer suggested US has the potential of delivering fluorescent dextran across an epithelial cell layer [46–48]. A suspension of MBs and dextran was delivered through a channel in the center of the transducer (similar to Fig. 9.13A). US and MBs promoted the delivery of fluorescent dextran across the cell layer more significantly than when dextran was applied alone. The transducers, including a delivery channel, were further miniaturized and implemented in tethered endoscopic capsules (Fig. 9.13C, [47]).

The capsules were then tested *ex vivo* and *in vivo*. *Ex vivo*, the capsule delivered a combination of US, MBs and quantum dots (QDs) to the small intestine of murine models. Laser scanning confocal microscopy showed the QDs lodged in the mucus layer, not penetrating into the underlying epithelia. For *in vivo* tests, the capsule was inserted through a stoma in the small intestine of terminally anaesthetized porcine models. US alone and QDs alone did not mark the lumen of the small intestine with fluorescence. However, a combination of US, MBs, and QDs facilitated visible retention of QDs in the intestinal wall, as seen in Fig. 9.14, in 80% of cases.

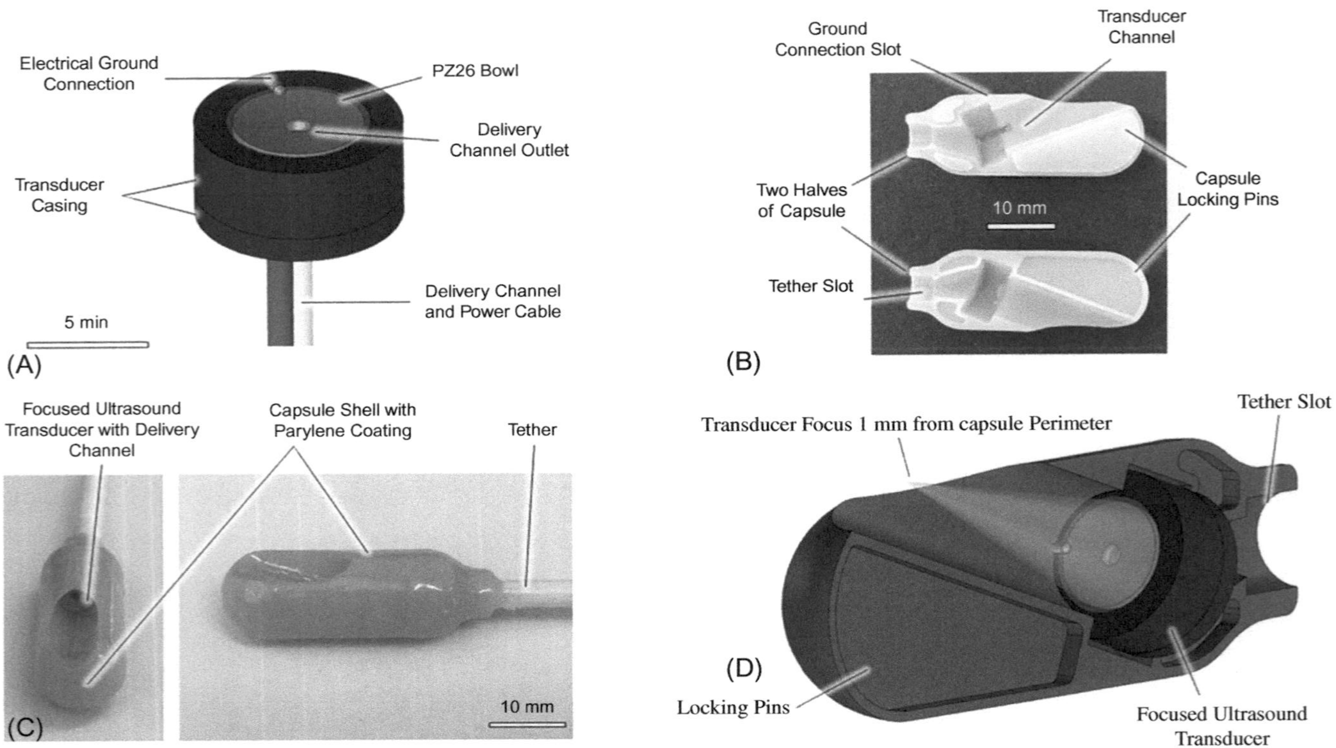

FIG. 9.13 (A) Schematic of ultrasound transducer used in delivery capsule. (B) Locking mechanism for aligning capsule. (C) External view of capsule for *in vivo* drug delivery. (D) Cross-sectional image of *in vivo* delivery capsule. *(From Stewart, F., Cummins, G., Turcanu, M.V. et al. Ultrasound mediated delivery of quantum dots from a proof of concept capsule endoscope to the gastrointestinal wall. Sci Rep 2021;11:2584. https://doi.org/10.1038/s41598-021-82240-1.)*

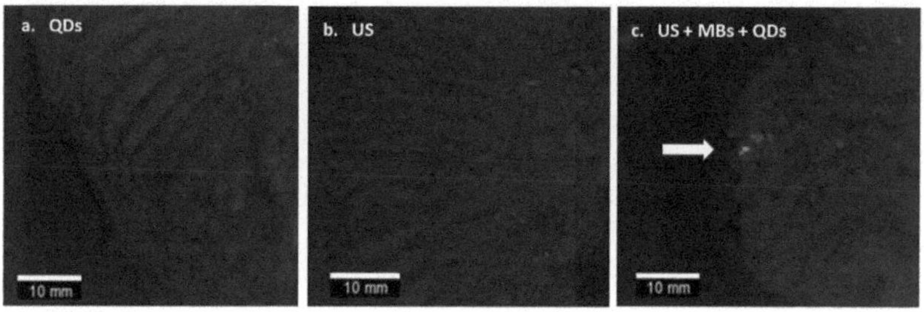

FIG. 9.14 Fluorescent QDs did not lodge into tissue (A) when they were released in the absence or (B) presence of US alone, but (C) they were retained when delivered combined with insonation and MBs. *(From M. V. Turcanu et al., "Ultrasound and Microbubbles Promote the Retention of Fluorescent Compounds in the Small Intestine," 2018 IEEE International Ultrasonics Symposium (IUS), 2018, pp. 1–4, https://doi.org/10.1109/ULTSYM.2018.8580036. With Permission.)*

9.4.3 Transducer arrays

Research into transducer arrays for USCE therapy is at an earlier stage, following the results demonstrating feasibility outlined in the previous section. However, work by Moldovan et al. on prototyping of arrays for intravascular sonoporation [8] and on design studies for intracorporeal sonoporation [49] suggests that the previous positive single-element results can be translated into array-based USCE devices with appropriate ancillary technologies.

9.5 Future developments

USCE has great potential for both diagnosis and therapy, with specific attributes that make robotics an attractive partner technology. Future developments are required in both clinical domains and may also be found in hybridization with other capsule endoscopy modalities.

9.5.1 Diagnosis

As noted in Section 9.2.1, μUS describes ultrasound scanning at frequencies higher than the upper limit of conventional ultrasound systems, in the range above 20–30 MHz. Its advantages include improving axial resolution from $\sim$200 μm to $<\sim$75 μm, enhancing the ability to detect changes in the spatial distribution of cells in a given tissue. μUS may also have the potential to detect accurately pathological changes in the intestinal epithelia [50].

Further developments may be expected in a range of ancillary technologies, including robotic manipulation by magnetic [27] or other means [25]; in miniature transducer arrays, of both piezoelectric [16] and CMUT types; in both the partitioning and design of the necessary electronic systems; in energy supply, from within the device and by remote delivery; and in wireless communications [51].

Once even early autonomous diagnostic USCE devices and systems have been engineered, it is likely that automated data interpretation will emerge as clinical applications are explored. As outlined in Section 9.2.2, USCE devices will produce very large amounts of data. Ordinarily, this would require a corresponding amount of time from a clinician for review. However, in early work, a learning-enabled computer-aided system has classified μUS B-scans of inflamed and healthy murine intestinal tissue with a high success rate [52]. This could reduce workload for clinicians and identify pathology that might otherwise be overlooked.

9.5.2 Therapy

Therapeutic US in the GI tract could deliver medication locally or systemically. Local delivery is relevant for diseases of the intestine, such as inflammatory bowel disease or bowel cancer, translating into reduced side effects since the amount of drug required will be much smaller than the regular dose administered systemically [53]. Systemic delivery is relevant for oral administration of any type of drug (antibodies, proteins, nanoparticles, and vaccines traditionally only delivered intravenously or intramuscularly). An US capsule ingested orally will reduce dependence on needle injections, increasing patient treatment adherence.

Preliminary data have suggested that MBs and US can facilitate the passage of large molecules across tissue layers [54]. There are three main mechanisms by which this may be possible: transcellularly, i.e., through cells; paracellularly, i.e., through intercellular spaces; and/or by sonoporation, i.e., through formation of pores in cell membranes [55]. Sonoporation has been suggested to be the main phenomenon for the passage of drugs through the epithelium as a result of acoustic cavitation [56–58]. It can also induce bioeffects, such as spatiotemporal changes of intracellular calcium concentration ([Ca^{2+}]) [59], reorganization of the actin cytoskeleton [60], and chemical dissociation of water leading to the formation of free radicals [61].

9.5.2.1 Nanodroplets rather than MBs

Nanodroplets (NDs) are liquid-filled compounds. Once they are exposed to US of a resonant frequency, they can undergo acoustic droplet vaporization (a phase shift from liquid core to gaseous core) and can further undergo cavitation [62]. According to the application and method used, one could choose between NDs or MBs. MBs have a net overall density lower than that of blood, rendering them buoyant [63], whereas NDs often have a net overall density higher than that of blood, enabling them to reside at the bottom of a liquid volume [64].

MBs are normally too large to extravasate and have a short half-life (<5 min). In contrast, NDs are small enough to extravasate and accumulate in tumors and have a longer half-life when injected in blood [65]. Although NDs are promising theranostic agents, they are yet to be clinically approved.

9.5.2.2 Fully autonomous therapeutic USCE

Autonomous therapeutic USCE devices operating *via* sonoporation are a promising approach for treatment of various bowel diseases such as Crohn's disease, ulcerative colitis, and cancer [22,66]. Compared to current procedures performed with tethered endoscopes, capsule treatment could be delivered for much longer periods (days vs minutes or hours) with less invasion. They will also facilitate access to the small bowel, which is difficult to reach with current endoscopic technologies [67].

Depending on the therapeutic applications, such as sonoporation, hyperthermia, or HIFU, design considerations for the US transducers and driving electronics vary in terms of frequency of operation, duty cycle, and electrical power. However, one of the most important milestones in the development of any kind of autonomous therapeutic capsule will be the removal of the tether. This is presently employed in all capsule prototypes to house the electrical interconnects between the ultrasonic transducer and the bench equipment, as well as, the channels used for delivery of therapeutic agents.

9.5.3 Hybridized multimodal USCE devices

Integrating multiple capabilities into one capsule could improve both diagnosis and therapy. Such capabilities include optical imaging, autofluorescence imaging (AFI), microwave ablation, pH sensing, and pressure sensing.

The lack of substantial morphological changes in early-stage cancerous tissue makes these lesions undetectable with white light imaging [68]. US may detect the slight thickening of the epithelia and enhanced vascularization, but AFI could differentiate between natural fluorescence of healthy tissue and diseased tissue [69]. Healthy areas emit green autofluorescence, whereas diseased areas appear as darker regions. This could be further combined with cancer-selective molecular fluorescence probes, which react with metabolites overexpressed in cancerous tissue, producing highly fluorescent products, stronger than AFI [70,71]. Hybridized with USCE, these would decrease the error rate of diagnosis of cancer in the GI tract.

Similar hybridization may be possible with a miniaturized fluorescent imager. One such device, suitable for integration into capsule endoscopy, has successfully distinguished between healthy and malignant tissue in freshly resected human bowel tissue [72]. This was achieved with an optical block and a 64×64 single-photon avalanche diode (SPAD) array which detected green fluorescence from both an endogenous autofluorescent fluorophore and a cancer-selective molecular probe that labels colorectal cancer cells.

9.6 Conclusions

Ultrasound capsule endoscopy is a promising technology for imaging that aims to increase the provision of diagnostic information while reducing the invasiveness of classical endoscopy and increasing patient comfort. Therapeutic capsule endoscopes can also be employed to deliver targeted drugs with enhanced delivery rates through the GI tract by means of sonoporation performed with the onboard ultrasonic transducers.

USCE prototypes tested on *ex vivo* and *in vivo* tissues have used a mechanically rotated single-element piezoelectric transducer to create B-mode images of the GI tract. This technique has been preferred so far due to its relative simplicity compared to US array implementation. However, arrays have the advantage of faster scanning and are not prone to movement errors. Future USCE devices are therefore expected to include axial or circumferential arrays. However, the necessity to include all electronics inside the small capsule volume is a difficult task, not only because of the small size of the circuits, but also because wireless communication and power delivery systems must be developed.

All capsule prototypes have been tethered so far because of the difficulty of embedding the electronics. However, ASIC technology can be developed in the future for miniature control circuits that can fit inside the capsule. Similarly, wireless power transfer technology has already been tested and demonstrated to be feasible and, with future development, it will be possible to supply enough power to drive all the electronics in both imaging and therapeutic ultrasonic ingestible devices.

Furthermore, μUS is the most promising approach for the development of USCE devices to image the wall of the GI tract because of the higher resolution it offers compared to conventional ultrasound. This higher resolution is more desirable in the case of GI tract imaging as it can differentiate better the small structures present in the gut and penetration depth is not an important issue because of the small thickness of the tissue to be imaged.

Therapeutic USCE is also promising. Relying partly on US and related technologies developed for other therapeutic purposes, its success may be motivated by the extreme convenience of oral drug delivery and the need for large-molecule drugs to be ingested, and by the potential for treatment of parts of the GI tract that are presently *terra incognita*.

References

[1] Iddan G, Meron G, Glukhovsky A, Swain P. Wireless capsule endoscopy. Nature 2000;405 (6785):417.

[2] Given Imaging [Internet]. Available from: www.givenimaging.com.

[3] Medtronic [Internet]. Capsule endoscopy. Available from: https://www.medtronic.com/covidien/en-us/products/capsule-endoscopy.html.

[4] Keuchel M, Hagenmüller F, Tajiri H, editors. Video capsule endoscopy [Internet]. Berlin, Heidelberg: Springer Berlin Heidelberg; 2014. [cited 28 December 2020]. Available from: http://link.springer.com/10.1007/978-3-662-44062-9.

[5] Szabo TL. Diagnostic ultrasound imaging: inside out. 2nd ed. Amsterdam; Boston: Elsevier/AP, Academic Press is an imprint of Elsevier; 2014. 806 p. (MATLAB examples).

[6] Moglia A, Menciassi A, Schurr MO, Dario P. Wireless capsule endoscopy: from diagnostic devices to multipurpose robotic systems. Biomed Microdevices 2007;9(2):235–43.

[7] Cobbold RSC. Foundations of biomedical ultrasound. Oxford; New York: Oxford University Press; 2007. 802 p.

[8] Moldovan AC, Qiu Z, Lines D, Gachagan A, Cochran S. Development of a 1-D linear phased ultrasonic array for intravascular sonoporation. In: 2019 IEEE international ultrasonics symposium (IUS); 2019. p. 2056–9.

[9] Vives AA, editor. Piezoelectric transducers and applications [Internet]. 2nd ed. Berlin Heidelberg: Springer-Verlag; 2008. [cited 18 December 2020]. Available from: https://www.springer.com/gp/book/9783540775072.

[10] Khuri-Yakub BT, Oralkan Ö. Capacitive micromachined ultrasonic transducers for medical imaging and therapy. J Micromech Microeng 2011;21(5):054004–14.

[11] Savoia AS, Caliano G. MEMS-based transducers (CMUT) for medical ultrasound imaging. In: Frontiers of medical imaging. World Scientific; 2014. p. 445–64. https://doi.org/10.1142/9789814611107_0021.

[12] Qiu Y, Huang Y, Zhang Z, Cox BF, Liu R, Hong J, et al. Ultrasound capsule endoscopy with a mechanically scanning micro-ultrasound: a porcine study. Ultrasound Med Biol 2020;46(3):796–804.

[13] O'Donnell M, Eberle MJ, Stephens DN, Litzza JL, Vicente KS, Shapo BM. Synthetic phased arrays for intraluminal imaging of coronary arteries. IEEE Trans Ultrason Ferroelectr Freq Control 1997;44(3):714–21.

[14] Szabo TL. Diagnostic ultrasound imaging: inside out. Academic Press; 2004.

[15] Tekes C, Karaman M, Degertekin FL. Optimizing circular ring arrays for forward-looking IVUS imaging. IEEE Trans Ultrason Ferroelectr Freq Control 2011;58(12):2596–607.

[16] Lay HS, Cummins G, Cox BF, Qiu Y, Turcanu MV, McPhillips R, et al. In-vivo evaluation of microultrasound and thermometric capsule endoscopes. IEEE Trans Biomed Eng 2019;66(3):632–9. https://doi.org/10.1109/TBME.2018.2852715.

[17] ter Haar G. The safe use of ultrasound in medical diagnosis [Internet]. London: British Institute of Radiology; 2012. [cited 28 December 2020]. Available from: http://www.birpublications.org/pb/assets/raw/Books/SUoU_3rdEd/Safe_Use_of_Ultrasound.pdf.

[18] Wu J, Nyborg WLM, editors. Emerging therapeutic ultrasound. Hackensack, NJ: World Scientific; 2006. 346 p.

[19] Kooiman K, Vos HJ, Versluis M, de Jong N. Acoustic behavior of microbubbles and implications for drug delivery. Adv Drug Deliv Rev 2014;72:28–48.

[20] Dijkmans P, Juffermans L, Musters R, Vanwamel A, Tencate F, Vangilst W, et al. Microbubbles and ultrasound: from diagnosis to therapy. Eur J Echocardiogr 2004;5(4):245–56.

[21] Lindner JR, Song J, Jayaweera AR, Sklenar J, Kaul S. Microvascular rheology of Definity microbubbles after intra-arterial and intravenous administration. J Am Soc Echocardiogr 2002;15(5):396–403.

[22] Schoellhammer CM, Schroeder A, Maa R, Lauwers GY, Swiston A, Zervas M, et al. Ultrasound-mediated gastrointestinal drug delivery. Sci Transl Med 2015;7(310), 310ra168.

[23] Kono K, Nakashima S, Kokuryo D, Aoki I, Shimomoto H, Aoshima S, et al. Multi-functional liposomes having temperature-triggered release and magnetic resonance imaging for tumor-specific chemotherapy. Biomaterials 2011;32(5):1387–95.

[24] Gourevich D, Dogadkin O, Volovick A, Wang L, Gnaim J, Cochran S, et al. Ultrasound-mediated targeted drug delivery with a novel cyclodextrin-based drug carrier by mechanical and thermal mechanisms. J Control Release 2013;170(3):316–24.

[25] Moglia A, Menciassi A, Dario P, Cuschieri A. Capsule endoscopy: progress update and challenges ahead. Nat Rev Gastroenterol Hepatol 2009;6(6):353–61.

[26] Ciuti G, Valdastri P, Menciassi A, Dario P. Robotic magnetic steering and locomotion of capsule endoscope for diagnostic and surgical endoluminal procedures. Robotica 2010;28(2): 199–207.

[27] Norton JC, Slawinski PR, Lay HS, Martin JW, Cox BF, Cummins G, et al. Intelligent magnetic manipulation for gastrointestinal ultrasound. Sci Robot 2019;4(31). https://doi.org/10.1126/scirobotics.aav7725.

[28] Yim S, Sitti M. 3-D localization method for a magnetically actuated soft capsule endoscope and its applications. IEEE Trans Robot 2013;29(5):1139–51.

[29] Rumack CM, Levine D. Diagnostic ultrasound. Elsevier; 2017. ISBN: 9780323401715.

[30] Media Backgrounders – Detail [Internet]. Default. [cited 17 December 2020]. Available from: https://www.asge.org/home/about-asge/newsroom/media-backgrounders-detail.

[31] Wang X, Seetohul V, Chen R, Zhang Z, Qian M, Shi Z, et al. Development of a mechanical scanning device with high-frequency ultrasound transducer for ultrasonic capsule endoscopy. IEEE Trans Med Imaging 2017;36(9):1922–9. https://doi.org/10.1109/TMI.2017.2699973.

[32] Cochran S. Sonopill: the future of capsule ultrasound. In: Europe – Korean Conference on Science and Technology; 2018. Glasgow.

[33] Chen Y, Zhou D, Lam KH, Cheung KF, Dai J, Chan HLW. Endoscopic ultrasound radial array transducers fabricated with PZT tube by a rotate-and-dice method. Sensors Actuators A Phys 2013;201:357–62.

[34] Cummins G, Desmulliez MPY, Hand DP, Wlodarczyk KL, Demore CEM, Lay H, et al. Sonopill pathfinder: rapid prototyping for ultrasound capsule endoscopy. In: Yang G-Z, Darzi A, editors. Proceedings of the Hamlyn symposium on medical robotics, 25–28 June 2016. Imperial College London and the Royal Geographical Society; 2016. p. 102–3.

[35] Anon. GF-UCT180 Curvilinear Array Ultrasound Gastrovideoscope | Olympus America | Medical [Internet]. [cited 16 December 2020]. Available from: https://medical.olympusamerica.com/products/curved-linear-array-gf-uct180.

[36] Kohli DR, Baillie J. 3—How endoscopes work. In: Chandrasekhara V, Elmunzer BJ, Khashab MA, Muthusamy VR, editors. Clinical gastrointestinal endoscopy. 3rd ed. Philadelphia: Elsevier; 2019. p. 24–31. [Internet]. [cited 16 December 2020]. e2. Available from: http://www.sciencedirect.com/science/article/pii/B9780323415095000037.

[37] Uchida D, Kato H, Matsumoto K, Ishihara Y, Matsumi A, Saragai Y, et al. Single-session esophagogastroduodenoscopy and endoscopic ultrasound using a forward-viewing radial scan ultrasonic endoscope. BMC Gastroenterol 2019;19(1):220. https://doi.org/10.1186/s12876-019-1141-7.

[38] Dietrich CF, Arcidiacono PG, Braden B, Burmeister S, Carrara S, Cui X, et al. What should be known prior to performing EUS exams? (Part II). Endosc Ultrasound 2019;8(6): 360–9.

[39] Zhou D, Cheung KF, Chen Y, Lau ST, Zhou Q, Shung KK, et al. Fabrication and performance of endoscopic ultrasound radial arrays based on PMN-PT single crystal/epoxy 1–3 composite. IEEE Trans Ultrason Ferroelectr Freq Control 2011;58(2):477–84.

[40] BS EN 60601-2-37:2008+A1:2015. Medical electrical equipment. Particular requirements for the basic safety and essential performance of ultrasonic medical diagnostic and monitoring equipment. BSI; 2008.
[41] Lee JH, Traverso G, Ibarra-Zarate D, Boning DS, Anthony BW. Ex vivo and in vivo imaging study of ultrasound capsule endoscopy. J Med Devices 2020;14(021005). https://doi.org/10.1115/1.4046352.
[42] Lee JH, Traverso G, Schoellhammer CM, Blankschtein D, Langer R, Thomenius KE, et al. Towards wireless capsule endoscopic ultrasound (WCEU). In: 2014 IEEE international ultrasonics symposium; 2014. p. 734–7.
[43] Lay HS, Cox BF, Seetohul V, Demore CEM, Cochran S. Design and simulation of a ring-shaped linear array for microultrasound capsule endoscopy. IEEE Trans Ultrason Ferroelectr Freq Control 2018;65(4):589–99.
[44] Memon F, Touma G, Wang J, Baltsavias S, Moini A, Chang C, et al. Capsule ultrasound device: further developments. In: 2016 IEEE international ultrasonics symposium (IUS); 2016. p. 1–4.
[45] Wang J, Memon F, Touma G, Baltsavias S, Jang JH, Chang C, et al. Capsule ultrasound device: characterization and testing results. In: 2017 IEEE international ultrasonics symposium (IUS); 2017. p. 1–4.
[46] Stewart F. Capsule-based ultrasound-mediated targeted drug delivery [Doctoral Thesis]. University of Dundee; 2018.
[47] Stewart F, Cummins G, Turcanu MV, Cox BF, Prescott A, Clutton E, et al. Ultrasound mediated delivery of quantum dots from a proof of concept capsule endoscope to the gastrointestinal wall. Sci Rep 2021;11:2584. https://doi.org/10.1038/s41598-021-82240-1.
[48] Turcanu MV, Stewart FR, Cox BF, Clutton RE, Mulvana H, Vllasaliu D, et al. Ultrasound and microbubbles promote the retention of fluorescent compounds in the small intestine. In: 2018 IEEE international ultrasonics symposium (IUS); 2018. p. 1–4.
[49] Moldovan AC, Qiu Z, Lines D, Gachagan A, Cochran S. Enhanced modelling of a 1-D phased ultrasonic array for intracorporeal sonoporation. In: 2020 IEEE international ultrasonics symposium (IUS). Las Vegas, NV, USA: IEEE; 2020. p. 1–4. https://doi.org/10.1109/IUS46767.2020.9251375.
[50] Ødegaard S, Nesje LB, Lærum OD, Kimmey MB. High-frequency ultrasonographic imaging of the gastrointestinal wall. Expert Rev Med Devices 2012;9(3):263–73.
[51] Faerber J, Gregson R, Clutton RE, Khan SR, Cochran S, Desmulliez MPY, et al. In vivo characterization of a wireless telemetry module for a capsule endoscopy system utilizing a conformal antenna. IEEE Trans Biomed Circuits Syst 2018;12(1):95–105.
[52] Yang S, Lemke C, Cox BF, Newton IP, Nathke I, Cochran S. A learning based microultrasound system for the detection of inflammation of the gastrointestinal tract. IEEE Trans Med Imaging 2021;40(1):38–47. https://doi.org/10.1109/TMI.2020.3021560.
[53] Stewart F, Mulvana H, Näthke I, Cochran S. Theranostics in the gut. Theranostics and image guided drug delivery. Royal Society of Chemistry; 2018.
[54] Stewart F, Verbeni A, Qiu Y, Cox BF, Vorstius J, Newton IP, et al. A prototype therapeutic capsule endoscope for ultrasound-mediated targeted drug delivery. J Med Robot Res 2018;03(02):1840001.
[55] Hersh DS, Wadajkar AS, Roberts NB, Perez JG, Connolly NP, Frenkel V, et al. Evolving drug delivery strategies to overcome the blood brain barrier. Curr Pharm Des 2016;22(9):1177–93. https://doi.org/10.2174/1381612822666151221150733.
[56] van Wamel A, Kooiman K, Harteveld M, Emmer M, ten Cate FJ, Versluis M, et al. Vibrating microbubbles poking individual cells: drug transfer into cells via sonoporation. J Control Release 2006;112(2):149–55.

[57] Meijering BDM, Juffermans LJM, van Wamel A, Henning RH, Zuhorn IS, Emmer M, et al. Ultrasound and microbubble-targeted delivery of macromolecules is regulated by induction of endocytosis and pore formation. Circ Res 2009;104(5):679–87.

[58] Kooiman K, Foppen-Harteveld M, van der Steen AFW, de Jong N. Sonoporation of endothelial cells by vibrating targeted microbubbles. J Control Release 2011;154(1):35–41.

[59] Fan Z, Kumon RE, Park J, Deng CX. Intracellular delivery and calcium transients generated in sonoporation facilitated by microbubbles. J Control Release 2010;142(1):31–9.

[60] Chen X, Leow RS, Hu Y, Wan JMF, Yu ACH. Single-site sonoporation disrupts actin cytoskeleton organization. J R Soc Interface 2014;11(95):20140071.

[61] Escoffre J-M, Campomanes P, Tarek M, Bouakaz A. New insights on the role of ROS in the mechanisms of sonoporation-mediated gene delivery. Ultrason Sonochem 2020;64:104998.

[62] Correas J-M, Quay SD. EchoGen emulsion: a new ultrasound contrast agent based on phase shift colloids. Clin Radiol 1996;51:11–4.

[63] Ja'afar F, Leow CH, Garbin V, Sennoga CA, Tang M-X, Seddon JM. Surface charge measurement of SonoVue, Definity and Optison: a comparison of laser Doppler electrophoresis and micro-electrophoresis. Ultrasound Med Biol 2015;41(11):2990–3000.

[64] Mannaris C, Yang C, Carugo D, Owen J, Lee JY, Nwokeoha S, et al. Acoustically responsive polydopamine nanodroplets: a novel theranostic agent. Ultrason Sonochem 2020;60:104782.

[65] Moyer L, Timbie K, Sheeran P, Price R, Miller W, Dayton P. Direct nanodroplet and microbubble comparison for high intensity focused ultrasound ablation enhancement and safety. J Ther Ultrasound 2015;3(S1):0–67. https://doi.org/10.1186/2050-5736-3-S1-O67.

[66] Stewart F, Cox B, Vorstius J, Verbeni A, Qiu Y, Cochran S. Capsule-based ultrasound-mediated targeted gastrointestinal drug delivery. In: 2015 IEEE international ultrasonics symposium (IUS); 2015. p. 1–4.

[67] Ciuti G, Menciassi A, Dario P. Capsule endoscopy: from current achievements to open challenges. IEEE Rev Biomed Eng 2011;4:59–72.

[68] Haringsma J, Tytgat GNJ, Yano H, Iishi H, Tatsuta M, Ogihara T, et al. Autofluorescence endoscopy: feasibility of detection of GI neoplasms unapparent to white light endoscopy with an evolving technology. Gastrointest Endosc 2001;53(6):642–50.

[69] Curvers WL, Singh R, Song L-MW-K, Wolfsen HC, Ragunath K, Wang K, et al. Endoscopic tri-modal imaging for detection of early neoplasia in Barrett's oesophagus: a multi-centre feasibility study using high-resolution endoscopy, autofluorescence imaging and narrow band imaging incorporated in one endoscopy system. Gut 2008;57(2):167–72.

[70] Mitsunaga M, Kosaka N, Choyke PL, Young MR, Dextras CR, Saud SM, et al. Fluorescence endoscopic detection of murine colitis-associated colon cancer by topically applied enzymatically rapid-activatable probe. Gut 2013;62(8):1179–86.

[71] Matsuzaki H, Kamiya M, Iwatate RJ, Asanuma D, Watanabe T, Urano Y. Novel hexosaminidase-targeting fluorescence probe for visualizing human colorectal cancer. Bioconjug Chem 2016;27(4):973–81.

[72] Melino G, Accarino C, Riehle M, Potter M, Fineron P, Annese VF, et al. Capsule endoscopy compatible fluorescence imager demonstrated using bowel cancer tumours. IEEE Sensors J 2020;20(17):9763–71.

Chapter 10

Modeling of capsule-intestine contact

Yang Liu[a], Jiyuan Tian[a], Bingyong Guo[a,b], and Shyam Prasad[c]
[a]*College of Engineering, Mathematics and Physical Sciences, University of Exeter, Exeter, United Kingdom,* [b]*School of Marine Science and Technology, Northwestern Polytechnical University, Xi'an, Shaanxi, China,* [c]*The Royal Devon and Exeter NHS Foundation Trust, Exeter, United Kingdom*

10.1 Introduction

Two major developments in recent years have advanced the diagnosis of small intestinal disorders. The first is the increasing resolution and sophistication of cross-sectional imaging, in particular computed tomography (CT) radiography and magnetic resonance imaging (MRI), which assess the anatomical and pathological structure of organs [1]. The second is the increasing use of capsule endoscopy (CE) for visualising the inner lining (mucosa) of the small intestine, an anatomical site previously considered to be inaccessible to clinicians. CE involves the ingestion of a device, typically 26 mm in length and 11 mm in diameter, which captures upward 100,000 photographic images over a number of hours, either storing them internally or transmitting them to an externally worn receiver. Software combines these images into video footage for clinicians to view. Since its inception twenty years ago [2], the only significant improvements in CE have been in image quality (typically 512 × 512 pixels), frame rate (2–6 frames per second), and angle of field [3]. All available CE systems continue to rely on peristalsis for passage through the gastrointestinal (GI) tract, and this leads to significant limitations, in particular due to the unpredictable and variable locomotion velocity. Significant abnormalities may be missed, due to intermittent high transit speed that leads to incomplete visualisation of the intestinal surface [3]. On the other hand, the often lengthy small-bowel transit time and resulting video (typically 2–8 h) make the procedure time-consuming for both patients and clinicians.

In theory, an active and controllable locomotive mechanism for CE could dramatically reduce the duration of the procedure, while at the same time

Endorobotics. https://doi.org/10.1016/B978-0-12-821750-4.00010-4

allowing the endoscopist to focus visualisation onto potential areas of interest. The ability to anchor the capsule at a specific position within the small intestine opens up the possibility of introducing new modalities, such as biopsy and targeted drug delivery. The vibro-impact system [4,5] is one such mechanism being explored by the Applied Dynamics and Control Lab at the University of Exeter to achieve these aims.

The performance of the vibro-impact system for CE significantly depends on the friction induced by capsule-intestine interaction, and, hence, it is critical to depict the capsule-intestine interaction during its passage through the GI tract in terms of intestinal friction and contact pressure. Recent studies concluded that the intestinal friction for a capsule moving on a flat specimen of the small intestine ranged from 10 to 200 mN, depending on capsule's shape, dimension, and instantaneous velocity [6,7]. Consequently, the friction coefficient ranged from 0.08 to 0.2. When a large friction is required in some special scenarios, e.g., anchoring the capsule for biopsy, the friction coefficient can be increased up to 0.49 by coating the capsule surface with the micro-patterned adhesives [8]. However, the capsule-intestine contact conditions may change according to the gesture of the capsule and GI peristalsis. Therefore, modeling methods should cover all possible scenarios during the passage of the capsule through the GI tract. In this chapter, three typical capsule-intestine contact modes will be depicted by analytical and finite element (FE) models to evaluate the intestinal resistance and pressure acting on the capsule.

The rest of this chapter is organised as follows. In Section 10.2, mathematical modeling of capsule-intestine contact is studied. In Section 10.3, FE models developed using ANSYS Workbench are introduced. In Section 10.4, FE results are presented and analysed, followed by the conclusions in Section 10.5.

10.2 Mathematical modeling of capsule-intestine contact

Since capsule-intestine contact conditions may vary over time according to the gesture of the capsule and intestinal peristalsis, this section will study three typical contact modes reported in Ref. [9] according to the real environment of the small intestine as presented in Fig. 10.1.

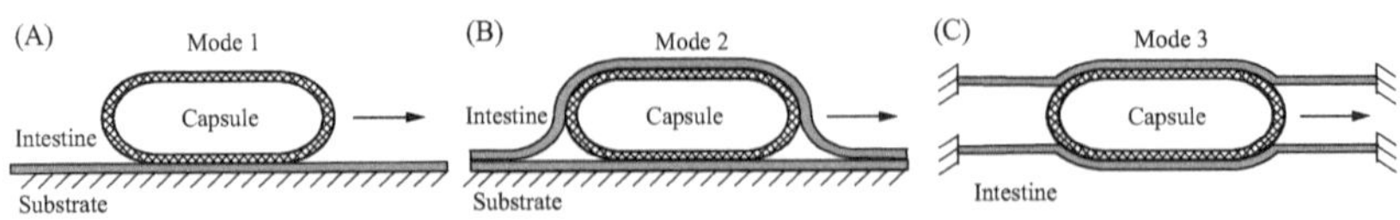

FIG. 10.1 Schematic diagrams of three contact modes: (A) Mode 1: the capsule moves on a cut-open intestinal surface; (B) Mode 2: the capsule slides inside a collapsing intestine; (C) Mode 3: the moving capsule is surrounded by the small intestine.

10.2.1 Mode 1

When the small intestine expands, the internal diameter of the intestine R_i becomes larger than the external diameter of the capsule R_c as shown in Fig. 10.1A. Assuming that $R_c \ll R_i$, capsule-intestine contact can be considered as the cylinder-cylinder contact case as illustrated in Fig. 10.2A, and the contact pressure P_{c1} is caused by capsule's gravity that can be written as

$$P_{c1} = P_c,$$

where P_c represents the contact pressure due to capsule's gravity. Correspondingly, the friction force can be written as

$$f_{c1} = f_c,$$

where $f_c = \mu M_c g$ represents the friction due to the capsule's gravity, μ, M_c and g are the friction coefficient between the capsule-intestine contact surface, the total mass of the capsule and the acceleration due to gravity, respectively.

When the capsule stays in stationary, the Hertz contact theory [10] can be applied to calculate the contact pressure between the capsule and the small intestine as shown in Fig. 10.2B, where $2a$ and L represent the width and length of the contact area, respectively, and r is the distance of the contact location from its central axis. Thus, the semicontact width can be expressed as

$$a = \sqrt{\frac{4FR_e}{\pi L E_e}},$$

where $F = M_c g$ is the load acting on the capsule, $R_e = \left(\frac{1}{R_c} + \frac{1}{R_i}\right)^{-1}$ is the equivalent radius, and $E_e = \left(\frac{1-\nu_c^2}{E_c} + \frac{1-\nu_i^2}{E_i}\right)^{-1}$ is the equivalent Young's modulus. E_c and E_i are Young's moduli, ν_c and ν_i are Poisson's ratios of the capsule and the small intestine, respectively.

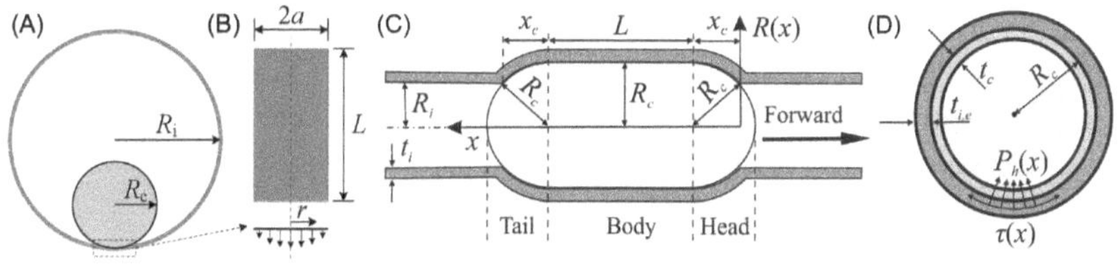

FIG. 10.2 (A) Sectional view of Mode 1, (B) top view of the contact area and the distribution of contact pressure, (C) geometric parameters of the capsule in the small intestine for Mode 3, and (D) sectional view at the center of capsule-intestine contact.

As the contact pressure distributes elliptically along the direction of the contact width, it can be written as

$$P = P_{\max}\sqrt{1-\left(\frac{r}{a}\right)^2},$$

where $P_{\max}$ is the maximum contact pressure along the central axis of the contact area given as

$$P_{\max} = \frac{2F}{\pi aL}.$$

Since it is difficult to measure the maximum contact pressure experimentally, the average contact pressure over the contact area,

$$P_{\text{avg}} = \frac{F}{2aL},$$

can be adopted and measured in experiments. Therefore, in practice, P_{avg} can be used as an analytical prediction of the capsule-intestine contact pressure for Mode 1.

In order to use the Hertz contact theory, the following four conditions must be satisfied: (1.a) the contact surface is continuous and nonconforming, $a \ll R_e$; (1.b) the strain is small, $a \ll R_e$; (1.c) each solid is considered as an elastic half-space, $a \ll R_i$, $a \ll R_c$, and $a \ll L$; and (1.d) the surface is frictionless. Since the friction coefficient is small in general, the friction-induced tangential traction has little influence on the normal pressure of the contact area [10], so the Hertz contact theory is still valid here.

10.2.2 Mode 2

When there is no peristalsis, the small intestine falls down naturally on the capsule, and the external radius of the capsule is smaller or equal to the internal radius of the intestine, $R_c \le R_i$. There are two contact surfaces, with one on the top and the other at the bottom of the capsule, as shown in Fig. 10.1B. Under this condition, in addition to capsule's gravity, the gravity of the small intestine also contributes to the contact pressure. Thus, the contact pressure P_{c2} can be written as

$$P_{c2} = P_c + P_i,$$

where P_i represents the contact pressure due to the gravity of the small intestine falling on the capsule. Consequently, the friction can be written as

$$f_{c2} = f_c + f_i,$$

where f_i represents the friction due to the gravity of the small intestine falling on the capsule. However, the analytical representations of P_i and f_i are difficult to be obtained since only a small varying part of the intestine acts on the capsule.

For this contact case, P_{c2} and f_{c2} can be numerically calculated through FE modeling, detailed in the next section.

10.2.3 Mode 3

When the intestine contracts, the capsule is surrounded by the intestine leading to $R_c > R_i$. The viscoelastic deformation of the intestinal wall may induce a high hoop pressure on the capsule as shown in Fig. 10.2C. Under this condition, the contact pressure includes three parts, the contact pressure due to capsule's gravity (Mode 1), the contact pressure due to the gravity of the small intestine falling on the capsule (Mode 2), and the hoop pressure caused by the hoop stress of the expanded intestine, which can be expressed as

$$P_{c3} = P_c + P_i + P_h,$$

where P_h represents the hoop pressure caused by the hoop stress of the expanded small intestine.

In order to model the hoop pressure analytically, the small intestine needs to be considered as a viscoelastic material [11,12], and the following assumptions are made: (2.a) the internal surface of the small intestine surrounds the capsule external surface when they are in contact; (2.b) the intestinal material is incompressible; and (2.c) the intestinal deformation is isotropical and symmetrical.

To model the hoop pressure, a local coordinate system [9,13] is built in Fig. 10.2C, with x and $R(x)$ representing the axial and radial directions, respectively. The capsule is divided into three segments, a semisphere head, a cylindrical body, and a semisphere tail. In the local coordinate, x_c is the distance from the contact point to the centre of the head in the axial direction.

Based on Assumption (2.a), the internal radius of the expanded intestine contacting the capsule body $R(x)$ is the same as the external radius of the capsule. x_c and $R(x)$ can be written as

$$x_c = \sqrt{R_c^2 - R_i^2}, \quad R(x) = \begin{cases} \sqrt{R_c^2 - (x_c - x)^2}, & x \in [0, x_c] \\ R_c, & x \in (x_c, x_c + L]; \\ \sqrt{R_c^2 - (x - L - x_c)^2}, & x \in (x_c + L, 2x_c + L]. \end{cases}$$

According to Assumption (2.b), the thickness of the intestinal wall attenuates when the capsule body expands the intestine radially. So, the thickness of the expanded intestine $t_{i,e}$ can be expressed as

$$t_{i,e} = \sqrt{(R_i + t_i)^2 + R^2(x) - R_i^2} - R(x),$$

where t_i is the original thickness of the intestinal wall. Then according to Assumption (2.c), the hoop strain can be calculated as

$$\epsilon = \frac{R_i - R(x)}{R_i}.$$

In Refs. [11,12], the viscoelastic property of the small intestinal lining was studied, and the stress-strain relationship of the small intestine was represented by the Maxwell model that includes three elastic springs and two viscous dampers [14]. In Ref. [9], since it fitted better with their stress relaxation test, the Maxwell model with two elastic springs and one viscous damper was used, which can be written as

$$\tau(t) = \epsilon\left(E_1 e^{-\frac{E_1}{\eta_1}t} + E_2\right),$$

where E_1, E_2 are the Young's moduli of the springs, and η_1 is the viscosity coefficient of the damper.

For a capsule endoscope moving in a constant velocity, the hoop stress can be calculated using

$$\tau(t) = \epsilon\left(E_1 e^{-\frac{E_1 x}{\eta_1 \nu}} + E_2\right).$$

The relationship between the hoop pressure and the hoop stress of Mode 3 is the same as the pressure vessels in [15], so it can be calculated as

$$P_h(x) = \frac{t_{i,e}}{R(x)}\tau(x).$$

For the semisphere head section ($0 < x \leq x_c$), the hoop stress acting on the head can be decomposed into two parts: (i) the hoop stress along the x-axis contributing to the resistance directly marked as f_{hr}, and (ii) the hoop stress along $R(x)$ contributing to the resistance based on the Coulomb friction marked as f_{hf}, which are given by

$$f_{hr} = \frac{2\pi}{R_c}\int_0^{x_c}(x_c - x)P_h(x)R(x)\sqrt{1 + R'^2(x)}\mathrm{d}x,$$

and

$$f_{hf} = \frac{2\pi\mu}{R_c}\int_0^{x_c}\sqrt{R_c^2 - (x_c - x)^2}P_h(x)R(x)\sqrt{1 + R'^2(x)}\mathrm{d}x,$$

where $R'(x)$ is the derivative of $R(x)$ with respect to x.

For the cylindrical body section ($x_c < x \leq x_c + L$), the hoop stress is normal to the capsule surface, and hence, only introduces the Coulomb friction that can be written as

$$f_{bf} = 2\pi\mu\int_{x_c}^{x_c + L} P_h(x)R(x)\sqrt{1 + R'^2(x)}\mathrm{d}x,$$

where $R(x) = R_c$ and $R'(x) = 0$.

For the semisphere tail section ($x_c+L < x \leq 2x_c+L$), the hoop-induced friction is similar to its counterpart of the head section, including the hoop resistance on the tail f_{tr} and the hoop stress induced Coulombic friction on the tail f_{tf} that can be expressed as

$$f_{tr} = \frac{2\pi}{R_c}\int_{x_c+L}^{2x_c+L}(x_c - x)P_h(x)R(x)\sqrt{1+R'^2(x)}\mathrm{d}x,$$

and

$$f_{tf} = \frac{2\pi\mu}{R_c}\int_{x_c+L}^{2x_c+L}\sqrt{R_c^2-(x_c-x)^2}P_h(x)R(x)\sqrt{1+R'^2(x)}\mathrm{d}x,$$

Finally, the total friction on the capsule can be written as

$$f_{c3} = \mu M_c g + f_{hr} + f_{hf} + f_{tr} + f_{tr} + f_{tf}.$$

The friction due to the small intestine's gravity is neglected here as its contribution to the friction is relatively small.

10.3 Finite element modeling of capsule-intestine contact

Three FE models were built using the ANSYS Workbench to study the contact pressures of Modes 1–3. The geometric parameters shown in Fig. 10.2 and the mechanical properties of the capsule and the intestine are given in Table 10.1, and the set-up for each model is detailed as follows.

It is worth noting that since the synthetic small intestine is synthesised from synthetic human tissue analogs and woven fibers, the stiffness of the small intestine is expressed as Young's modulus of human tissue analogs $E_i = 25$ kPa used in Modes 1 and 2, and the strain-stress relationship in Mode 3 is represented by the Maxwell model with two elastic springs (E_1 and E_2) and one viscous damper (η_1) obtained through stress relaxation experiments [9].

10.3.1 Mode 1

The FE set-up for Mode 1 ($R_1 = \infty$ mm) is presented in Fig. 10.3A. When the small intestine expands, the upper intestinal wall is far away from the capsule. Therefore, the capsule slides on the intestine lower surface subject to gravity to the entire model. In this case, the capsule-intestine interaction can be modeled as a capsule moves on a flat-open intestine with a total length of 200 mm supported by a steel substrate. To satisfy the deformation requirement, two types of elements were used to mesh the model. The 8-node solid element (SOLID185) was used for the capsule and the substrate. The 20-node solid element (SOLID186) that exhibits quadratic displacement behaviour, having 3 DOFs at each node, was used to simulate the viscoelastic intestine. The restraint

TABLE 10.1 Geometric parameters and mechanical properties of the capsule and the synthetic small intestine.

Parameter	Value	Unit
R_c	5.50	mm
L	15.00	mm
E_c	0.11	GPa
E_i	25 (Modes 1 and 2)	kPa
E_1	196.43 (Mode 3)	kPa
E_2	757.48 (Mode 3)	kPa
R_1	∞	mm
R_2	13.00	mm
R_3	5.00	mm
t_c	0.50	mm
ρ_c	0.95	g/mm^3
ρ_i	1	g/mm^3
t_i	0.69	mm
η_1	5.36 (Mode 3)	MPas
ν_i	0.49	
ν_c	0.42	
μ	0.2293	

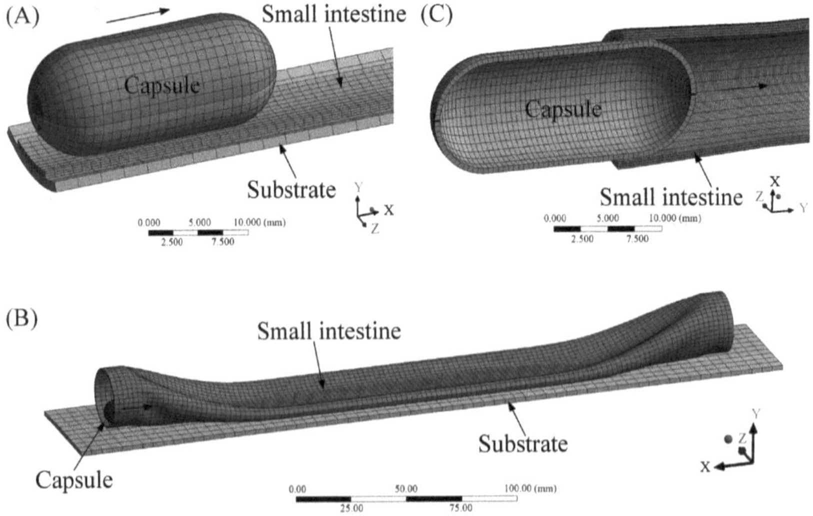

FIG. 10.3 FE set-ups for (A) Mode 1—the capsule moves on a cut-open small intestinal surface; (B) Mode 2—the capsule slides inside a collapsing intestine; (C) Mode 3—the moving capsule is surrounded by the small intestine.

and loading conditions were applied to the capsule-intestine-substrate construction in three steps. In the first step, the substrate was fixed at the lower surface to support the intestine and the moving capsule. Secondly, gravity $g = 9.81\,\text{m/s}^2$ was applied to the whole model. For the capsule, constrained DOFs are UZ (translation in the *Z* direction), ROTX (rotation around the *X*-axis), and ROTY (rotation around the *Y*-axis). That is, the DOFs of UX (translation in the *X* direction), UY (translation in the *Y* direction), and ROTZ (rotation around the *Z*-axis) are set free. At last, the capsule was exerted a displacement of 150mm along the *X* direction with progression speed varying from 0.5 to 10mm/s.

10.3.2 Mode 2

As the intestine relaxes, the upper part of the intestine falls naturally on the upper surface of the moving capsule due to gravity as shown in Fig. 10.3B. Before the collapsing of the small intestine, the intestine with an inner diameter of 26mm and a length of 250mm was laid on the steel substrate, and the capsule was set at the left entry of the small intestine as the starting position. The contacting attribute between the intestine and the supporting substrate was set as rough frictional contact, so there is no sliding between these two surfaces, but they can be separated and fit freely while the intestine falls. Also, the element type and the constraint of the capsule used in Mode 1 were used in Mode 2. As a supporting structure, the DOF of the substrate was fully constrained on the bottom surface. The inlet and outlet of the small intestine were fixed, and these two positions always maintained their initial shapes, while the middle position of the small intestine was set as free, so it can fall down due to the gravity. Mode 2 loading includes two steps: (1) The gravity is applied uniformly to the entire model for two seconds; (2) the capsule starts to move in a constant speed along the collapsing intestine.

10.3.3 Mode 3

Due to intestinal spontaneous movement, the intestine can produce a certain length of contraction ring. For Mode 3 as presented in Fig. 10.3C, as the diameter of the capsule is always greater than the diameter of a contracting intestine ring, the small intestine is expanded and the capsule is surrounded by the intestine. In the FE model, the capsule's outer diameter was 11mm. In order to reduce the computational time, Mode 3 was simplified to a 2-dimensional and axisymmetrical model, which includes a trumpet-shaped inlet and outlet and a straight section in the middle with an inner diameter smaller than the outer diameter of the capsule. The total length of the small intestine was configured as 250mm, and four campaigns of FE simulation were conducted with the inner diameter of the middle section set as 8.8, 9.2, 9.6, and 10mm, respectively. In order to get a better convergence, the pure penalty algorithm was used to describe the capsule-intestine contact and the normal contact stiffness factor was set as 1. Again, the same element type used in Mode 1 was applied for this case. On the left boundary of the small intestine, the ROTZ was set as free, while the other DOFs were restricted. On the right boundary, the small intestine was only set as free to UY, so the small intestine can be elongated when the capsule moves inside. This configuration was similar to the experiments in Ref. [9], which can prevent the occurrence of stress singularity and program error causing by the distortion of the small intestine. The capsule's loading adopts displacement along the *Y* direction with various constant speeds in the contractive intestine, and UX, UZ, ROTX, ROTY, and RTOZ were restricted. Finally, the FE results from the middle intestine which was far away from both ends of the intestine were used according to the Saint Venant's Principle.

10.4 Results and analysis

Results will focus on the contact pressures and frictional forces between the capsules and the intestines in Modes 1–3. Fig. 10.4 reveals the behaviour of the contact pressure and the resistance results of Modes 1 and 2 with the capsule moving speed varying for a wide range. As it can be seen from Fig. 10.4A, the capsule speed has little effect on the average contact pressure under the set-up of Model 1, and the average contact pressure is around 0.223 kPa. Consequently, the resistance is around 7.8 mN, which shows a good agreement with the experimental results in Ref. [16]. In Fig. 10.4B, as the speed increases, the average contact pressure and resistance decrease slightly, but their average values are around 0.134 kPa and 15.2 mN, respectively. The average contact pressure of Mode 2 is lower than that of Mode 1, but the resistance of Mode 2 is nearly twice that of Mode 1, which is expected that as the collapsing intestine increased, the normal force increased slightly and the contact area increased significantly.

Fig. 10.5 presents the average contact pressure and the resistance tendencies of different intestine inner diameters by varying the capsule moving speed in Mode 3. Taking the intestinal inner diameter of 10 mm as an example, the average contact pressure increases slightly from 9.95 to 10.91 kPa when the capsule moving speeds increase from 0.5 to 10 mm/s, which is shown in Fig. 10.5A. In Fig. 10.5B, the resistance between the capsule and the small intestine increases slightly from 1.54 to 1.67 N when the capsule moving speed increases. On the other hand, the inner diameter of the small intestine has a significant effect on the capsule-intestine interaction. A smaller inner diameter of the intestine leads to a greater contact pressure and resistance. Therefore, the magnitude of intestinal contraction determines the values of the contact pressure and the resistance, which is one of the factors that needs to be considered for capsule design.

Taking the capsule's moving speed of 0.5 mm/s as an example, the average contact pressures, maximum contact pressures, and frictional forces of Modes 1–3 are present in Table 10.2. It can be seen that the maximum contact pressure of Mode 2 is greater than that of Mode 1 and the friction of Mode 2 is about

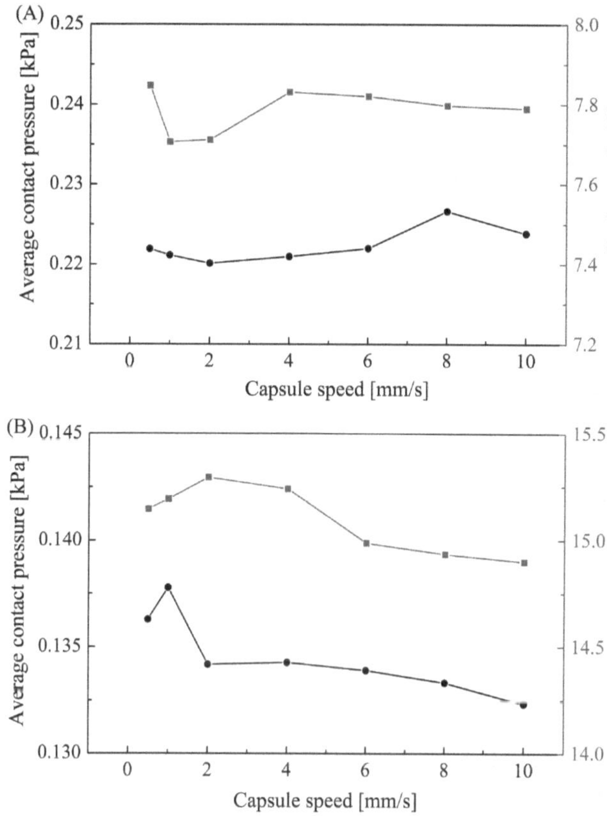

FIG. 10.4 FE average contact pressure *(black dots)* and resistance [*blue squares* (*gray* in print version)] on the capsule with various progression speeds for (A) Mode 1, and (B) Mode 2.

twice that of Mode 1. In Mode 3, the intestine's inner diameter is 10 mm. The capsule enlarged the small intestine in a uniform circumferential direction during its movement. Since the contact pressure of Mode 3 mainly comes from the hoop pressure, the result of Mode 3 is much greater than those of Modes 1 and 2.

The distribution of contact pressure in each case is shown in Fig. 10.6. As shown in Fig. 10.3A, the rectangular area enclosed by the blue area (gray in print version), about 15 mm in length and 2.6 mm in width, is the capsule-intestine contact area. In Fig. 10.6B, due to the upper half of the intestine dropping down and pressing on the capsule's upper surface, the maximum contact pressure in Mode 2 is higher than that in Mode 1. The contact pressure mainly distributes at the centre of the lower half surface of the capsule and at both ends of the upper half surface. Fig. 10.6C presents the distribution of the hoop pressure on the capsule induced by the contracting small intestine. The highest pressure situates at both ends of the capsule cylindrical surface. In addition, the

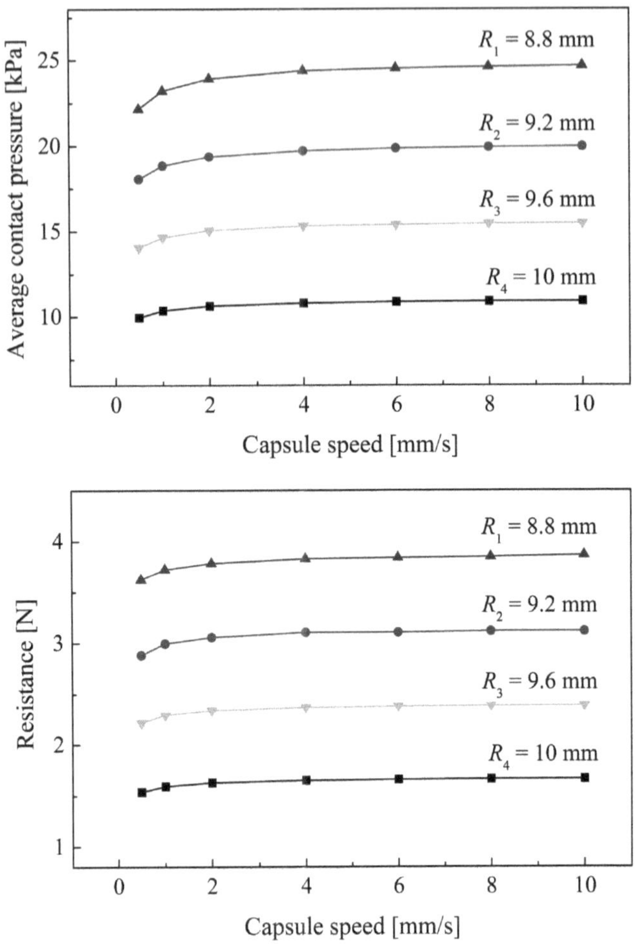

FIG. 10.5 (A) FE average contact pressure and (B) resistance applying on the capsule with a wide range of progression speeds in Mode 3.

TABLE 10.2 FE contact pressure and frictional force in each mode.

Mode	Speed (mm/s)	Average contact pressure (kPa)	Maximum contact pressure (kPa)	Frictional force (N)
1	0.5	0.22193	1.9609	0.007847
2	0.5	0.13628	2.5355	0.015146
3	0.5	9.9538	43.3229	1.53728

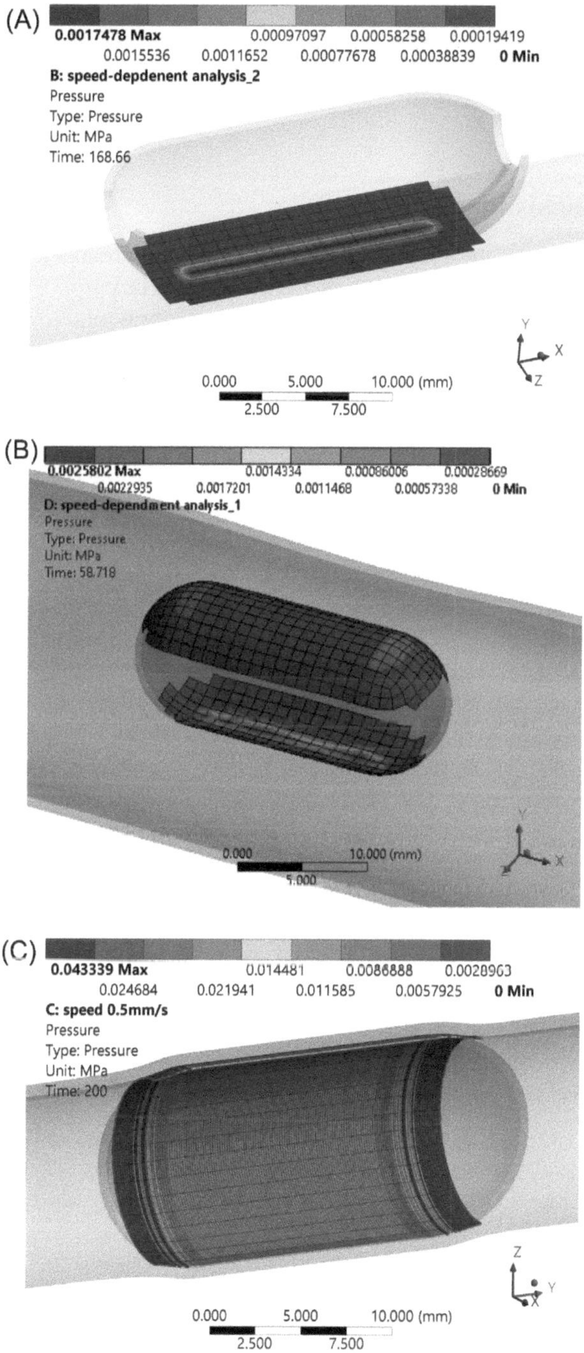

FIG. 10.6 Distribution of contact pressure in (A) Mode 1, (B) Mode 2, and (C) Mode 3 of FE simulations.

maximum value of the contact pressure on the left tail is larger than that of the right head, which could be interpreted by the stress relaxation of synthetic material of the small intestine.

10.5 Conclusions

This chapter studies the dynamic model of an endoscopic capsule moving in the small intestine through analytical and FE modeling. The purpose of this work is to develop a small-bowel simulator to understand capsule-intestine interactions during an endoscopic procedure to provide guidance on new endoscopic tool design. By taking intestinal peristalsis into account, three typical capsule-intestine contact models were studied: Mode 1, the capsule moves on a cut-open intestinal surface; Mode 2, the capsule slides inside a collapsing intestine; Mode 3, the moving capsule is surrounded by the small intestine. These simulations were based on the prototype parameters used in Ref. [4]. According to the FE results, they show that the capsule's speed has little effect on the average contact pressure of the intestine in Model 1. The average contact pressure is about 0.223 kPa, and the resistance on the capsule is about 7.8 mN, which has a high agreement with the experimental results in Ref. [16]. Due to the collapsing intestine in Mode 2, which can increase the normal force on the capsule slightly but the contact area significantly, the average contact pressure of Mode 2 is lower than that of Mode 1, but the resistance of Mode 2 is nearly twice that of Mode 1. As the capsule moves faster, both the average contact pressure and the resistance drop. For Mode 3, both the average contact pressure and the resistance increase as the capsule's speed increases. In particular, the inner diameter of the small intestine has a significant effect on the capsule-intestine interaction. The intestine with a small inner diameter may cause a large contact pressure and resistance on the capsule. Therefore, the magnitude of intestinal contraction can determine the contact pressure and the resistance of the capsule-intestine contact, which is one of the key factors that needs to be considered in capsule design.

References

[1] Hodler J, Ubik-Huch RA, von Schulthess GK. Diseases of the abdomen and pelvis 2018-2021: diagnostic imaging—IDKD Book (IDKD Springer Series). 1st ed. Springer Open; 2018.

[2] Iddan G, Meron G, Glukhovsky A, Swain P. Wireless capsule endoscopy. Nature 2000; 405:417.

[3] Vasilakakis M, Koulaouzidis A, Yung DE, Plevris JN, Toth E, Iakovidis DK. Follow-up on: optimizing lesion detection in small bowel capsule endoscopy and beyond: from present problems to future solutions. Expert Rev Gastroenterol Hepatol 2019;13:129–41.

[4] Liu Y, Páez Chávez J, Zhang J, Tian J, Guo B, Prasad S. The vibro-impact capsule system in millimetre scale: numerical optimisation and experimental verification. Meccanica 2020; 55(10):1885–902.

[5] Guo B, Liu Y, Birler R, Prasad S. Self-propelled capsule endoscopy for small-bowel examination: proof-of-concept and model verification. Int J Mech Sci 2020;74:105506.
[6] Kim JS, Sung IH, Kim YT, Kwon EY, Kim DE, Jang YH. Experimental investigation of frictional and viscoelastic properties of intestine for microendoscope application. Tribol Lett 2006;22(2):143–9.
[7] Wang X, Meng M. An experimental study of resistant properties of the small intestine for an active capsule endoscope. Proc Inst Mech Eng H 2010;224(1):107–18.
[8] Zhang H, Yan Y, Gu Z, Wang Y, Sun T. Friction enhancement between microscopically patterned polydimethylsiloxane and rabbit small intestinal tract based on different lubrication mechanisms. ACS Biomater Sci Eng 2016;2(6):900–7.
[9] Guo B, Liu Y, Prasad S. Modelling of capsule-intestine contact for a self-propelled capsule robot via experimental and numerical investigation. Nonlinear Dyn 2019;98:3155–67.
[10] Johnson KL. Contact mechanics. Cambridge, UK: Cambridge University Press; 1987.
[11] Kim JS, Sung IH, Kim YT, Kim DE, Jang YH. Analytical model development for the prediction of the frictional resistance of a capsule endoscope inside an intestine. Proc Inst Mech Eng H 2007;221:837–45.
[12] Zhang C, Liu H, Tan R, Li H. Modeling of velocity-dependent frictional resistance of a capsule robot inside an intestine. Tribol Lett 2012;47:295–301.
[13] Yan Y, Liu Y, Manfredi L, Prasad S. Modelling of the self-propelled vibro-impact capsule in small intestine. Nonlinear Dyn 2019;96:123–44.
[14] McCrum NG, Buckley CP, Bucknall CB. Principles of polymer engineering. Oxford, UK: Oxford University Press; 2003.
[15] Gere JM, Goodno BJ. Mechanics of materials. Boston, USA: Cengage Learning; 2012.
[16] Guo B, Ley E, Tian J, Zhang J, Liu Y, Prasad S. Experimental and numerical studies of intestinal frictions for propulsive force optimisation of a vibro-impact capsule system. Nonlinear Dyn 2020;101:65–83.

Chapter 11

Haptic interfaces

Peter P. Pott
Institute of Medical Device Technology, University of Stuttgart, Stuttgart, Germany

11.1 Human haptic perception

11.1.1 Introduction

Haptic describes the sensory and motor skills in the skin, joints, muscles, tendons, and ligaments.

Tactile describes the mechanical interaction with the skin. Tactile perception is the perception of exclusively mechanical interaction: normal and shear forces, moments, (angular) acceleration, (angular) speed, and heat.

Kinesthetic refers to both actuator and sensory properties of muscles and joints canthus considering forces and moments, positions and angles, and movements. As a result, every external kinesthetic interaction has a tactile component.

A haptic device (Fig. 11.1) is a system that provides mechanical output, which can be perceived haptically. It thus has at least one output, but not necessarily an input. The markings on the letters F and J of a keyboard are tactile patterns that transport information on the (right) position of the index finger. These simple keys are already tactile devices determined by their surface. Going further into detail of the key, it becomes clear that the characteristic curve of the spring holding the key in place has a second feature. It is producing the force felt by the user when he or she presses the key. This is designed in a way that a clear sensation of the fact that the key has been pressed is available. This information is propagated kinesthetically by the skin, muscles, and joints of the user. Thus, these simple keys are already a haptic device.

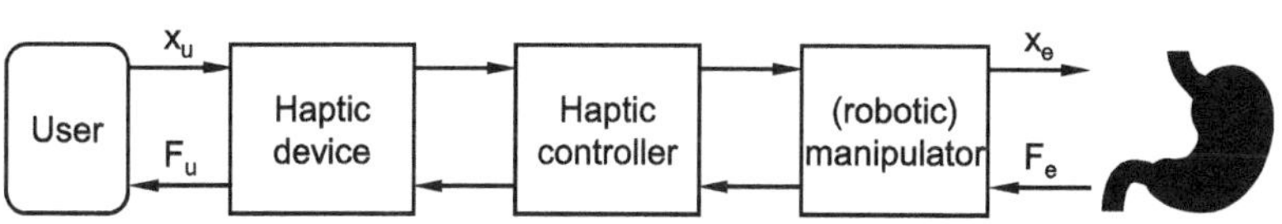

FIG. 11.1 System structure of a haptic display and input device in a telerobotics setup.

Endorobotics. https://doi.org/10.1016/B978-0-12-821750-4.00011-6

11.1.2 Anatomic and physiologic background

11.1.2.1 Human senses

The humans have various senses to perceive their environment (Fig. 11.2). These comprise the following:

- Thermal receptors providing information on (basically) temperature changes in the skin
- Chemical receptors that react to taste and smell
- Optical receptors that provide information on the intensity of electromagnetic waves between 450 and 700 nm wavelength—the visible range
- Pain receptors specialized on sensing physical damage (nociceptor)
- Mechanical receptors that provide information on strain and tension in the skin, tendons, or muscles

It has to be noted that these senses provide information on the change of a value only. Absolute measurements or even static haptic sensation are almost impossible, i.e., we are not able to perceive our clothes on the skin as soon "as we got used" to them and even a scratchy woolen pullover cannot be felt as long as we do not move.

The channel capacity of these receptors has a rather wide range. While optical receptors are able to record some 10 million bit/s, the sense of touch records roughly 1 million bit/s and auditive channel only approximately 100 kbit/s [1]. The processing of the receptor data is achieved in the cerebral cortex. This has several functional brain areas. The primary motor cortex is the physical location of signal processing for the sense of touch.

11.1.2.2 The haptic sense

The haptic sense is the most important within the sensorimotor system. It comprises a group of mechanical receptors that detect the deformation of tissues resulting from forces within the skin, joints, and muscles. Haptic perception thus comprises the signals from a large number of measuring points in and

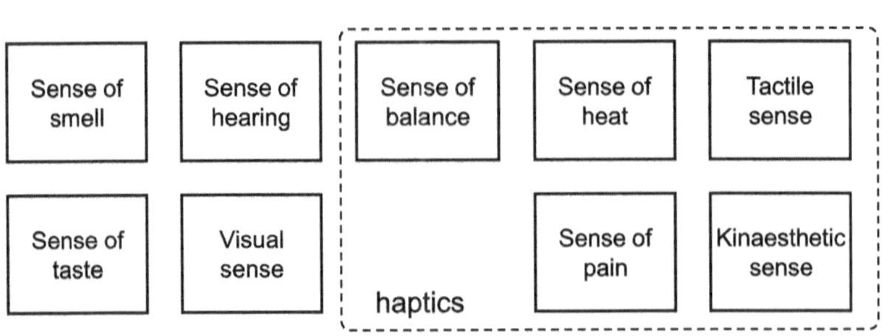

FIG. 11.2 Schematic overview of the human senses.

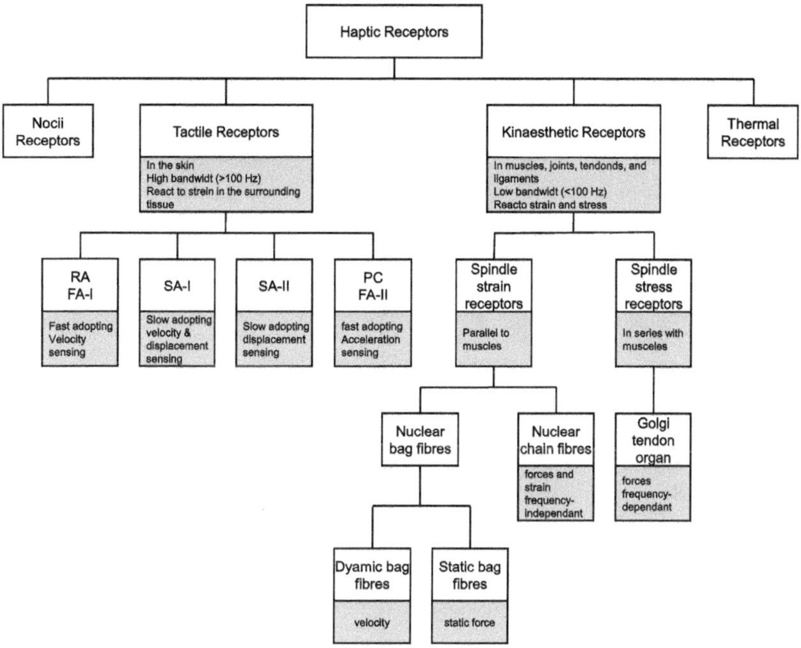

FIG. 11.3 Overview on mechanoreceptors in the skin (following [2] from [3]).

on the body by at least six types of receptors. These receptors are divided into two basic groups: tactile and kinesthetic receptors (Fig. 11.3).

Tactile receptors can be found in the outer layers of the skin (Fig. 11.4). Their density is strongly depending on the specific area of the skin. There are numerous receptors in the fingertips but a far smaller number on the back of the hand. The receptors react to strain in the skin and differ in the property of responding to displacement, velocity, or acceleration. In neurophysiology, the following types of tactile receptors are known [2,5]:

- Rapid-adaption or fast-adaption (RA or FA-I) Meissner corpuscles—velocity-proportional reception.
- Slow-adapting (SA-I and SA-II) Merkel cells and Ruffini corpuscles—velocity-proportional and displacement-proportional receptions, but with lower dynamics compared to Meissner corpuscles.
- Fast-adaption (FA-II) Pacinian corpuscles—acceleration-proportional reception.

It becomes clear that these receptors basically measure dynamic properties (velocity, acceleration). Thus, the haptic perception is a dynamic property. To gain a sensation of the haptic properties of a surface we usually manipulate it dynamically. We slide or sweep the surface with the fingertips to allow

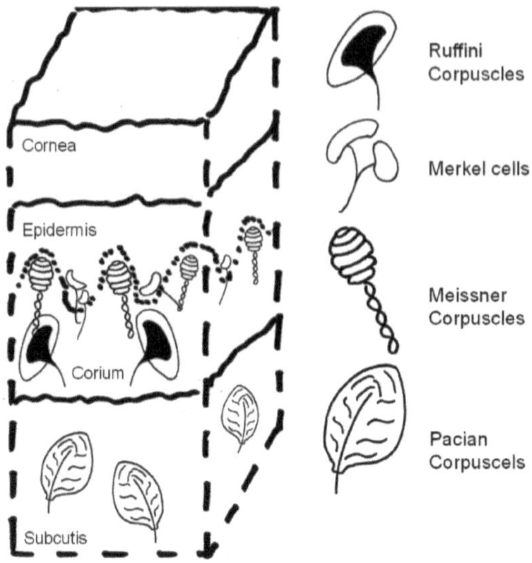

FIG. 11.4 Schematic cut through the skin showing the arrangement of mechanoreceptors (following [4] as shown in [3]).

interaction of our skin surface (fingerprints) with the surface roughness. This leads to characteristic frequencies that are perceived and allow us to distinguish haptic properties.

In contrast to tactile receptors, kinesthetic receptors are located inside the muscles, joints, and tendons. They detect forces that act on entire extremities. Due to the inherent low-pass filtering characteristics of the limbs' mass, stiffness, and damping, the requirements on dynamics are low. However, the requirements for the resolution of the lowest possible perceptible forces relative to the maximum acting force (amplitude dynamics) are comparable to the tactile receptors. Kinesthetic receptors are divided into two groups:

- Spindle-stretch receptors (dynamic bag fibers and static bag fibers)—in parallel arrangement to the muscle strains
- Spindel-tension receptors (Golgi tendon organ)—in serial arrangement to the muscle strains

11.1.3 Quantifying senses

For the engineer, it is important to see which physical values—in terms of displacement, vibration forces, torques, and pressure—can be perceived and which resolution during this perception can be achieved by the human user of a haptic device. The following tables provide a brief overview of the average achievable values [3] (Tables 11.1 and 11.2).

TABLE 11.1 Parameters of perception at the human hand.

Basic property	Parameter	Part of the body	Value
Static displacement, position	Displacement of the skin, absolute value	Fingertip (tactile)	10 μm
	Two-point discrimination (spatial resolution)	Fingertip (tactile)	2–3 mm
		Palm of the hand (tactile)	10–11 mm
	Position resolution, differential threshold	Interdigital joint (kinesthetic)	2.5°
		Wrist (kinesthetic)	2°
Dynamic displacement (vibration)	Frequency	Finger (tactile)	5–10 kHz
	Frequency	Body (kinesthetic)	20–30 Hz
	Maximum sensitivity	Fingertip	at 200–300 Hz
	Absolute amplitude	Fingertip	0.1–0.2 μm at 200–300 Hz
	Amplitude resolution, differential threshold	Fingertip (tactile)	10%–25%
	Frequency resolution, differential threshold	Fingertip (tactile)	8%–10%
Force and pressure	Force, absolute value	Fingertip (tactile)	0.8 mN
		Palm of the hand (tactile)	1.5 mN
	Force, differential threshold	Body (kinesthetic)	5%–10%
	Absolute pressure	Finger	2 mN/mm^2
	Pressure, differential threshold	Wrist (kinesthetic)	4%–19%
Torque	Differential threshold	Thumb and index finger (kinesthetic)	12.7%
Compliance	Differential threshold	Thumb and index finger (kinesthetic)	5%–15%

(Adapted from Hatzfeld C, Kern TA. Engineering haptic devices. Springer Berlin Heidelberg; 2016.)

TABLE 11.2 Output or actuation parameters of the human hand.

Basic property	Parameter	Part of the body	Value
Velocity	Maximum velocity	Finger (most proximal joint)	17 rad/s
		Finger (interdigital joint)	18 rad/s
Movement	Bandwidth (displacement and force)	Finger	5–10 Hz
		Arm (unexpected movement)	1–2 Hz
		Arm (periodic movement)	2–5 Hz
		Trained trajectory	5 Hz
		Reflex	10 Hz
Force	Maximum force	Index finger	40–50 N
		Thumb	85–100 N
		Wrist	35–65 N
		Fingertip grip	45–65 N
		Key grip	76–109 N
	Precision	Fingertip (with visual feedback)	40 mN
		Fingertip (without visual feedback)	11%–15%
Torque	Maximum torque	Finger (most proximal joint)	3.7–5 Nm
		Finger (interdigital joint)	1.2–2.9 Nm
		Finger (distal joint)	0.4–0.85 Nm

(Adapted from Hatzfeld C, Kern TA. Engineering haptic devices. Springer Berlin Heidelberg; 2016.)

It is important to understand that these values are determined through different experiments by a large number of authors. Thus, they might not be comparable and apply to certain groups of subjects only. Above all, these are average values that can be far higher and lower for an individual. In addition, there is a strong dependency on frequency for the dynamic parameters. Fig. 11.5 shows the minimum perceivable amplitude of a tactile stimulus over frequency.

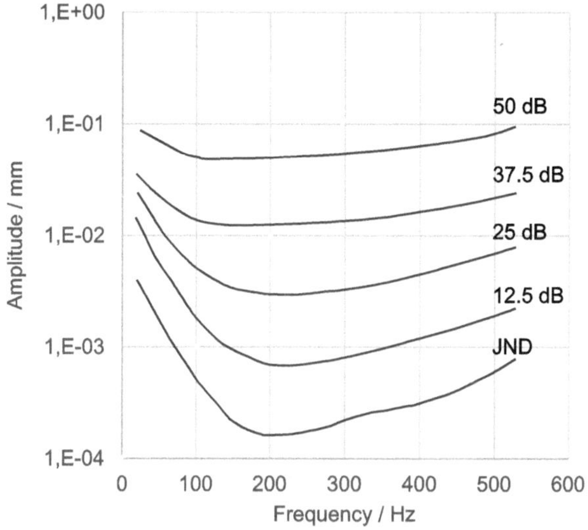

FIG. 11.5 Perception of a haptic stimulus depending on frequency and amplitude. *(After Hugony A. Über die Empfindung von Schwingungen mittels des Tastsinnes. Z Biol 1935; 96:548–553.)*

The lower the stimulus, the more pronounced the frequency dependence becomes. The just noticeable difference (JND) has its extremum at around 200 Hz. Table 11.3 sums up the amplitude and frequency range of the three main senses of the humans.

TABLE 11.3 Range of values for the three main human senses.

Parameter	Visual sense	Sense of hearing	Tactile sense
Dynamic range	140 dB	130 dB	80 dB
Reference value	10^{-6} cd/m^2	20 μPa	0.25 mN
Frequency range	450–800 nm	16 Hz – 16 kHz	0–10,000 Hz
Repetition rate	25 Hz	20 Hz	20 Hz
Differentiation In theory Practically	2×10^7 Colors 80 Shades of gray	4×10^5 tones	>0.8 mm laterally >0.4 mm vertically
	Max. 12 colors practically distinguishable	84 Keys on the piano keyboard	–

11.2 Engineering

11.2.1 Impedance

The mechanical impedance Z describes the quotient of force F and velocity $v = \dot{x}$ or torque T and angular velocity ω.

$$\overline{Z} = \frac{F}{\dot{x}} = \frac{T}{\omega} = \frac{1}{\overline{Y}}$$

Mechanical admittance Y is the reciprocal value of impedance. Both can be used to describe dynamic technical systems. A high impedance is exhibited by a "stiff" or "inertial" system. A low impedance indicates that a system is "light" and/or "soft" or "gliding". The concept of impedance is applied to haptic systems whether a haptic display (input device) or haptic manipulator (output device) is considered. For an input device, the impedance Z_D refers to the impedance that a system exhibits at its mechanical output (e.g., the user interface) when moved. Comparable to the technical system, the impedance is also suitable for a simplified description of the user and his or her mechanical influence on the technical system. The user impedance Z_U is used for this purpose.

The user impedance or stiffness can be influenced voluntarily in some range. For example, a handshake can be either hard or soft. The scale of this effect is frequency-depending. Thus, even involuntarily, the same hand is softer at low frequencies compared to high frequencies, which is caused basically by the moved mass of the hand.

11.2.2 Haptic transparency

The principle of haptic transparency [6] is basically applied to consider stability in the area of control-related problems and should be around ±3 dB. Transparency is a simple well-established, frequency-dependent, characteristic measure to describe haptic interfaces. In most cases, only the real part is considered. Transparency is the quotient of input and output impedance:

$$\overline{Z} = \frac{\overline{Z}_{in}}{\overline{Z}_{out}}$$

and is a complex number. A transparency in the range of 1 shows that the input impedance, i.e., the mass of a moving object or resistance of a structure, is not influenced by the actual technical system. The user manipulating the haptic device (one end of the transmission chain) experiences the mechanical properties at the input device (e.g., robot, other end of the transmission chain) without any misrepresentation. The principle of transparency can be applied to telemanipulation systems as well as to haptic simulators (interaction with virtual objects only). As $\overline{Z}_{in}$ and $\overline{Z}_{out}$ are frequency-dependent values, haptic transparency is also frequency-dependent.

In addition to haptic transparency, some authors [7] also use transparency width (Z-width) of a haptic system as a performance measure:

$$\overline{Z}\text{-width} = \overline{Z}_{max} - \overline{Z}_{min}$$

Z-width is the difference of the impedance of full blocked haptic display (e.g., in contact with a maximum stiff body, without considering friction and inertia) and the same display freely moving in space (considering friction and inertia). This measure describes the performance of individual devices and allows a comparison between devices after changes to the technical setup or the control algorithm.

11.2.3 One kHz as the engineering goal of a haptic system

Haptic perception extends into frequency ranges up to 10 kHz, with the most important frequency well below 1 kHz. The perception of microstructures of surfaces and the precise recognition of highly dynamic events, e.g., the moment of contact with a hard surface, is achieved by this high bandwidth. A simple model of a material could be the parallel combination of a mass m and a spring of stiffness k the 3 dB cutoff frequency f_c of a hard contact can be estimated by

$$f_c = \frac{1}{2\pi}\sqrt{\frac{k}{m}} = \frac{1}{2\pi}\sqrt{\frac{k}{V \cdot \rho}}$$

taking into account a virtual volume of 1 and individual density ρ. Young's modulus E determines k, and for most materials, the relation of E and ρ is more or less constant. Thus, with the exception of very soft materials such as rubber, biological tissue, and lead, the hard contact with most materials will lead to cutoff frequencies above 700 Hz. Thus, for a high-quality perception of these materials, both the signal processing and the mechanical system should be able to least exhibit dynamics in this range, since humans are sensitive here.

11.3 Haptic input devices

11.3.1 History and state of the art

Haptic input devices were originally designed and built for interaction with the virtual world. With the broad availability of powerful computers at the end of the 1990s, first simulations of virtual objects in real time could be achieved. This led to the necessity to make the physical properties of such objects (geometry, stiffness, mass) tangible.

Haptic displays can be designed by a serial or parallel kinematic. The first group has a large workspace but rather large moving masses reducing the available frequency bandwidth. Parallel kinematic setups are the opposite: their workspace is rather small but actuators are mounted to the base. This reduces moving masses to a minimum and allows high mechanical bandwidth. Of course, combinations—leading to hybrid kinematic setups—are also possible.

Haptic displays can be controlled considering their impedance or the admittance. Admittance-controlled haptic interfaces have advantages over impedance-controlled haptic interfaces during interaction with high impedance (virtual) environments. However, the performance of admittance-controlled haptic interfaces is often limited by the lower boundary of the impedance that can be achieved without causing issues with stability. Especially, the low inertia in an admittance model can cause unstable interaction [8].

Finally, considering the actual application, the number of active degrees of freedom can vary from one (a rotating knob, active accelerator pedal, or single-axis joystick) to six for full control of, e.g., a telemanipulator. Doubtlessly, any combination of active and passive degrees of freedom is possible.

In the following, various devices will be briefly introduced. A good overview for systems deployed in the medical context is provided by Escoba-Castillejos [9] while for more specialized applications, e.g., for haptic gloves, Perret provides nice review [10]. Finally, the aspect of haptic man–machine interaction is also considered for industrial applications [11].

11.3.2 Haptic feedback systems

11.3.2.1 Three to six axis displays

Haptic displays are used to provide the user with a haptic perception of a (virtual) object. The display is also used to create an input signal (Fig. 11.1). Thus, it not only produces a force but also senses the position, velocity, and acceleration of the user input. To achieve this, a more or less sturdy mechanical setup is needed (Figs. 11.6, 11.7). The size varies with the specification derived from the task. Especially when interacting with large (virtual) objects, rather large

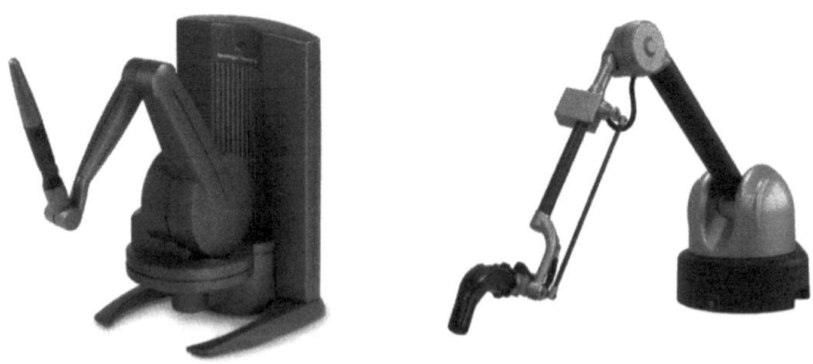

(A) (B)

FIG. 11.6 Serial kinematics haptic input devices. (A) Shows the Touch X by 3D Systems, Rock Hill, NC, USA with three passive and three active degrees of freedom and impedance control; (B) Shows the Virtuose 6D35–45 by Haption, Soulgé-sur-Ouette, France, device with six active degrees of freedom. *((A) Courtesy of 3D Systems; (B) Courtesy of Haption.)*

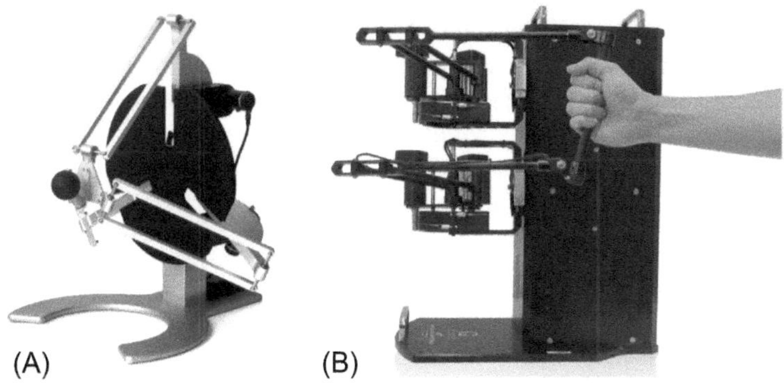

FIG. 11.7 Parallel kinematics input devices with five active and one passive degree of freedom. (A) Delta 3; (B) HD^2 High-definition haptic device. *((A) By Force Dimension, Nyon, CH; Courtesy of Force Dimensions, CH; (B) By Quanser, Markham, Canada; Courtesy of Quanser.)*

FIG. 11.8 Cable-robotics-based haptic input devices with six degrees of freedom. (A) IPAnema 3 mini by Fraunhofer IPA, Stuttgart, Germany [12]; (B) device by Haption, Soulgé-sur-Ouette, France. *((A) Courtesy of Fraunhofer IPA, Stuttgart, Germany; (B) Copyright Groupe PSA, Paris, France.)*

displays may be needed (Fig. 11.8B). On the opposite, while interacting with tissue in a surgical setting usually desktop-sized systems (Figs. 11.6A and B, 11.7) would be used.

Depending on the desired dynamics and workspace, parallel kinematics (Figs. 11.6A, 11.7B, 11.8) or serial kinematics (Figs. 11.6B and C, 11.7A) will be used.

Rope-based or cable-driven haptic input devices (Fig. 11.8) are a special case of parallel kinematics. Here, instead of struts and discrete joints (Fig. 11.6A), ropes and winches are used. This leads to very lightweight systems as ropes can transmit comparably large pulling forces, while a strut—transmitting also pushing forces—has to withstand buckling and thus would always be larger in cross-section and higher in weight [12,13]. The main drawback of rope-based systems is that they always need to be built around the interface point (and thus—more or less—the user) and at least one more rope than degrees of freedom is needed. Most systems with six degrees of freedom use eight ropes and winches for symmetry reasons. Systems are available from small systems like the SPIDAR-G 6 degrees of freedom (DOFs) haptic input device [14] to be used on a desktop to medium-sized input device (Fig. 11.8A) and large room-filling systems for virtual reality applications (Fig. 11.8B).

11.3.2.2 One- and two-axis displays

Haptic displays with one or 2 DOFs are quite common. Simple examples are input devices such as the BMW iDrive, which use a DC motor to provide software-defined indents or viscous damping allowing intuitive interaction with driver assistance systems in the car. Also, the accelerator pedal in the car can be equipped with an actuator creating a haptic display that suggests, e.g., less sportive driving style to the driver. Finally, fly-by-wire systems in planes provide a haptic feedback to the pilot to let him or her feel rudder response as it would be possible in conventional mechanically actuated control surfaces.

However, more sophisticated systems are also under investigation and available on the market. For example, the MagLev system [15] uses electro- and permanent magnets to create forces that can be felt by the user during interaction with (virtual) objects.

11.3.3 Tactile feedback systems

11.3.3.1 Vibrotactile

The vibrotactile display is one of the most common and simple. The simplest setups are found in smart phones to produce a haptic sensation following the activation of virtual key being pressed. The user percepts are a mechanical response to his or her action. This is achieved by using an inertial linear shaker or rotating system with an eccentric mass. More sophisticated systems use piezoelectric activation at varying frequencies [16] or electroactive polymers [17]. For car applications, systems exist that shake to whole screen to allow the driver to distinguish different virtual key blindly [18].

11.3.3.2 Surface modification

Using actuators, the surface touched by the user can be modified to create a roughness or pattern. The most prominent systems are used for the blind to create Braille letters. This can be accomplished by either piezoelectric [19,20] (Fig. 11.9), electromagnetic, thermal actuators or by using electroactive polymers for example.

FIG. 11.9 STReSS^2: Array of piezoelectric bending actuators to modify surface properties [19]. *(Photograph taken by Jerome Pasquero.)*

11.3.3.3 Surface vibration

Using high-frequency actuators, the surface of a system can be modified producing standing waves. This can be accomplished by, e.g., a piezoelectric plate with structured electrodes that are driven individually creating different surface patterns by using bending modes at different frequencies that then can be perceived by the user (Fig. 11.10).

11.3.3.4 Electro tactile

The mechanoreceptors in the skin (see Fig. 11.3) produce tiny electric signals that are transmitted by the nerves. This can be stimulated directly using electric fields. Thus, electrotactile displays produce small—spatially distributed—electric fields that interact with the receptors and nerves in the skin and lead to a tactile perception. Arranging the necessary electrodes in an array (Fig. 11.11) could make it possible to actually feel an image. But, also single-axis virtual key can be represented in this way.

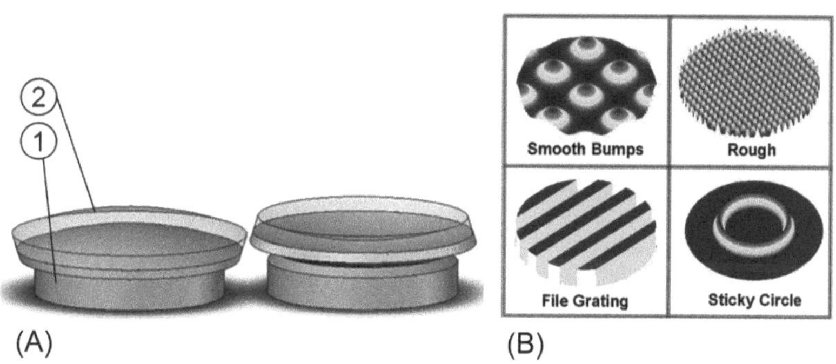

FIG. 11.10 Surface vibration tactile feedback device. Standing waves on the surface can be felt as surface patterns or roughness. Piezodisc (1) bonded to a glass substrate (2). (A) Mechanical setup; (B) possible surface structures [20]. *(Courtesy of IEEE.)*

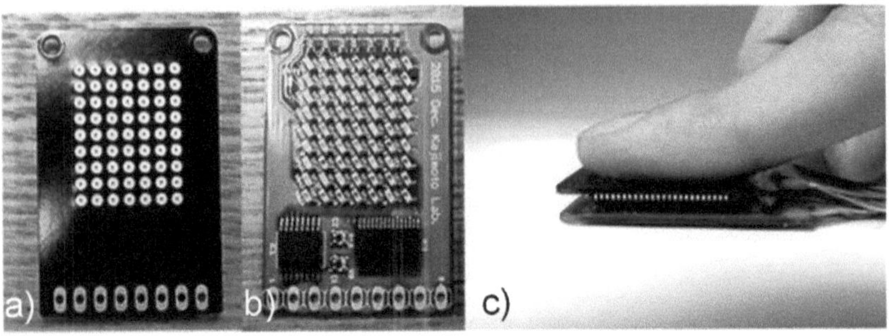

FIG. 11.11 Electrotactile display. (A) Electrode pattern; (B) backside of the PCB with integrated electronics; (C) representation of graphic content [21]. *(Courtesy of SpringerNature.)*

11.3.3.5 Ultrasonic

Acoustic waves produce periodic pressure changes in air. By creating standing waves of short wavelength and high pressure, a tactile sensation in the skin can be achieved. Thus, ultrasound in range between 40 and 100kHz is either focused on the skin using an acoustic lens or arrays of small transducers are driven to create a phased array that focuses sound pressure on the skin (Fig. 11.12). Very pronounced and sharp sensations can be achieved this way.

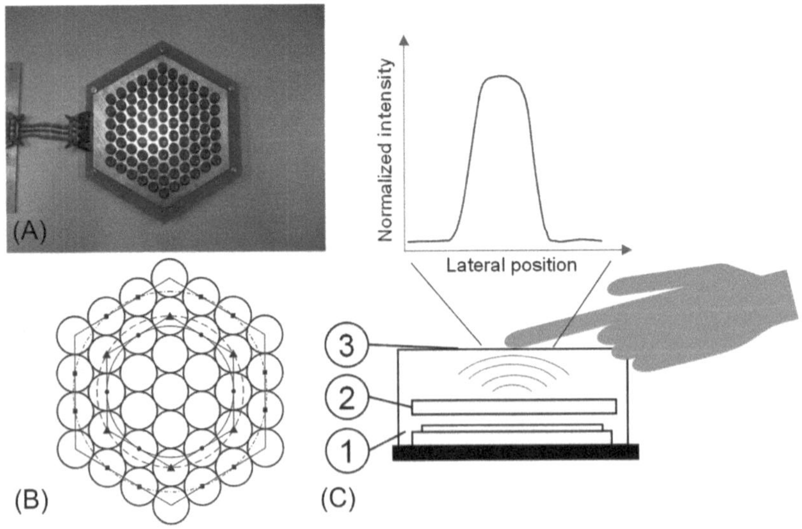

FIG. 11.12 Ultrasonic tactile display. (A) System with fluidic coupling medium and a spherical ultrasonic lens; (B) array of transducers to produce a sound pressure field in air [22], (C) principal function: (1) ultrasonic transducer array in housing, (2) ultrasonic lens, (3) finger on surface for tactile sensation. *((A and B) Courtesy of SpringerNature.)*

11.3.4 Summary

Haptic input devices exist in a wide range of variety. Applications such as mechatronic applications in cars but also feedback systems in airplanes are quite common. On the contrary, robotic surgical interaction of today's available systems relies on visual feedback only. Still, systems for multi-axis haptic interaction exist commercially but at rather high cost.

11.4 Pseudohaptics

Haptic feedback can be derived also by indirect methods. While *direct* would mean that interaction forces are measured and displayed to the user by systems described earlier (over a preferably large frequency band), *indirect* means that other physical values are measured and displayed. Two mayor approaches are described in the following.

11.4.1 Acceleration mapping

The system described by the group of Kuchenbecker [23] uses MEMS-based acceleration sensors attached to the laparoscopic instruments of a surgical robot to measure accelerations that occur during tissue contact. After processing, the derived signals are fed back to the user by means of small vibration motors, inertial shakers, and also by making the audible. The user receives a certain feeling for the surgical situation by correlating the haptic perception with the visual impression. The measurement principle does not allow to measure static forces (e.g., during pulling a thread while suturing or retracting tissue) but enable the surgeon to feel dynamic forces (such as ripping of tissue during excessive force application) only. As the skin is not sensitive to the direction of forces applied, the direction of the measured accelerations is of no importance.

This approach is promising as simple acceleration sensors are used on situs, which are rather small compared to force sensors that would have to be integrated into the flux of forces through the instrument and robot. Also, the haptic display consists of simple and easy-to-integrate parts that do not alter the system's structure. While the actual haptic feedback is limited by the earlier described reasons, the approach allows rather precise interpretation of the actual situation.

11.4.2 Displacement-force mapping

Another form of pseudohaptic feedback was introduced by Lécuyer [8] to simulate a haptic perception while interacting with objects in a virtual environment. The approach produces a haptic illusion by combining the visual feedback of actively seeing, e.g., a gripper closing while exerting force to an input device, which does not move at all. So, in contrast to conventional input devices where the user produces a displacement x_u and receives a force F_u, here the user produces a force F_u and sees the displacement of the end effector x_e. Combined with experience, the user receives a perception of the stiffness of the manipulated object [24].

By defining his coupling characteristics between input- and output device, the behavior—which is also strongly depending on the latency of the visual feedback—can be defined between wide boundaries. Combining a changing coupling characteristic between input force F_u and end effector displacement x_e depending on the actual interaction forces F_e and under consideration of the visual feedback will lead to pseudohaptic feedback (Fig. 11.13).

The relation of end effector displacement x_e is a function of input force F_u and interaction force F_g:

$$x_e = f(F_u, F_e)$$

In order to achieve a pseudohaptic feedback, F_u and F_e determine each other. While increasing F_g leads to an increase of x_e, (i.e., closing a gripper) the interaction force F_e must have a decreasing effect on the displacement x_e.

The most straightforward relation between F_g and resulting x_e is the subtraction of the forces scaled by a factor s ($[s] = N^{-1}$):

$$x_e = s \cdot (F_u - F_e)$$

In the case of no interaction with the environment ($F_e = 0$), the coupling is defined by the absolute value of s only. As soon as a contact with an object is achieved, the coupling curve changes depending on the interaction force F_e. A simple linear stiffness ($F_e = x_e/n$) leads to a reduction of the slope of the coupling curve (Fig. 11.14). So, a larger user input force F_u is required for the same—visually perceived—displacement x_e of the end effector. This force difference can be perceived as haptic feedback.

Displacement-force mapping requires a force measurement in the situs but reduces the need of active force-feedback systems. It thus leads to simpler—and

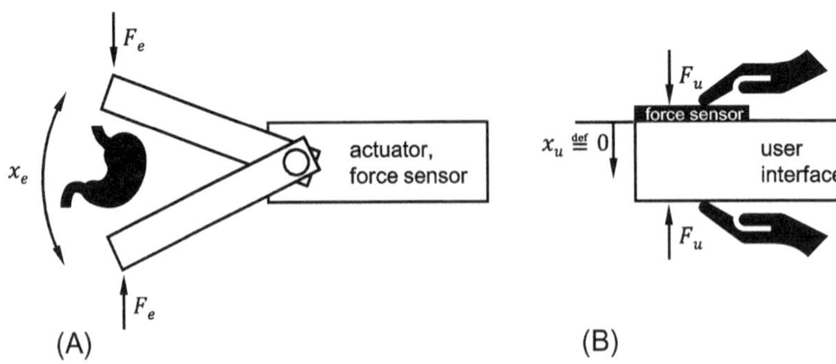

FIG. 11.13 Principle of an isometric user interface for measuring gripping forces (B) and an end effector with variable branch displacement and measurement of the interaction force (A).

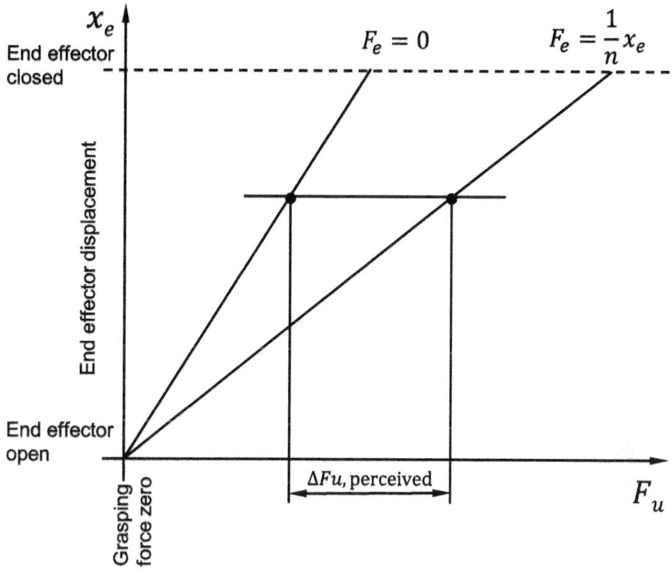

FIG. 11.14 Exemplary coupling relations for grasping force and end effector displacement: idle case (*left*), linear compliance n (*right*). The difference between exerted gripping forces for reaching identical displacements can be perceived as haptic feedback $\Delta F_{u,\ perceived}$. *(After Hatzfeld C, Kern TA. Engineering haptic devices. Springer Berlin Heidelberg; 2016.)*

cheaper—input devices. The frequency bandwidth of the perceived haptic feedback is limited to the actively achievable movement. It is thus limited to gripping movements or very slow displacements of a robot.

References

[1] Jungnickel U. Miniaturisierte Positioniersysteme mit mehreren Freiheitsgraden auf der Basis monolithischer Strukturen. Disseratation, Germany: TU Darmstadt; 2004.

[2] Smith CUM. Biology of sensory systems. 2nd ed. Chichester, UK, Hoboken, NJ: John Wiley & Sons; 2008.

[3] Hatzfeld C, Kern TA. Engineering haptic devices. Berlin Heidelberg: Springer; 2016.

[4] Doerrer C. Entwurf eines elektromechanischen Systems für flexibel konfigurierbare Ein- gabefelder mit haptischer Rückmeldung. Dissertation, Germany: TU Darmstadt; 2003.

[5] Schmidt RF, Schaible H-G. Neuro-und sinnesphysiologie. Berlin Heidelberg: Springer; 2006.

[6] Lawrence DA. Stability and transparency in bilateral teleoperation. IEEE Trans Robot Automat 1993;9(5):624–37.

[7] Colgate JE, Brown JM. Factors affecting the Z-width of a haptic display robotics and automation. In: 94 International conference. Los Alamitos: IEEE Computer Society Press; 1994. S3205–3210.

[8] Nabeel M, Jafari A, Ryu J-H. Realizing low-impedance rendering in admittance-type haptic interfaces using the input-to-state stable approach IROS Vancouver 2017. In: IEEE, Piscataway, NJ, S914–919; 2017.

[9] Escobar-Castillejos D, Noguez J, Neri L, Magana A, Benes B. A review of simulators with haptic devices for medical training. J Med Syst 2016;40(4):104.

[10] Perret J, Vander Poorten E. Touching virtual reality: a review of haptic gloves ACTUATOR 2018. In: 16th International conference on new actuators: 25–27 June 2018. Frankfurt am Main: VDE; 2018.

[11] Aziz FA, Mousavi M. A review of haptic feedback in virtual reality for manufacturing industry. J Mech Eng 2009;40(1):68–71.

[12] Pott A. Cable-driven parallel Robots. Theory and application. Heidelberg: Springer; 2018.

[13] Ho WY, Kraus W, Mangold A, Pott A. Haptic interaction with a cable-driven parallel robot using admittance control cable-driven parallel robots. Springer; 2015. S 201–212.

[14] Kim S, Hasegawa S, Koike Y, Sato M. Tension based 7-DOF force feedback device: SPIDAR-G. In: Loftin B, editor. IEEE virtual reality 2002. Proceedings; 24–28 March 2002, Orlando, Florida. IEEE Computer Soc, Los Alamitos, S 283–284; 2002.

[15] Berkelman PJ, Butler ZJ, Hollis RL. Design of a hemispherical magnetic levitation haptic interface device. In: Proceedings of the ASME international mechanical engineering congress and exposition, Bd 58, S 483–488; 1996.

[16] Debus T, Becker T, Dupont P, Jang T-J, Howe RD. Multichannel vibrotactile display for sensory substitution during teleoperation. In: Stein MR, editor. Telemanipulator and telepresence technologies VIII. SPIE, S 42–49; 2002.

[17] Matysek M, Lotz P, Winterstein T, Schlaak HF. Dielectric elastomer actuators for tactile displays World haptics 2009. In: Third Joint EuroHaptics Conference, 2009 and Symposium on Haptic Interfaces for Virtual Environment and Teleoperator Systems; WHC 2009; 18–20 March 2009, Salt Lake City, UT, USA. IEEE, Piscataway, NJ, S 290–295; 2009.

[18] Tzudnowski I. Improved safety and convenience in your car: continental presents touch display with active haptic feedback; 2015. Continental AG. Zugegriffen: 22. Oktober 2020.

[19] Levesque V, Pasquero J, Hayward V. Braille display by lateral skin deformation with the STReSS2 tactile transducer world haptics 2007. Los Alamitos, CA: IEEE Computer Society; 2007. S 115–120.

[20] Winfield L, Glassmire J, Colgate JE, Peshkin M. T-PaD: tactile pattern display through variable friction reduction world haptics 2007. Los Alamitos, CA: IEEE Computer Society; 2007. S 421–426.

[21] Uematsu H, Suzuki M, Kanno Y, Kajimoto H. Tactile vision substitution with tablet and electro-tactile display. In: Bello F, Kajimoto H, Visell Y, editors. Haptics. Perception, devices, control, and applications: 10th international conference, EuroHaptics 2016, London, UK, July 4-7, 2016; 2016. Proceedings, Bd 9774. Springer, Cham, Heidelberg, S 503–511.

[22] Iwamoto T, Tatezono M, Shinoda H. Non-contact method for producing tactile sensation using airborne ultrasound. In: Ferre M, editor. Haptics: perception, devices and scenarios. 6th international conference, EuroHaptics 2008 Madrid, Spain, June 10-13, 2008; 2008. Proceedings, Bd 5024. Springer, Berlin, Heidelberg, S 504–513.

[23] McMahan W, Gewirtz J, Standish D, Martin P, Kunkel JA, Lilavois M, Wedmid A, Lee DI, Kuchenbecker KJ. Tool contact acceleration feedback for telerobotic surgery. IEEE Trans Haptics 2011;4(3):210–20. https://doi.org/10.1109/TOH.2011.31.

[24] Neupert C, Matich S, Scherping N, Kupnik M, Werthschützky R, Hatzfeld C. Pseudo-haptic feedback in teleoperation. IEEE Trans Haptics 2016;9(3):397–408.

Chapter 12

Case study of vision systems: Optimized compression architecture for wireless endorobots

Abdelkrim Zitouni[a], **Nedra Jarray**[b], **and Majdi Elhajji**[c]
[a]*Department of Physics, College of Science & Humanities - Jubail, Imam Abdulrahman Bin Faisal University, Jubail Industrial, Saudi Arabia,* [b]*Electronics and Microelectronics Laboratory, Faculty of Sciences of Monastir, University of Monastir, Monastir, Tunisia,* [c]*Department of Electrical Engineering, College of Engineering, Shaqra University, Dawadmi, Saudi Arabia*

12.1 Introduction

Many fatal diseases such as tumor, cancer, and ulcer can occur in different regions of the gastrointestinal path without any specific symptoms. Among the symptoms of these diseases, we distinguish constipation, vomiting, diarrhea, and blood in stool. These symptoms indicate only the presence of a given disease but cannot clearly specify its nature and its location in the gastrointestinal (GI) tract. To avoid these problems, some novel endoscopic investigation techniques (ultrasound and wireless methods) have been presented [1].

Among them, the capsule designed by Iddan [2] is the first wireless capsule that is presented rather than traditional wire-based endoscopy. It needs to be small sufficiently in order to be taken by patients easily. The main advantage of the capsule module is its direct examination of the disease without using any anesthesia [3,4]. Thus, several radio frequency (RF) capsules have been designed in the 433/868/915-GHz frequency bands [5] or in the 2.4-GHz frequency bands [6].

In 2000 [7], the first commercialized capsule has been designed. An extended version of this capsule has been designed in 2006 [8]. A set of other societies (Siemens, Pentax, and Olympus) have also proposed other comparable capsules [8,9].

Fig. 12.1 shows a standard wireless capsule that is composed by the electronic capsule, a workstation, and a wireless link. Two LEDs are implemented

Endorobotics. https://doi.org/10.1016/B978-0-12-821750-4.00012-8

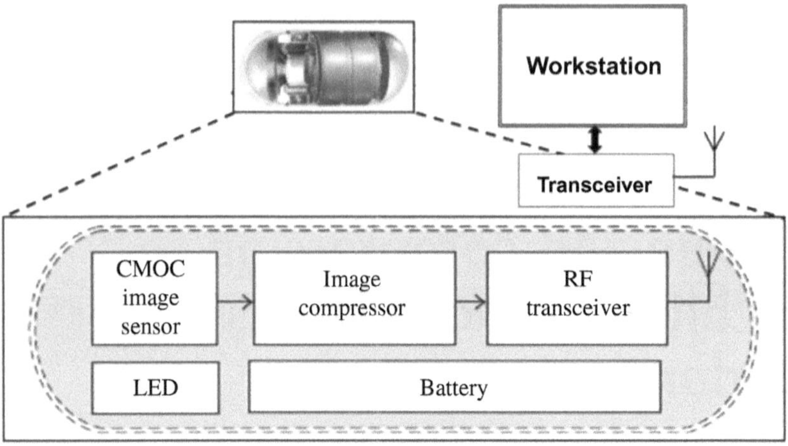

FIG. 12.1 Block diagram of a wireless endoscopic capsule.

inside the capsule to assure illumination. The generated images that are captured by a CMOS sensor are transmitted by a wireless radiofrequency module to the external world. These images are stored in a sophisticated memory for a future diagnosis after being reconstructed. A control module is used for operating the compression steps and orders.

A miniscule wireless camera is integrated in the capsule in order to take pictures of the gastrointestinal path. This camera can take between 50,000 and 60,000 images to be examined by the doctor. In order to swallow easily through human gastrointestinal path, it is necessary that the capsule has a small dimension. In general, one day is required for it to travel from entrance to evacuation. Captured images are submitted to the outside of the body through a wireless RF transmitter and then stored in the workstation [8]. The transferred image data dissipates about 90% of the entire power in the endoscopy capsule [8]. For this reason, it is necessary to compress this data while reducing power consumption and enlarging the communication bandwidth. However, the resolution of the constructed images needs to be saved. In fact, these images can be perturbed when zooming is done by a physician in order to well analyze the diseases. Though, power dissipation in the RF broadcaster is greatly amplified due to the increase in image resolution.

The chapter is organized as follows. Section 12.2 presents and discusses the related works. Section 12.3 describes all steps and methods of the architecture design. Section 12.4 shows the experimental results, studies these results, and compares them with related works. Finally, the conclusions are reported in Section 12.5.

12.2 Related works

The principal concern during the design of a wireless endoscopy system is to maximize the performance (power, area, computational speed, etc.). For this reason, we have interested principally on the optimization of the main part (image compression) of the endoscopy system by adopting first the discrete cosine transform (DCT) based technique that has been used by many works [10–15]. In Ref. [14], a high-performance-oriented hardware design of an 8×8 transform is proposed by authors. The design targets pipelined calculation architecture and memory resources optimization. The authors in Ref. [15] present an field programmable gate array (FPGA) parallel architecture that minimizes the real-time resources constraints. Adaptive block size (for 4×4 and 8×8 transforms) hardware architectures are also studied in Ref. [12]. In Ref. [10], Loffler presents an optimized DCT algorithm that performs 29 additions and 11 multiplications for computing the DCT of a vector with 8 components. This algorithm usually performs arithmetic operations by using the floating point technique, so it consumes a lot of memory and power. In the past few years, a simplified DCT process that necessitates 38 additions and 16 shifting operations has been discussed in Refs. [16,17].

While minimizing the complexity, we have proposed an optimized integer Cordic-DCT architecture based on modified carry look ahead adder (MCLA) technique that reduces the number of additions from 38 to 30 operations [18]. In this architecture, the multiplication by constants that is used in floating point-based methods is frequently realized by a succession of shifts and additions. This implementation is characterized by a reduced silicon area and dissipated power.

The disadvantage of the DCT-based technique resides in the appearance of the false shape and blocs artifacts, which may causes the loose of information at higher compression speed [19]. In order to avoid this problem, another technique for image compression called discrete wavelet transform (DWT) is also used in some other works [20–25]. The disadvantage of the DWT technique resides in the fact that it needs great computation resources rather than the DCT one. For this reason, we have proposed to explore the positive points of the two techniques by merging them into a unique hybrid DCT/DWT design.

Concerning the image coding part, which is performed after the compression step, many techniques can be used inside endoscopy systems (Huffman, Lempel-Ziv, and Entropy Coding). Lempel-Ziv technique necessitates an addressable memory in order to store the coder dictionary [26,27]. Huffman technique [28] needs also a memory in order to store Huffman probabilities table [29]. Two kinds of lossless entropy compression schemes can also be used: context-based binary arithmetic coding (CABAC) and context-adaptive variable length coding (CAVLC). CABAC algorithm [30] needs large amount of processing for decoding step compared to similar algorithms. CAVLC [31] is used when the high quality of coded image or video is not required. Due to its efficiency than the other techniques, we have chosen the CABAC technique during the coding step of the endoscopy system image. In fact, in comparison with CAVLC, CABAC algorithm is more effective and achieves an average more important between 9% and 14% of bit rate saving.

The embedded modules (image compression, image coding, data transmitter, etc.) inside an endoscopy system may inevitably need to communicate and share the common resources. The majority of the works undertaken by several researchers, referring to the integration of the communication within an endoscopic capsule system, have focused on the study of point-to-point interconnection [32–35]. This type of communication architecture leads sometimes to a system which can include an increased area due to the intercomponents wires routing. Other researchers are interested on the improvement of the I2C (inter-integrated circuit) bus or specific parallel bus protocols in order to accentuate the performances and to facilitate the interfacing of the communicating modules inside endoscopy systems [36,37]. These works propose communication architectures based on an I2C bus-equipped by a bus inverter and a bus encoder/decoder for interfacing the used memories and the image sensors. Since the buffers of the used memories and the image sensors are 8-bits, these approaches use a serial/parallel converters and 16/32-bits cyclic redundancy check (CRC) correction schemes, which lead to silicon area increasing. In Ref. [38], a micro-ball and multiple cameras master-slave system based on an 8-bits specific parallel bus have been presented. This architecture can lead to a silicon area increasing since they are based on distributed and hybrid memory scheme for each slave device. In order to accentuate the communication performances, we have developed a bridge advanced microcontroller bus architecture (AMBA)/I2C bus-based architecture.

12.3 Architecture design

The proposed endoscopic design integrates three main parts (compression, encoding and communication) in a manner targeting the optimization of the performance criteria.

We focus our work on the optimization of the digital part, including compression and communication modules. So the output image data are assumed to be handled by a CMOS image sensor equipped by a color filter array (CFA) of Bayer format [39]. At the beginning, the CFA data lines extracted from the sensor are filtered and stored in a SRAM (static random-access memory). The data compression process is then performed by a hybrid 2D-DWT-DCT-based architecture.

12.3.1 Compression step

The compression task is based on a hybrid 2D-DWT-DCT architecture that allows the application of the DWT and the DCT techniques sequentially as illustrated by Fig. 12.2.

First, the original frame, for example, 256×256, is considered to be divided into 32×32 blocs in the sensor SRAM. The input data are serially read from the SRAM. Then, a serial-parallel conversion is carried out. These operations are

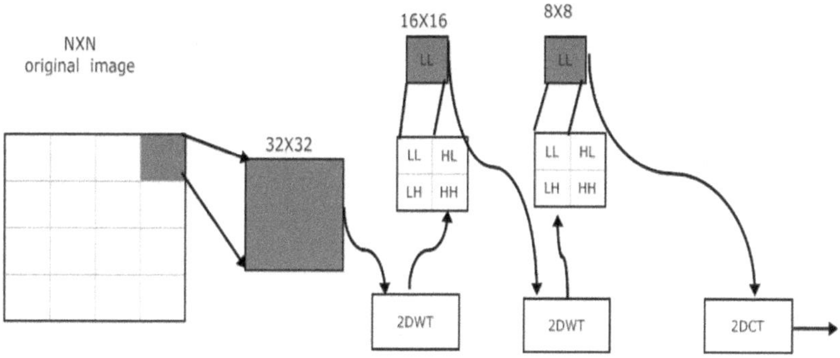

FIG. 12.2 Flow of the 2D-DWT-DCT compression technique.

performed by the proposed I2C/AMBA communication architecture. Each bloc is then decomposed by using the 2D-DWT technique. After applying the 2D-DWT to the specified sequences, a group of LL, LH, HL, and HH coefficients is handled. The LL coefficients represent the low-frequency components. Due to their much reduced values, the LH, HL, and HH coefficients are ignored in order to guarantee greater compression. In fact, during DWT regeneration, entered data is reached in multiple resolutions [40] by extra decomposition of the LL coefficients for different stages. During the reconstruction step of the generated data outputs, a 2-factor-based up-sampling is performed to the compressed data. The signal is navigated many times over the same group of high- and low-pass filters mutually in lines and columns. As example, while attaining about 50% of compression proportion, the HL and HH coefficients need to be lengthened with zeros.

The regenerated data are turned from the input value with very lesser error due to the compression process as demonstrated in Ref. [41]. So, 75% of data are compacted at this stage, and the LL coefficients are pushed to the next level when they are decomposed using the 2D DWT. Besides, the 8-DCT points are used for these coefficients.

The procedure is executed by adding one more level in each step, and then a constant scaling factor (SF) is used for dividing the coefficients to attain the preferred compression ratio. Lastly, for data regeneration, a rescaling and a zeros padding step are performed to the data before its passing via the DWT filters.

To attain further compression, a JPEG-based quantization algorithm is applied. In this level, most of the coefficients that have the highest frequency are smoothed to zero. A scaling factor (SF) is further applied for scaling the quantized coefficients.

In order to prove the efficiency of the proposed hybrid algorithm, the 256×256 "Lena" standard image is studied as example. The transformation/reconstruction of this image is performed by the proposed hybrid and standalone

DCT and DWT algorithms in order to verify the results. The fully subsampling DWT method is used for the DWT algorithm where all the LL, HL, LH, and HH components are considered.

The compression of the image is performed for all the three algorithms based on a compression ratio (CR) of about 96%. The compression ratio represents the ratio between the rejected data and the original one. So it represents the amount of data coefficient degradation and is expressed by $CR = \frac{\text{Discarded data}}{\text{Original data}}$.

For a CR of 96% and when the standalone DCT is used for image reconstruction, the PSNR is about 28.66 dB. On the other hand, when the reconstruction is performed by the DWT algorithm, the PSNR is about 25.49 dB. Nevertheless, a PSNR of about 30.01 dB is measured when the proposed hybrid algorithm is used for image reconstruction. So this PSNR is reasonably higher than the PSNR obtained by the other algorithms.

The efficiency of our hybrid algorithm has also been compared with that of the standard JPEG and JPEG 2000 [42,43] for 256×256 "Lena" and "Barbara" images at a compression rate of around 32%. When the images are reconstructed using the standard JPEG, the PSNR is 32.49 dB and 27.77 dB for "Lena" and "Barbara" images, respectively. For the images reconstructed using the standard JPEG 2000, the PSNR is 34.15 and 28.89 dB for "Lena" and "Barbara" images, respectively. However, when the images are reconstructed using the proposed hybrid algorithm, the PSNR is 36.98 and 31.46 dB for "Lena" and "Barbara" images, respectively. From these results, we notice that the proposed hybrid algorithm is characterized by a higher performance relative to the standard JPEG and JPEG 2000 algorithms for "Lena" and "Barbara" images. We notice also the independence of the image from the effects of the blocking artifacts and the false contouring.

This study shows that the proposed hybrid technique ameliorates PSNR results relatively to the others DCT and DWT algorithms due to the combination of the DCT and the DWT advantages.

2D-DWT part: Traditional DWT architectures are based on convolutions [44,45]. These architectures are characterized by higher power computation complexity and also require more memory. To overcome these problems, second-generation DWT lifting-based algorithms are proposed in Refs. [20,46]. Nevertheless, the direct hardware implementation of these algorithms causes an extended data path and low efficacy. To more optimize these kinds of architectures, a new method that allows the reduction of the DWT computation complexity is proposed in Ref. [25]. This method that is chosen during the design of our architecture allows also logic and memory units minimization without any degradation of the throughput. This method is also called DWT-Gall 5/3 that is constituted by the three main steps: (i) Split, (ii) Predict, and (iii) Update [47,48] (Fig. 12.3).

The proposed modification concerning the 2D-DWT is constituted by three units: line computation unit, transposition unit, and column computation unit as presented in Fig. 12.4. The first part interests on the computation of the (LH)

$$Xodd = X(2i+1)$$

$$Xeven = X(2i)$$

$$H(m) = X(2i+1) + \alpha(X(2i) + X(2i+2))$$

$$L(m) = X(2i) + \beta(H(m) + H(m-1))$$

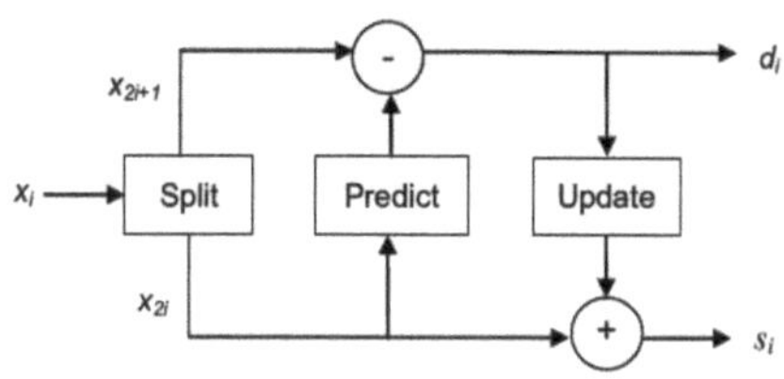

FIG. 12.3 Formulation of the Gall 5/3 DWT steps.

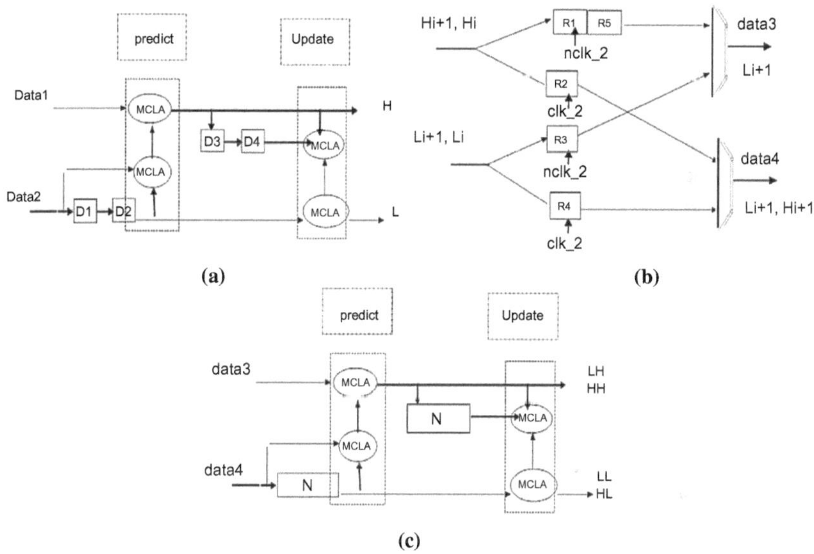

FIG. 12.4 Modules of the modified 2D-DWT architecture: (A) line computation, (B) transposition, and (C) column computation.

coefficients. Then, the transposition unit performs the management of the existent coefficients of the line computation unit in order to facilitate the calculation of the details and the approximation coefficients by the line computation unit.

Since the proposed architecture belongs to the lifting scheme with (5/3) wavelet filter approaches and uses a computational complexity reduction technique [25], the on-chip memory is hardly reduced. In fact, as explained in Ref. [25], only two temporal dual-port buffers RAM1 and RAM2 with a depth of 2N ($N \times N$ represents the size of each image) is used. For storing the data of raw/column filters and the transposition between them, a set of about 36 buffers and registers are also used.

2D-DCT part: The traditional DCT methods, even the most optimized [10], that are based on the floating point scheme, need more memory and require multiplication circuits. So they are less suitable for portable devices like wireless endoscopy capsule where the power consumption should be limited. For this

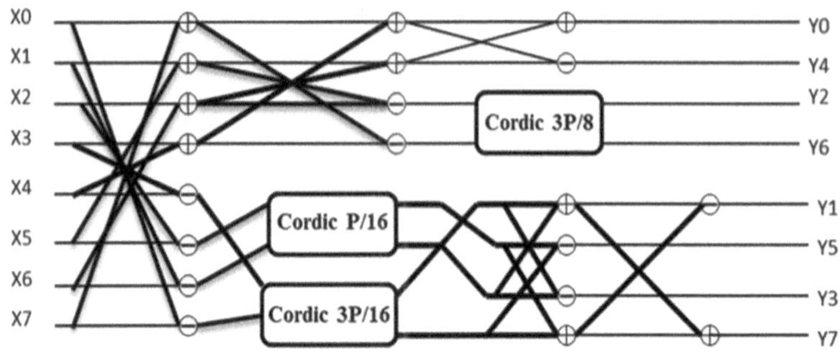

FIG. 12.5 Architecture of the proposed 2D-DCT algorithm.

reason, we have adopted an optimized integer-based Cordic DCT technique [16] that eliminates the multiplication and uses only adders and shifters. The performed optimization is based on reducing the number of addition and shifting operations and integrating the MCLA adders rather than the traditional CLA ones. The details concerning the optimization process are described in our last work published in Ref. [18]. The aim is to more reduce the dissipated power and improve the computation speed.

The DCT architecture part (Fig. 12.5) is composed by 8 2D-DCT modules addressing the pixels and generating the columns quantization matrix in parallel. This architecture is merged with a quantization block in a parallel manner in order to allow the quantification of the coefficients. The merged DCT and the quantification modules allow the simultaneous quantification of 8 coefficients in each clock cycle. So each column is treated in one click cycle. Fig. 12.6A and B illustrate the DCT-Q architecture and the quantization part, respectively. The quantization part is formed by 3-binary modules and a shifting module. The aim of this architecture is to consider all multiplication factors. In order to reduce the switching activity that minimizes the dynamic power consumption, a gated clock technique is used. This technique is based on a finite state machine controller (Fig. 12.6C).

MCLA part: The computation tasks in both 2D-DCT and 2D-DWT parts of the hybrid compression architecture are performed by a MCLA technique. This technique is used due to its computation rapidity and reduced surface. Rather than the AND and the NOT gates used in traditional CLA [49,50], we have implemented the NAND gates in order to increase the computation speed and to minimize the cost factor of the CLA adders.

12.3.2 Coding step

The coding step is performed by a CABAC coder that is composed by three stages (binarizer, context modeler, and arithmetic coder). The context modeler [51] reads the binary chain and produces a context value based on the top and the left macro-blocs neighbor data. A probability distribution model is allocated to the specified symbols in the modeling steps. This model allows the driving of

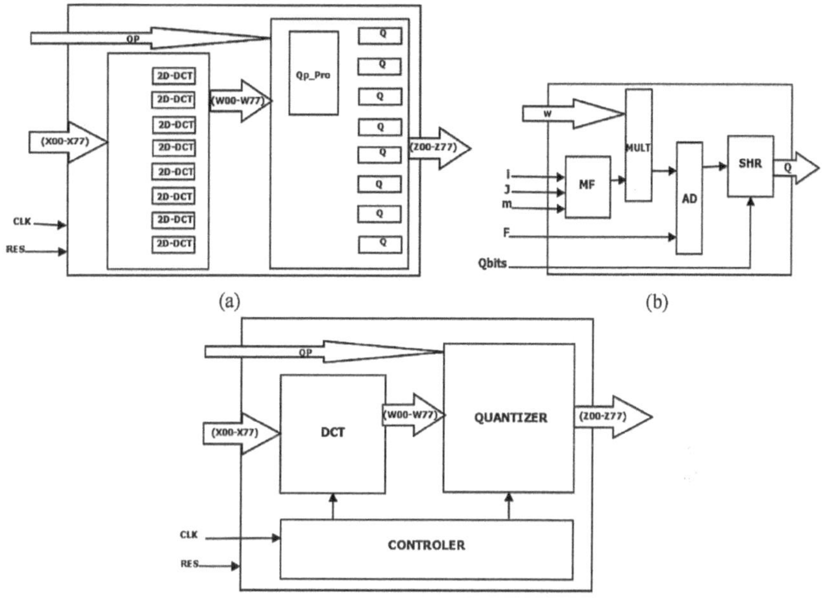

FIG. 12.6 DCT-Q architecture (A), block diagram of quantization (B), and gated clock controller based architecture (C).

the actual device to create an arrangement of bit as a coded illustration of the symbol according to the coding stage modeling distribution. The first type of CABAC design types includes a context pattern with up to neighboring syntax elements in the former of the actual syntax element to encode. The second type of context models is defined for the syntax element. Finally, the 3rd and the 4th types of the context model are used for residual coefficients.

The arithmetic encoder block is composed by three coding steps: regular, by-pass, and termination. The binarization method consists of mapping the integer value of (SE) into a sequence of bits (bin string). This process allows the reduction of the coded symbol size. Thus, it simplifies the costs of context modeling as well as the calculus of arithmetic coding. To implement the binary process, CABAC defines four methods (unary, truncated unary, fixed length, unary/Kth Exp-Golomb). The binarization architecture (Fig. 12.7) is composed by a demultiplexer, a multiplexer, the binarization module, and a gating clock module. The gating module is attached to each other block to stop it whenever it is not under running.

12.3.3 Communication step

The use of a bridge in the capsule endoscopy system is necessary when the adaptation between two or several bus must be operated in order to offer the locally

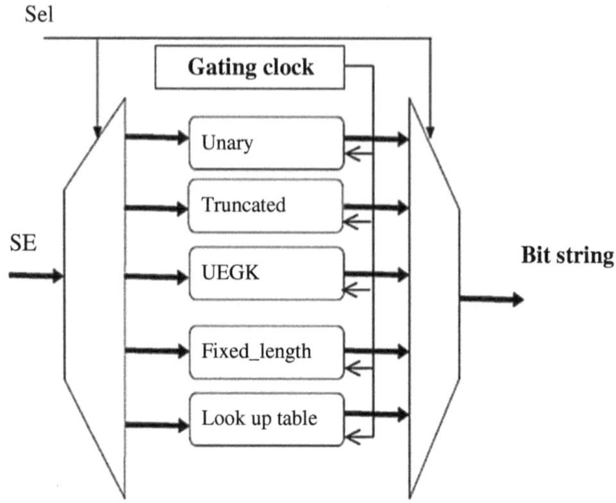

FIG. 12.7 Binarization architecture.

needed performances. We have used the AMBA bus for controlling the endoscopy system and assuring its communication with the outside world such as sensors and actuators. In fact, these modules can be assembled to an endorobot in order to perform several tasks such as stopping of the device during the exploration phase, tissue sampling, pH sensing, etc. Due to the scalability offered by I2C bus, it prevents data corruption and doesn't occupy a lot of silicon area. So we have used this bus for assuring endoscopy modules communication.

Based on architectural exploration as presented in Ref. [52], we have allocated the system modules that have the same affinities in terms of communication on the same bus (Fig. 12.8). The designed bridge makes it possible to adapt two resources of communication belonging to the same bus or to the two busses by guaranteeing a given bandwidth. To begin a transfer, the initiator submits a request and the address of the target module (e.g., channel coding module) with which it needs to communicate. To uncouple the synchrony from the AMBA/I2C busses, two buffer memories (FIFO) have been used (FIFO I and FIFO O). These FIFOs allow the management of the transfers between the two modules that have different speeds and support the bandwidths suitable for the AMBA/I2C busses (Fig. 12.9). A control unit and two multiplexers/demultiplexers are also used.

12.4 Experimental results

The endoscopic system design has been specified by using the VHDL language for the hardware part and with Matlab language for the software part. Simulation step is performed by using the Modelsim tool. Synthesis steps are proved

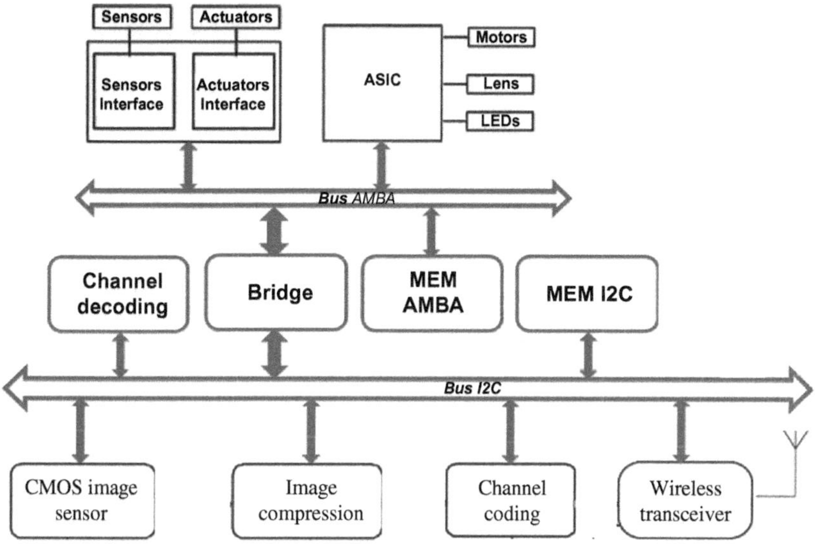

FIG. 12.8 Endoscopy communication architecture.

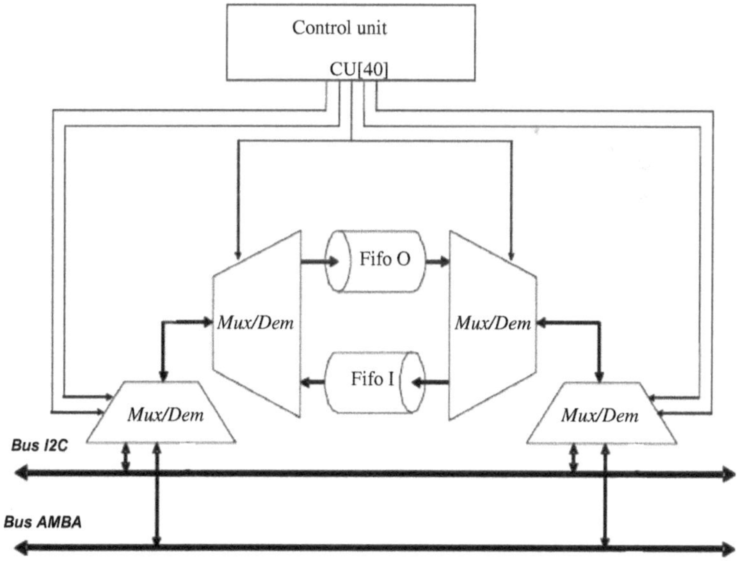

FIG. 12.9 Bus bridge architecture.

using Synplify and the ISE 13.1 tools targeting an implementation into a Virtex 5 Xilinx FPGA. Power consumption is estimated with XPower tool. Images qualities are evaluated by measuring the PSNR.

Concerning the hybrid compression 2D-DWT-DCT architecture and in order to study the effect of using the MCLA, we have decomposed the original image into 4 blocs LL, LH, HL, and HH and started by modeling the lifting equations. Then, we have inserted the proposed MCLA equations.

12.4.1 PSNR and CR results relative to standard algorithms

We have measured the PSNR and the compression ratio obtained by our architecture and the standard algorithms (DCT, DWT) in the case of 10 endoscopy images downloaded from the source [53]. Fig. 12.10 presents the compared PSNR (peak signal-to-noise ratio) values for the DCT, the DWT, and the proposed 2D-DWT/DCT algorithms. We notice that the PSNR values of the images 1, 5, and 9 are sharply increased. This can be explained by the fact that these images are bright in color, with fine details and textures. It is also noticed that the proposed hybrid algorithm is better than the DCT algorithm and the DWT algorithm and that the DWT algorithm has an average PSNR of about 32 dB. For this reason, diverse applications are performed for the variation of the CR at that PSNR value. The obtained drawing is shown in Fig. 12.11. We notice that the proposed hybrid algorithm has performed higher CR when it is compared by the DCT and the DWT standard algorithms at a fixed PNSR.

Fig. 12.12 shows the original and the reconstructed images for the image number 2. The PSNR values for DCT, DWT, and proposed 2D-DWT/DCT algorithms are 31, 29, and 36.5 dB, respectively. It is noticed that the proposed hybrid algorithm has better reconstruction quality. The DCT algorithm suffers from blocking and false contouring effects, and the image reconstructed based on DWT algorithm is unreadable and has least PSNR value.

12.4.2 PSNR and CR results relative to related works

Compared to the works from those that are based on DCT, DWT, DPCM, and JPEG methods [34,54–59], the proposed algorithm has viable overages CR and

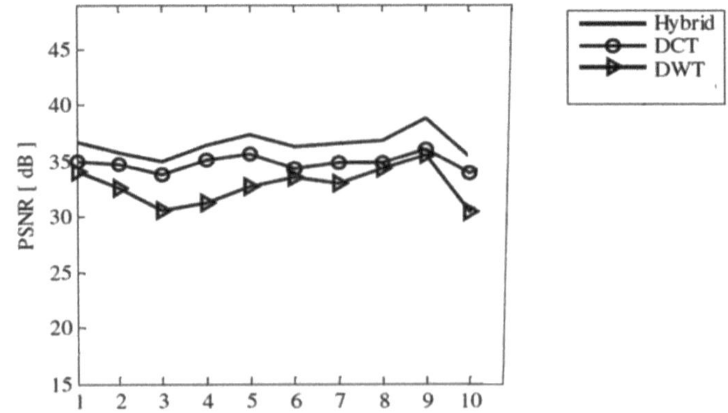

FIG. 12.10 PSNR for average compression ratio of 96%.

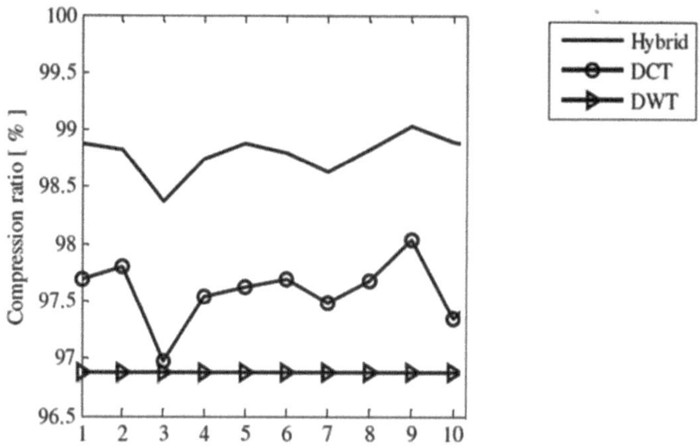

FIG. 12.11 Compression ratio for average PSNR of 32dB.

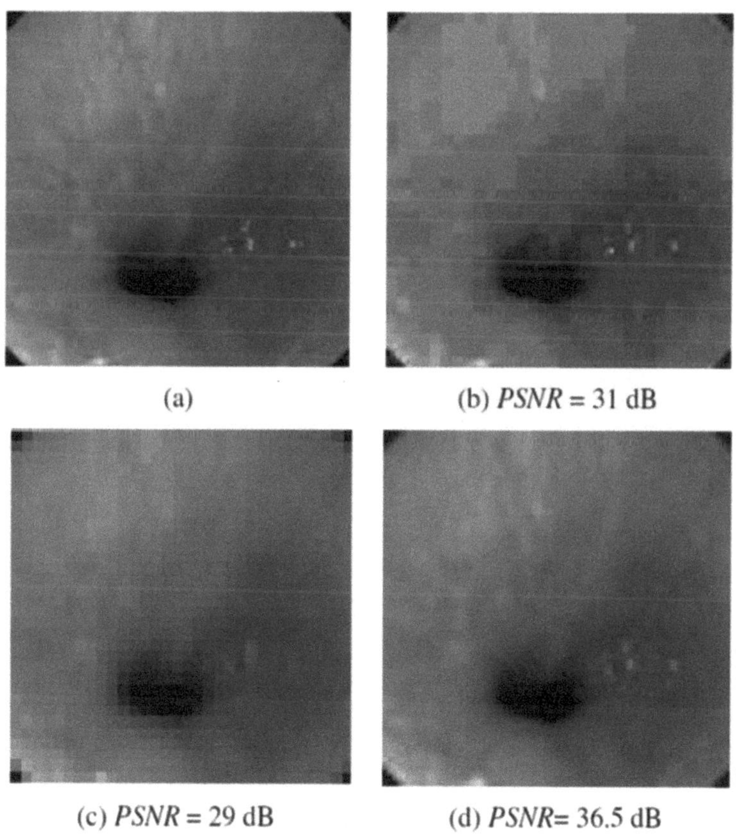

FIG. 12.12 (A) Original (image no. 2), reconstructed image using (B) DCT, (C) DWT, (D) Hybrid DWT-DCT.

TABLE 12.1 PSNR and CR comparison with other capsule prototype works.

	Compression algorithm	Average CR %	Overall PSNR (dB)
Chen et al. [56]	JPEG-LS	56.70	46.43
Wahid et al. [34]	DCT	87.13	32.95
Turcza and Duplaga [54]	DCT	91.20	34.7
Lin et al. [55]	DCT	79.65	32.51
Hu et al. [57]	DWT	71.98	39.64
Hasan-Khan and Wahid [58]	DPCM	80.35	45.51
Nithin-Varma and Surender-Kumar [59]	FT-DPCM	60.90	∞
Proposed	Hybrid 2D-DCT-DWT	96.00	37.50

PSNR (Table 12.1). It can be observed that the majority of these designs are based on the DCT algorithms. This may lead to a very expensive computation (logarithm operation) and need extra buffer memory to memorize the overall image rows. Furthermore, blocking artifacts can alter the reconstructed images due to the intrinsic nature of DCT-based algorithms. Hence, since the proposed architecture realizes a compromise between CR and PSNR, it is characterized by efficient results. In fact, we notice that only Nithin-Varma and Surender-Kumar [59], Hasan-Khan and Wahid [58], and Chen et al. [56] that use FT-DPCM (fixed threshold DPCM), DPCM, and JPEG-LS architectures, respectively, are characterized by PSNR that are greater than the PSNR of our architecture. Since the algorithm proposed in Nithin-Varma and Surender-Kumar [59] is near-lossless, the obtained PSNR is considered as infinity. The highest PSNR (34.7 dB) for the DCT-based work is obtained by Turcza and Duplaga [54]. The PSNR obtained by the work of the same author in Ref. [60] is about 34.7 dB. At the same time, these works are characterized by CR that are less than the CR of our architecture. But if we consider that previous research [61,62] for image compression stipules that the suitable PSNR for medical images would be equal or larger than 35 dB, it is noticed that, even for very larger CR of around 96%, the PSNR of our algorithm is greater than 35 dB. Concerning natural image that have in general sharp transitions and edges, the PSNR values of all architectures may decrease relatively to endoscopy or medical images.

12.4.3 FPGA synthesis results

The FPGA synthesis results of the proposed hybrid architecture are compared to two other typical FPGA-based architectures for endoscopy compressor [36,62].

TABLE 12.2 FPGA synthesis results comparison with other capsule prototype works.

	Khan [58]	Cavallotti [36]	Proposed
Number of LUT	957	744	635
Memory (RAM + Registers)	1610	1562	1447
Max. CLK frequency (MHz)	32.5	39	520
Power consumption relative to the proper frequency (mW)	12.42	10	150
Power consumption relative to 520-MHz frequency (mW)	198.72	133.33	150

These results (Table 12.2) show that the proposed architecture contains about 635 LUT and 1447 memory count (RAM + Registers). The 1562 memory count of Cavallotti [36] represents 25 RAM blocks of the used FPGA. The 1610 memory count of Hu et al. [57] represents some of the registers and RAM locations of the used FPGA.

This study shows that the proposed architecture integrates the smallest LUT count, which is 33.6% smaller than that of Hu et al. [57] and 14.6% smaller than that of Cavallotti [36]. These results also show that the proposed hybrid architecture contains the lowest memory count, which is 10.12% lower than that of Khan [62] and 7.36% lower than that of Cavallotti [36].

The proposed design operates with a 520-MHz running frequency. This frequency is 16 times higher than that of Khan [62] and 13 times higher than that of Cavallotti [36]. At this maximal executing frequency, the power consumption of our architecture is around 150 mw. This result shows that the proposed architecture consumes a power, which is lower than that of Khan [62] and only slightly higher than that of Cavallotti [36]. From raw four of the table, it can be thought that the architecture of Khan [62] consumes low power than the proposed architecture. But by converting the power dissipation of this architecture relatively to 520 MHz (greatest executing frequency of the proposed architecture) based on the equation: $\frac{\text{Power(P)} \times 520}{\text{Freq(P)}}$, where Power (P) and Freq (P) characterize the consumed power and the frequency of Khan [62] architecture, we notice that this architecture consumes about 198.72 mw as shown in line 5 of Table 12.2. In fact, as presented in the equation, power is proportional to the frequency. Thus, our architecture dissipates about 24.51% less power than Khan [62] architecture and only about 11.11% more than Cavallotti [36] architecture.

The optimized architecture of the proposed hybrid 2D DCT-DWT algorithm explains the reduced LUT and memory counts. Since the majority of the other architectures use the DCT techniques, we make a comparison between the

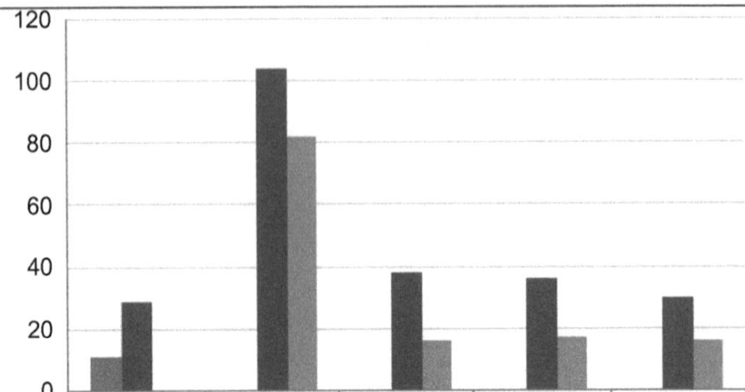

	Loeffler DCT	Cordic DCT	Cordic-based Loeffler DCT	Bin DCT	Proposed
Operation Multiply	11	0	0	0	0
Add	29	104	38	36	30
Shift	0	82	16	17	16

FIG. 12.13 Performances comparison of the hybrid 2D-DWT-DCT algorithm.

proposed technique and different versions of the typical DCT architectures. Fig. 12.13 recapitulates the number of processes for diverse DCT architectures. It can be noticed that our architecture needs only 30 adding and 16 shifting steps when it is compared to the traditional Cordic-based Loeffler DCT [16,17] that needs 38 adding operations, and 16 shifting operations. Also, it is characterized by a low calculation complexity relative to the binDCT algorithm that requires 36 adding operations and 17 shifting operations [63]. Thus, our architecture has higher performance and reduced hardware complexity relative to Cordic-Loeffler DCT and binDCT algorithms. So, it is more proper for a low-power and a high-quality-oriented Codecs, particularly for embedded systems.

12.5 Conclusions

This chapter describes a low-complexity/power-oriented architecture for wireless capsule endoscopy. The compression task is based on a hybrid 2D-DWT-DCT architecture that allows the application of the DWT and the DCT techniques sequentially. The MCLA adder is used to implement the arithmetic operations due to its efficiency than the traditional CSA techniques.

During the coding step of the system images, we have chosen the CABAC technique. In fact, this technique is more efficient in comparison with the other techniques such as CAVLC and CABAC since it achieves an average more important between 9% and 14% of bit rate saving CAVLC.

In order to emphasize the communication performances, we have developed a bridge AMBA/I2C bus-based architecture. In order to uncouple the synchrony from the two busses, two buffer memories (FIFO) have been used inside this architecture.

The proposed architecture has higher performance, reduced hardware complexity, and low power consumption than the previous techniques presented in related works. So it is more suitable for a low-power and high-quality-oriented codecs, particularly for the embedded systems.

References

[1] Chi B, Yao J, Han S. A 2.4GHz low power wireless transceiver analog front-end for endoscopy capsule system. IEEE Int Symp Circuits Syst 2006;4907–10.

[2] Iddan G, Meron G, Glukhovsky A. Wireless capsule endoscopy. Nature 2000;405:417–8.

[3] Glukhovsky A. Wireless capsule endoscopy. Sens Rev 2003;23:128–33.

[4] Wang L, Drysdale T, Cumming D. In situ characterization of two wireless transmission schemes for ingestible capsules. IEEE Trans Biomed Eng 2007;54:2020–7.

[5] Quinlan P. A multimode 0.3-200-kb/s transceiver for the 433/868/915-MHz bands in 0.25-/spl mu/m CMOS. IEEE J Solid State Circuits 2004;39:2297–310.

[6] Fakher A, Deen M, Bruin H. Low-voltage, low-power and low phase noise 2.4GHz VCO for medical wireless telemetry. In: Proceedings of Canadian conference on electrical and computer engineering; 2004. p. 1321–4.

[7] Pillcam. Given imaging. Available at: http://www.givenimaging.com/en-us; 2011.

[8] Moglia A, Menciassi A, Dario P. Recent patents on wireless capsule endoscopy. Recent Patents Biomed Eng 2008;1:24–33.

[9] McCaffrey C, Chevalerias O, Mathuna C, et al. Swallowable-capsule technology. IEEE Pervasive Comput 2008;7:23–9.

[10] Loeffler C, Lightenberg A, Moschytz G. Practical fast 1-D DCT algorithms with 11-multiplications. In: Proceedings of ICASSP, vol. 3; 1989. p. 988–91.

[11] Li J, Lu S. Low power design of two-dimensional DCT. In: *Proceedings of IEEE* conference *on ASIC and exhibit*; 1996. p. 309–12.

[12] Park J, Kwon S, Roy K. Low power reconfigurable DCT design based on sharing multiplication. In: *Proceedings of IEEE* conference *on acoustics, speech, and signal processing*, vol. 3; 2002. p. 3116–9.

[13] Shams A, Pan W, Chidanandan A, et al. A low power high performance distributed DCT architecture. In: Proceedings of IEEE computer society annual symposium on VLSI; 2002. p. 21–7.

[14] Amer I, Badawy W, Jullien G. High performance hardware implementation of the H.264 simplified transform and quantization. In: *Proceedings of international conference on acoustics*, speech and signal, vol. 2; 2005. p. 1137–40.

[15] Park J, Ogunfunmi T. A new hardware implementation of the H.264 8x8 transform and quantization. In: Proceedings of international conference on ICASSP; 2009. p. 585–8.

[16] Prithivi M, Kumar G, et al. VLSI implementation and area efficient cordic based integer DCT architectures for HEVC. Int J Innov Res Sci Eng Technol 2015;4:3210–5.

[17] Sun C, Ruan J, Heyne B, et al. Low-power and high-quality Cordic-based Loeffler DCT for signal processing. IET Circuits Devices Syst 2006;1:453–61.

[18] Jarray N, Elhaji M, Zitouni A. Low complexity and efficient architecture of 1D-DCT based Cordic-Loeffler for wireless endoscopy capsule. In: Proceedings of 12th international multi-conference on systems, signals & devices SSD; 2015. p. 1–5.

[19] Aziz S, Pham R. Efficient parallel architecture for multi-level forward discrete wavelet transform processors. J Comput Electr Eng 2012;38:1325–35.

[20] Daubechies I, Sweldens W. Factoring wavelet transform into lifting steps. J Fourier Anal Appl 1998;4:245–67.

[21] Shi G, Liu W, Zhang L. An efficient folded architecture for lifting based discrete wavelet transform. IEEE Trans Circuits Syst II Express Briefs 2009;56:290–4.
[22] Wu B, Lin C. A high-performance and memory-efficient pipeline architecture for the 5/3 and 9/7 discrete wavelet transform of JPEG2000 codec. J IEEE Trans Circuits Syst Video Technol 2005;15:1615–28.
[23] Lai Y, Chen L, Shih Y. A high-performance and memory-efficient VLSI architecture with parallel scanning method for 2-D lifting-based discrete wavelet transform. J IEEE Trans Consum Electron 2009;55:400–7.
[24] Huang C, Tseng P, Chen L. Flipping structure: an efficient VLSI architecture for lifting-based discrete wavelet transform. J IEEE Trans Signal Process 2004;52:1080–9.
[25] Zhang W, Jiang Z, Gao Z, et al. An efficient VLSI architecture for lifting-based discrete wavelet transform. IEEE Trans Circuits Syst II Express Briefs 2012;59:158–62.
[26] Ziv J, Lempel A. A universal algorithm for sequential data compression. J IEEE Trans Inform Theory 1977;23:337–43.
[27] Tajallipour R, Wahid K. Efficient data encoder for low-power capsule endoscopy application. In: *Proceedings of international conference on information sciences signal processing*; 2010. p. 512–5.
[28] Turgis D, Puers R. Image compression in video radio transmission for capsule endoscopy. Sensors Actuators A Phys 2005;123:129–36.
[29] Rigler S, Bishop W, Kennings A. FPGA-based lossless data compression using Huffman and LZ77 algorithms. In: *Proc. IEEE* Canadian conference of electrical and computer engineering; 2007. p. 1235–8.
[30] Chen J, Lin Y. A high-performance hardwired CABAC decoder for ultra-high resolution video. J IEEE Trans Consum Electron 2009;55:1614–22.
[31] Bjontegaard G, Lillevold K. Context-adaptive VLC (CAVLC) coding of coefficients. In: *Proc. 3rd meeting JVT-C028, Fairfax, VA*; 2002. p. 6–10.
[32] Al-Rawhani M, Chitnis D, Beeley J, et al. Design and implementation of a wireless capsule suitable for auto-fluorescence intensity detection in biological tissues. J IEEE Trans Biomed Eng 2013;60:55–61.
[33] Hasan-Khan T, Wahid K. Low power and low complexity compressor for video capsule endoscopy. J IEEE Trans Circuits Syst Video Technol 2011;21:1534–46.
[34] Wahid K, Bum-Ko S, Teng-Andre D. Efficient hardware implementation of an image compressor for wireless capsule endoscopy applications. In: Proceedings of the international joint conference on neural networks (IJCNN); 2008. p. 2761–5.
[35] Zargari F, Azimi E. An efficient hardware accelerator for M-coder of H.264 video coding standard. Eur J Sci Res 2009;35:615–22.
[36] Cavallotti C, Merlino P, Vatteroni M, et al. An FPGA-based versatile development system for endoscopic capsule design optimization. Sensors Actuators A Phys 2011;172:301–7.
[37] Xie X, Li G, Chen X, et al. A low-power digital IC design inside the wireless endoscopic capsule. IEEE J Solid State Circuits 2011;41:2390–400.
[38] Gu Y, Xie X, Wang Z. The design and implementation of a chipset for the endoscopic microball. In: Proceedings of international conference on ISCAS; 2012. p. 2633–6.
[39] Bayer B. Color imaging array. *U.S. Patent*, 3 971 065; 1976.
[40] Mallat S. A theory for multiresolution signal decomposition: the wavelet representation. IEEE Trans Pattern Anal Mach Intell 1989;11:674–93.
[41] Shrestha S. Hybrid DWT-DCT algorithm for image and video compression applications [PhD thesis]. Saskatchewan, Canada: Department of Electrical and Computer Engineering, University of Saskatchewan Saskatoon; 2010.

[42] Pennebaker MW. JPEG still image data compression standard. 3rd ed. New York: Springer; 1993.

[43] Thomos N, Boulgouris N, Strintzis M. Optimized transmission of JPEG2000 streams over wireless channels. IEEE Trans Image Process 2006;15:54–67.

[44] Parhi K, Nishitani T. VLSI architecture for discrete wavelet transforms. J IEEE Trans Very Large Scale Integration (VLSI) Syst 1993;1:191–202.

[45] Wu P, Chen L. An efficient architecture for two-dimensional discrete wavelet transform. J IEEE Trans Circuits Syst Video Technol 2001;11:536–45.

[46] Sweldens W. The new philosophy in biorthogonal wavelet constructions. Proc SPIE 1995;25:68–79.

[47] Darji A, Chandorkar A, Merchant S. Memory efficient and low power VLSI architecture for 2-D lifting based DWT with dual data scan technique. In: Proc. recent researches in circuits, systems and signal; 2012. p. 62–7.

[48] Babu P, Prasad K. Image interpolation using 5/3 lifting scheme approach. Signal Image Process 2011;2:71–80.

[49] Cheng F, Unger S, Theobald M, et al. Delay-insensitive carry-lookahead adders. In: Proc. VLSI design; 1997. p. 322–8.

[50] Nagendra C, Irwin M, Owens R. Area-time power tradeoffs in parallel adders. J IEEE Trans Circuits Syst II 1997;43:689–702.

[51] Marpe D, Schwarz H, Wiegand T. Context-based adaptive binary arithmetic coding in the H.264/AVC video compression standard. Proc IEEE Trans Circuit Syst Video Tech 2003;620–36.

[52] Poujet J. Test of system on chip: ordering and architectural solutions space exploration [PhD thesis]. France: University of Montpellier II: Sciences and Techniques of Languedoc; 2002.

[53] Anon. Gastro-lab-the gastrointestinal site. Available at: www.gastrolab.net; 2010.

[54] Turcza P, Duplaga M. Hardware-efficient low-power image processing system for wireless capsule endoscopy. IEEE J Biomed Health Inf 2013;17:1046–56.

[55] Lin M, Dung L, Weng P. An ultra-low-power image compressor for capsule endoscope. J Biomed Eng Online 2006;5:1–8.

[56] Chen X, Zhang X, Zhang L, et al. A wireless capsule endoscope system with low power controlling and processing ASIC. J IEEE Trans Biomed Circuit Syst 2009;3:11–22.

[57] Hu C, Meng M, Liu L, et al. Image representation and compression for capsule endoscope robot. In: Proc. IEEE international conference on information and automation; 2009. p. 506–11.

[58] Hasan-Khan T, Wahid A. Subsample-based image compression for capsule endoscopy. J Real-Time Image Proc 2013;8:5–19.

[59] Nithin-Varma M, Surender-Kumar S. A near lossless and low complexity image compression algorithm based on fixed threshold DPCM for capsule endoscopy. Multimed Tools Appl 2020;79:8145–60.

[60] Turcza P, Duplaga M. Energy-efficient image compression algorithm for high-frame rate multi-view wireless capsule endoscopy. J Real-Time Image Proc 2019;16:1425–37.

[61] Garawi S, Istepanian H, Abu-Rgheff A. 3 G wireless communications for mobile robotic tele-ultrasonography systems. IEEE Commun Mag 2006;44:91–6.

[62] Hasan-Khan T, Shrestha R, Khan A, et al. Modular and programmable development platform for capsule endoscopy system. J Med Syst 2014;38:57.

[63] Tran D. The BinDCT: fast multiplierless approximation of the DCT. IEEE Signal Process Lett 2000;7:141–4.

Part 3

Ethics, regulation, and project management

Chapter 13

Regulating endorobots in the European Union: An overview of the ethical and legal framework

Federico Costantini[a] and Fabio Balducci Romano[b]
[a]*University of Udine, Udine, Italy,* [b]*University of Rome 'Tor Vergata', Roma, Italy*

13.1 Introduction

It can be said that technology, law, and medicine have been deeply connected throughout the history of humankind. Indeed, in ancient times, drugs were administered and surgery was performed during rituals in which the devices used had little or no importance compared to that of spells and gestures, and no liability could be linked to healers—witches or shamans of any sort—who performed them according to some kind of belief. With civilization, surgeons obtained the right to be rewarded for their services, and conversely, they began to be exposed to liability for their mistakes. For example, in the Babylonian Code of Hammurabi (1972–50 BC)—not the first but the most renowned ancient collection of laws—the physician could claim a remuneration whose value depended on the social status of the patient—free man, servant, or slave—and on the difficulty of the intervention, which was qualified as surgery by the use of a "bronze lancet" [1]. In a case of malpractice, the punishment could be quite harsh: the surgeon not only had to pay a sum of money but also had a hand cut-off (paragraph 218, Code of Hammurabi). It could be said that medicine has always had its risks, but today the concern at least does not affect the physical integrity of surgeons.

In recent years, technology, medicine, and law have become more deeply intertwined than before, since healthcare is currently a thriving industry where crucial interests require accurate amalgamation. The individual patient expects to be cured with the best of medical knowledge and fair treatment. The public sector requires the assessment of the validity of the methods and the security of the devices employed to maintain fair competition among players and also

Endorobotics. https://doi.org/10.1016/B978-0-12-821750-4.00013-X

containment of the social costs. Manufacturers are concerned about recovering costs from the huge investments made in research, development, testing, and marketing their products. Professionals—surgeons, of course, but also paramedics and technicians—are worried about the expenses required to acquire the skills and competencies needed to use medical devices and the danger of being sued for malpractice. There are also many other players indirectly involved: above all, insurance companies, exposed to risks that are difficult to foresee and to assess. There are no surgeons' hands at stake, as in ancient Babylon, but healthcare has never stopped being a tough business. Thus, the need for a legal framework that—due to such numerous conflicting interests—cannot be other than complicated.

Compared to the millennial evolution of medicine, the deployment of information technologies in this field is very recent, within the past 50 years, yet an exponential progress has been made, and incredible results have been achieved through a pervasive automation of tasks and procedures. Besides the booming adoption of artificial intelligence techniques in healthcare, we have been witnessing the implementation of robotics in many different fields: diagnosis, surgery, assistance, personal care, and prosthetics. The progress in endoscopic robotics can be observed, from the pioneering deployments of industrial robots in surgery to the market uptake of sophisticated systems suitable for performing many tasks and even complex procedures [2–8]. Of course, nobody can foresee the future, and it is difficult to make previsions in uncertain times such as those we are living in. Yet, we can argue that robotic technologies in healthcare are ready to enter what has been called the "Slope of Enlightenment" [9].

Healthcare is considered one of the most promising frontiers of robotics, which in the past has usually been limited to fields where tasks to be performed were qualified by the three "Ds" (dull, dirty, and dangerous) [10]. In this regard, it is remarkable that in the standard ISO 8373:2012, which provides a technical definition of "robot"—namely, an "actuated mechanism programmable in two or more axes (4.3) with a degree of autonomy (2.2), moving within its environment, to perform intended tasks" (2.6), the devices used in hospitals are specifically named as examples for the category of the "professional service robot," which includes "service robot for professional use" and "service robot (2.10) used for a commercial task, usually operated by a properly trained operator (2.17)." This observation is relevant for two reasons. First, it clarifies a conceptual difference between the robots deployed in hospitals and those that are used in production plants. Considering the fact that PUMA560, the first robot to be used in a surgery room, was indeed an industrial robot, this is an important step forward, and it implies not only that the technical standards have to be different but also that the ecosystem should be specific [11]. Thus, the design should comply with peculiar requirements. Second, it states that robots cannot be left working alone, but they require the presence of a human who has the potential and ability to intervene. In this regard, it is useful to emphasize that healthcare robots fall into the definition of "collaborative robots" "designed for direct interaction with a human" (2.26). In this sense, endorobots are quite peculiar since, to be precise, they interact with humans in a twofold manner: with the operator, of course, and with the patient whose body the device is moving in. The combination of these two factors could be considered the empirical basis underpinning the entire ethical and legal framework.

Since in healthcare robotic technologies have reached the level of advanced automation, yet full autonomy is still far from being achieved, the interaction with the human operator is the most critical issue. In this regard, adopting a classification based on six levels of automation has been proposed—from "level 0" (*no automation*) to "level 5" (*full autonomy*)—from the SAE J3016 taxonomy [12] applied to autonomous vehicles [13]. The fact is that such categorization is currently disputed even in its original field, where experts prefer a more complex conceptual framework [14]. To be challenged, in particular, is level 3, "conditional driving automation," whereby "automation performs the complete dynamic driving task (DDT), but not DDT fallback, within a limited operational design domain (ODD)." In a scenario in which a precise and quick reaction is most needed—an obstacle on the road, a jaywalking pedestrian, an oncoming car, it is not clear how and when exactly the transition in control-authority could pass from the human driver to the artificial pilot, or vice versa [15]. In robotic surgery, for example, the human surgeon could be alerted by the robot with a "request to intervene" when it is already too late, as the surgeon is overwhelmed or confused by the tasks and decisions to make in a very short time. The result would be that the surgeon could be left not with a real chance of saving the patient, but just with the ultimate responsibility for the failure of a hopeless attempt. In this scenario, increasing the automation in devices would mean the unacceptable rise in probability that the human surgeons will be found liable.

It can be argued that the combination of tasks and procedures performed jointly by humans and robots are quite complicated and hardly fit into just six levels of automation because many different concomitant tasks are performed by the operator and the machine—some jointly, others separately. Hence here, as generally in the field of "cobots" (collaborative robots), the concept of control-authority is not binary—based on a "yes" or "no" answer—but depends on the levels of complexity of the human-robot interaction. In short, human-machine interaction has many nuances and intermediate degrees, depending on the function that is left to the machine, as confirmed by the many conceptual frameworks proposed [16,17]. Such technological complexity inevitably is mirrored by the attempts made by ethicists and laymen to draw moral guidance and outline a legal framework in the field of robotics. Such debates, of course, have repercussions as well for the development of endorobots.

This chapter provides a primer on the ethical and legal issues emerging from the deployment of endorobots, focusing on the design of such devices and limiting its scope to the legal system of the European Union. Therefore, here we cannot tackle other topics that could be relevant as well, such as the bioethical concerns expressed by the "Oviedo Convention" of 1997, or the legal liability of the human operator, since they rely mostly on international law—such as the Council of Europe convention already mentioned—on the one hand, and on legislation and jurisprudence at a national level, on the other. In our explanation, sometimes we use robotic surgery as a paramount example of the technological advancements in this field, but our arguments can be extended to any other kinds of tools used for other purposes, such as diagnostics or treatment.

In the first part, we offer an overview of the ethical issues. After briefly explaining the different theoretical perspectives on innovation and qualifying endorobots as "social robots," namely, human artifacts that embody the power of technology and change the fabric of social interactions, we provide an overview of the ethical theories that are relevant in healthcare robotics and then focus on "engineering ethics". Finally, we offer a synthesis of ethical principles that can be taken into consideration in designing and manufacturing endorobots.

In the second part, we outline the legal framework in the European Union concerning endorobots, which, as previously anticipated, is aimed at fostering market growth, innovation, and, of course, health protection, thus coordinating product safety and product liability. In other words, the EU establishes a two-level regulation providing public control on the marketing of devices, permitting claims for damages received, and setting common standards in healthcare. In each part, we provide a short overview on two documents that pave the way for deep changes in the legal system, namely, the Resolution adopted by the EU Parliament on the October 20, 2020, which includes a proposal for Regulation on liability for the operation of Artificial Intelligence-systems [18] and the proposal of EU Regulation "Artificial Intelligence Act" presented on April 21, 2021 [19].

We conclude by making some final and comprehensive remarks we deem to be useful in the ever- and fast-changing scenario of endoscopic robotics.

13.2 Ethics and technology: Endorobots as "social robots"

Contemporary debate on ethics and technology is very lively. In order to provide a simple introduction, two opposite positions can be isolated. On one side, there is an overall approach that can be denoted as "technological solutionism" [20]. According to this view, every human concern—natural, social, or personal—could be tackled with an appropriate technology. In other words, there is no problem that cannot be fixed by a tool, and conversely, there is a kind of equipment for any difficulty. Contrary to this vision stands a consolidated philosophical tradition that challenges such claims of omnipotence, revealing its philosophical roots in the rise of modern rationalism [21] and its purpose in filling the existential void of contemporary nihilism [22]. Below the surface of the current celebration of the extraordinary achievements of technological innovation—and of the biases that afflict public debate—flows a consistent discussion challenging the purpose of the race for progress and the hidden prejudice that innovation could—or should—be pursued at any cost [23].

Before addressing the topic of this section, an important clarification has to be made regarding the social impact of technological devices. Since the first pioneering studies by cyberneticians [24], it has been agreed that the advent of advanced automation would deeply change human communities worldwide and it would conflict with preexisting cultures, thus requiring a huge effort in order to achieve a shared ethical understanding. Robots were particularly addressed with concern, since they embodied—as they actually still do—both ideals and nightmares of technology. In this regard, it is remarkable how the debate of the first half of the last century is reflected in science fiction literature, and particularly in the works of Isaac Asimov, whose formulation of the famous "Laws of Robotics" is widely used even today as a starting point for the discussion in the field.

In the past 50 years, scholars have developed many theories that allow us to consider human artifacts not only as material objects but also as media of social interaction, similar to Information Communication Technologies (ICTs) in the digital realm. In this sense, "service robots"—as previously defined—have been qualified as "social robots" for their ability to shape the human ecosystem in which they operate [25]. According to these theories, due to the adoption of "social robots," we, as individuals, reframe the meaning attributed to places and circumstances, modifying our behavior in front of other people and, more importantly, shaping our personal beliefs.

In general, to consider healthcare robots as "social robots" is important in order to set the current debate in a wider theoretical perspective, based on a deeper understanding—technological, economic, social, and psychological—of the different levels of complexity in the relationship between technological innovation and ethics: how we personally, in our inner feelings, perceive hospitals, surgical rooms, and medical clinics with an advanced level of automation, how we, as professionals, patients, or even indirectly involved as next-of-kin, act toward others and react to other conduct, what we expect from society in terms of clinical results, medical care, and even personal assistance.

13.2.1 Roboethics: Between "machine ethics" and "engineering ethics"

Despite the fact that, as said before, ethical issues regarding autonomous systems have been addressed since the first studies in cybernetics, and the application of a moral philosophy in the field of computer and artificial intelligence has been attempted since 50 years ago [26], the coinage of the term "roboethics" [27] and the growth of a specific discipline is quite recent, dating from the beginning of the new millennium [28,29]. According to a useful simplification, the scope of "roboethics" includes three different kinds of issues: safety for single users and the surrounding workplace; liability for any kind of damage, including personal data breaches; and social consequences, especially the impact on the labor market [10]. In this regard, we can detect two very different approaches: one, called "machine ethics," investigates whether, how, and under whose guidance an artificial agent may be able to carry out a moral decision [30]; and the other, named "engineering ethics," questions the set of assessments that should be required in designing and implementing technological artifacts [31]. It is widely admitted that the term "ethics" is improperly used in addressing robots, and we can agree that here, it is invested with a different meaning in each case.

In the first case, ethics refers to a set of rules addressed directly to the artificial agents, mimicking human moral values. This kind of issue pertains to artificial intelligence—or, better, "autonomous and intelligent systems" [32]—where the tremendous progress of Machine Learning techniques is challenging the current understanding of the difference between artificial agents and natural beings. Indeed, such agents are able to change their behavior depending on the stimuli received, thus becoming unpredictable and exposing their environment to negative consequences. The fact that there are moral values involved does not necessarily mean that there is a choice of sorts—a free will, philosophically sound—behind the conduct of the artificial agent, but it just qualifies the performed action as compliant with a given moral model. On this topic, the debate has been ongoing since the beginning of the discipline of artificial intelligence [33] involving also the moral assessment of human decisions toward machines [34,35].

"Engineering ethics," instead, refers to principles set for a specific category of humans, those in charge of the design and implementation of technological artifacts, as an arrangement of general values in a particular field or as an enactment of special duties that are suitable for carrying out only in a specific sector [36]. In this regard, there are three different approaches that are relevant in healthcare and remarkably suitable for adoption in the case of endorobots: the first is a broader perspective envisioned by the "ethics of information", the second is given by the "value-sensitive design", and the third refers to "responsible research innovation".

13.2.2 Three perspectives in "engineering ethics"

In a wider perspective, the "ethics of information" is an attempt to provide a comprehensive "metaethical" framework suitable for offering guidance to both human and artificial agents [37]. Consequently, it challenges the current approach to innovation, which is focused on the mere "empowerment" of the human operator. The key concern in such a view is that underlying the celebration of power achieved by the individual vested with control of devices is either the claim of unlimited freedom in the exploitation of human possibilities, or an alarming paternalism that can be exercised by the owner of control's authority, whether it is a single doctor, a specific institution or company, or the healthcare system in general [38]. It is evident that there are risks in both perspectives: in the first, in the attempt to exploit her/his freedom, the individual could end up harming herself/himself, other individuals, or the whole society. In the second, the patient could become the victim of an abusive exercise of technological power, such as unethical experimentation, which could result in a form of torture or other violations of human dignity. On the contrary, framing the idea of "companionship" between humans and artificial agents, it is possible to directly link technological devices to the essential scope of medical practice, which is the cure of the patient, and to include technology as a third beneficial component in the relationship with doctors and with the entire healthcare system at the national and international levels [39,40]. This second perspective is suitable for application to endorobots, since there are many joint activities performed together by humans and machines, and the workspace is also located inside the body of the patient, so the use of technology is particularly invasive.

"Value-sensitive design" (VSD) has remarkably grown in recent years. This discipline is aimed at configuring processes and interfaces in order to move the use of technological artifacts toward the respect of certain principles, which result as integrated into the operating domain of the devices [41–43]. It adopts a tripartite methodology concerning the analysis of conceptual aspects—the detection of stakeholders and their conflicting values—of empirical findings—the observation of social contexts in which artifacts are deployed—and of the technical requirements—the effective design required for respecting the values set as binding. One famous example of this can be observed in Article 25 of the Regulation (UE) 2016/679 (henceforth GDPR) [44], which implements "privacy by design" and "privacy by default" principles in the field of personal data protection. Such a perspective has been applied in the healthcare field in the "care-centered value-sensitive design" (CCVSD) that provides principles, criteria, methodologies, and tools for developing ethically grounded healthcare robots and evaluating the impact of their adoption [45]. It is remarkable that this approach—together with its further implementations [46]—offers an analytical model aimed at providing substantial grounds for the ethical framework to be incorporated into the devices. In addition, as in the previous perspective, the essential allegiance between patient and doctor—in general, the robot operator—and, in particular, its beneficial purpose shapes the adoption of the artifacts as an activity that should not be other than favorable.

A deeper embedding of values into technological innovation is proposed by "responsible research and innovation" (RRI) [47–50]. Indeed, according to this perspective, ethical concerns are not only focused on the design of specific devices, but they are also included as building elements throughout the many phases of the innovative process. Consequently, this kind of methodology is aimed at offering direct support and constant cooperation to innovators in the attempt to balance two opposing needs. On the one hand, without any previous piloting guidance, the impact of design alone can hardly be effective, and it is very limited by such a case-by-case approach. On the other hand, the intrinsic risk in VSD is that it could result in a mere imposition of predefined assumptions, regardless of the peculiarity of a specific context. It is important to add that this approach has been adopted by the European Union in fostering innovation since the Horizon2020 program.

Focusing on endorobotics, we observe that the issues concerning "engineering ethics" can be divided into the two main groups, depending on whether they pertain generally to human operators of the devices or specifically to their designers.

In the first case, given the continuous and fast technological progress, the need to standardize training programs emerges to validate achieved competencies and to monitor performances in human operators [51–53]. In endorobotics especially, but not exclusively in surgery, the issue is currently tackled only by coaching sessions or simulation environments provided by the manufacturers or by the sharing of best practices among the communities of users. Instead, in order to gain public trust at an international level regarding these new tools and procedures, it is crucial to build an institutional framework that is suitable for assessing even nontechnical skills [54]. This is a prerequisite for nurturing the birth of a new interdisciplinary professional figure in healthcare and thus for the creation of a set of ethical duties and, eventually, the establishment of specific criteria for legal liability.

The same concerns can be expressed as regards the designers of such devices, although there are some further remarks to be made in light of recent discussions [55–58]. Since the design of technological devices is deemed as crucial in order to guarantee the respect of ethical principles in their use, as CCVSD and RRI are proposing, the most convenient solution seems to include them as requirements for the marked introduction, adopting the certification model as a tool for the regulation of the market of industrial products, as well known worldwide in many other sectors (e.g., energy consumption of electric devices), possibly in combination with a "code of ethical conduct" for robot engineers, as suggested for the first time in a Resolution regarding the civil regulations on "Smart Robots" of the European Parliament [59].

Both perspectives have been considered in the governance model that is included in the proposal of Regulation approved by the EU Parliament on the 21st of April 2021 [17,19]. According to the draft under discussion, every "AI system"—any artificial agent which falls under the definition provided by Article 3(1) and detailed in the Annex I—will be classified as "prohibited practice" (Article 5) or "high-risk" system (Article 6) depending on the risks posed. In this latter qualification, the artificial agent will be required to abide by a set of strict specifications including the adoption of a risk management system (Article 9), a policy to ensure data quality (Article 10), the release of detailed technical documentation (Article 11), the automatic recording of events (Article 12), a high level of transparency (Article 13), human oversight (Article 14) and security measures (Article 15). Furthermore, providers, manufacturers, importers, distributors, users and even third parties will be required to comply with many obligations, among which it is noteworthy the cooperation with competent authorities (Article 23) and the submission for conformity assessment procedure in order to obtain a declaration of conformity (Article 48) and a CE marking of conformity (Article 49), which will permit the distribution of their product or service in the EU digital market. In conclusion, the European Union is preparing to build a comprehensive framework in which ethical design would become a key component of technological innovation, being embedded both into the innovation process and into the political governance.

13.2.3 Final remarks on "roboethics"

Concluding this part, we can observe that "roboethics" represents a new discipline on which the debate is articulated on several different levels of complexity (technological and philosophical, academical and institutional, policy and business). It is remarkable that, while the discussion is still ongoing, the European Union is preparing to adopt a piece of legislation, which, if enacted, would establish in the sector a governance model which has proven, until today, to be a flexible and adaptable balance between market demands and social expectations. Yet this solution, which as of today seems the most sustainable, pairs with significant drawbacks since certification could determine, for instance, increased cost and delays for innovators, developers, and designers. It is quite unlikely that they, which are the direct recipients of such provisions, would have a real chance to spend the last word.

The adoption of a certification model in this context raises potentially a suggestive paradox. Indeed, the fact that moral philosopher's conundrums could be delegated to an impartial institution and then processed as a matter of industrial standards could even dispel the nightmares of cyberneticians, yet it could summon other, and more powerful, demons. Provided that, in general, the formalization of a problem into a procedure brings it nearer to an automated solution, therefore opening the possibility to be managed by an autonomous agent, it is likely that also the process for obtaining the ethical certification would be highly automated, thus requiring for itself to be assessed. Two opposite options here seem to occur: to allow the agency to certify itself or to empower of such duty another institution, albeit having the same nature and structure. In the first case, the effect is to create a self-referential power; in the second case, to appeal to an infinite regress. Only the future can tell how the European Union could avoid falling into this difficult dilemma.

13.3 Legal framework of endorobots in the European Union

Endorobots and, in general, surgical robots are currently regulated by EU legislation on medical devices [60]. As is known, the "medical device" notion, provided for by the new Medical Devices Regulation 2017/745 (MDR), includes "any instrument, apparatus, appliance, software, implant, reagent, material or other article intended by the manufacturer to be used, alone or in combination, for human beings for one or more of the following specific medical purposes: diagnosis, prevention, monitoring, prediction, prognosis, treatment or alleviation of disease, diagnosis, monitoring, treatment, alleviation of, or compensation for, an injury or disability, investigation, replacement or modification of the anatomy or of a physiological or pathological process or state, providing information by means of *in vitro* examination of specimens derived from the human body, including organ, blood and tissue donations, and which does not achieve its principal intended action by pharmacological, immunological or metabolic means, in or on the human body, but which may be assisted in its function by such means" [61].

Such a notion may be referred to robots operating under human control, i.e., tele-operated devices. Robots have been used for years in various surgical treatments: for example, the da Vinci system, which is remotely operated by surgeons, is classified as a medical device [62]. The same definition, however, might also apply to the more recent "smart" robot, i.e., devices that may perform surgery with a certain degree of autonomy.

As some authors observed, there are manifold definitions of "robot" in legal literature [63]. It is to be noted, though, that currently, no such definition is mentioned in EU law. In fact, the European legal system does not provide any general discipline either for surgical robots or for artificial intelligence, despite the numerous recent initiatives in this area [64]. However, it is possible to identify some legal principles specifically applicable to the operation of these cutting-edge technological devices [63].

13.3.1 EU regulation on medical devices

The traditional goal of European legislation in the medical sector is to implement the internal market, establishing common rules for the marketing and servicing of medical devices, thus allowing enterprises and consumers to rely on the right to free circulation of goods. In fact, the legal basis for such legislation is provided for by Art. 114 TFEU, formerly Art. 95 TEC, on the approximation of laws [65]. The right to free circulation of goods has to be balanced with the requirement of a high level of protection of health for patients and users, which is now protected as a fundamental right of the EU under Art. 35 of the CFREU [66]. The protection of such a right involves the establishment of high standards of quality and safety for medical devices. Moreover, an additional legal basis for the MDR is provided for by Art. 168(4)(c) TFEU that includes the power of the EU to set high standards of quality and safety for medical devices. Some authors offered an interesting comparative perspective of the different objectives of EU and American legislation [67,68].

Compliance with technical and safety standards is certified by the CE mark (*Conformité Européenne*). EU legislation classifies medical devices according to risk, from Class I to Class III, introducing a progressive system of controls, depending on the potential risk inherent in the various devices [69]. Currently, da Vinci surgical robots belong to Class IIb [70].

The procedure for the assessment of conformity of a medical device varies according to its classification. For Class I medical devices, manufacturers must notify their domicile and give a description of the device concerned to the Competent Authority of the Member State where they have their registered place of business. Manufacturers established in non-European countries must appoint an authorized representative based in the Union. For Classes IIa, IIb, and III medical devices, manufacturers must also involve an independent notified body that must issue a certificate stating the conformity; in these cases, the CE mark is followed by the identifying number of the notified body (NB). These NBs have to meet the same high-quality standards throughout the EU and must have qualified staff to perform conformity assessments and carry on inspections. Assessments of certain high risk devices (e.g., implants) might also involve EU-level panels of independent experts.

Thus, for example, a manufacturer of a Class I medical device (e.g., sterile gloves or body liquid collectors), or his relevant authorized representative in the European Union designated by him, must notify their domicile and give a description of the devices concerned to the Competent Authority of the Member State where they have their registered place of business. For medical devices of classes IIa, IIb, and III, Member States may, in addition, request to be informed of all data allowing for the identification of such devices, together with the label and the instructions for use, when such devices are put into service within their territory.

As observed [71], the EU regulatory scheme for medical devices does not contemplate a single regulatory body with nationwide scope similar to the US Food and Drug Administration (FDA), but rather a multifaceted system with roles for the European Commission, the private standards organizations, the Member States' competent authorities, and the private or public NBs, designated to carry out conformity assessment functions. Thus, approval decisions (made by NBs) and enforcement (implemented by Member States with or without assistance from the NBs) are highly decentralized [71].

The new MDR [61] will replace the Medical Devices Directive 93/42/EEC (MDD) and the Active Implantable Medical Devices Directive 90/385/EEC (AIMDD). The date of entry into force of the provisions of the MDR, initially scheduled for May 26, 2020, has been postponed by 1 year due to Reg. 2020/561, in consideration of the health emergency connected with COVID-19 and the consequent need to guarantee priority to patient care. It is important to note, though, that a number of transitional provisions are established.

However, in order to ensure the availability of the devices, until May 2025, products certified under the MDR and under the MDD will coexist on the market. Both will have equal status under EU legislation, and every discrimination in public tenders is forbidden.

Compared with the MDD, the MDR clarifies the respective obligations of manufacturers, authorized representatives, importers, and distributors (arts. 10–16 MDR). In general, the rules aim at extending the application of EU safety and security standards to medical devices manufactured in third countries. In fact, the assessment of conformity, certified by the CE mark, is a requirement that must be fulfilled before the medical devices are commercialized. Moreover, such an objective may explain why manufacturers established outside the EU must appoint an authorized representative based in the EU.

The MDR also clarifies the distinction between vigilance and postmarket surveillance. The first includes identifying and reporting serious incidents and conducting corrective actions; it requires a direct and efficient cooperation between health professionals, health institutions, national competent authorities, and manufacturers. The second is an obligation of the manufacturer and involves monitoring the available information through follow-up plans in order to confirm that the medical devices remain safe.

13.3.2 The challenging definition of medical devices

It should be noted that the definition of medical devices contained in European legislation has evolved over time in order to adapt to the development of technology.

The first interesting evolution tends to include software in the notion of medical devices when it satisfies two cumulative conditions, relating (i) to the purpose pursued and (ii) to the action produced (MDD, Art. 1.2 lett. a). In particular, software is considered in itself a medical device when it is specifically intended by the manufacturer to be used for one or more of the medical purposes set out in the definition of medical devices (Dir. 2007/47, recital 6). On the other hand, for the purpose of qualification as a medical device, the fact that the software acts directly or indirectly on the human body is considered irrelevant.

According to jurisprudence referred to the MDD, the notion of medical devices does not include software that does not perform any action on data or whose action is limited to storage, archiving, lossless compression or, finally, to simple research, i.e., software that has a numerical library function, and that allows you to find information from metadata, without modifying or interpreting it. For this reason, the Court of Justice has stated that software that, among other functions, allows the use of a patient's personal data, particularly in order to detect contraindications, interactions between medicinal products and excessive doses, constitutes a medical device, as far as this functionality is concerned, even if the said software does not act directly in or on the human body. In order to be placed on the market and circulate freely in the EU, such software must compulsorily obtain and bear the CE marking of conformity [72].

The MDR acknowledged the Court's interpretation. The new Regulation also added that the qualification of software, both as a device and as an accessory, is independent of the location of the software or the type of interconnection between the software and a device. In addition, software for purposes associated with lifestyle and well-being is not considered a medical device [73]. Moreover, devices or services sold via the internet are now explicitly covered by the MDR.

The innovations introduced make the legal notion of medical devices sufficiently elastic to adapt to the technological development of the next few years, including in the artificial intelligence sector.

13.3.3 The allocation of liability for endorobots

Liability for damage resulting from manufacturing defects of the medical devices and of endorobots is also governed by European law. The MDR in fact establishes the general rule of manufacturers' liability, referring to Directive 85/374/EEC on liability for defective products [74]. Thus, as some scholars observed [63], a strict liability regime is likely to apply to a defective surgical robot. In fact, under Art. 4 of Dir. 85/374/EEC, the injured person shall be required to prove the damage, the defect, and the causal relationship between defect and damage; on the contrary, proof of negligence is not necessary. In its case-law, the Court of Justice confirmed that it is for the national legal order of each Member State to establish the ways in which evidence is to be elicited, what evidence is to be admissible before the appropriate national court, or the principles governing that court's assessment of the probative value of the evidence adduced before it, as well as the level of proof required. However, the principle of effectiveness, as a general principle of EU law, requires, in terms of the detailed procedural rules governing actions for damages, that those rules should not render practically impossible or excessively difficult the right to compensation conferred by Dir. 85/374/EEC [75].

As regards the specific point of national evidentiary rules, it must be noted that often such rules do not require the victim to produce, in all circumstances, certain and irrefutable evidence of a defect in the product and of a causal link between the defect and the damage suffered; rather, national provisions often authorize the judge to conclude that a defect has been proven to exist on the basis of a set of evidence, the seriousness, specificity, and consistency of which allows him or her to consider, with a sufficiently high degree of probability, that such a conclusion corresponds to the reality of the situation. Examples of such evidentiary regimes are offered by Art. 2729 of the Italian Civil Code and Art. 1386 of the French Civil Code, as interpreted by national case-law; such rules authorize the national judge to ascertain a causal link between the defect in the product and the damage suffered by the person injured on the basis of serious, specific, and consistent presumptions.

Such an evidentiary regime of probable cause is of particular relevance when there is no scientific consensus supporting a causal relationship between a given fact and a harmful event.

In this respect, some authors suggest that surgical robots should contain a "black box" that would record data relating to the movements of the robot, environmental data detected by the robot, the commands given by the operator, and so on [63].

The first subject liable for damages resulting from defects is the manufacturer of the medical device. However, if the manufacturer is not established in a Member State and has not complied with his or her obligations, the authorized representative shall be legally liable for defective devices on the same basis as, and jointly and severally with, the manufacturer. Such joint and several liability ensures a higher level of protection of the victims. As observed by Bouquet and Fouassier, in the case of a defective medical device, the damaged persons will generally be able to claim compensation, suing a subject, the manufacturer or its authorized representative based in the EU territory [76].

In parallel, notified bodies take responsibility for negligence related to their conformity assessment or inspection tasks. Such cases arise, for example, when NBs fail to withdraw, restrict, or suspend certificates, and they require a specific insurance coverage, as discussed below.

However, recent case-law shows that such rules prove insufficient, especially when the manufacturer or its representative is insolvent or ceased their business [77].

After having experienced serious adverse events (e.g., the *Poly Implant Prothèse* case), the MDR intended to increase the guarantees established by EU law, obliging manufacturers to provide sufficient financial coverage in respect of their potential liability.

In particular, Art. 10 of the MDR states that manufacturers shall take measures proportionate to the risk class, type of device, and size of the enterprise to provide sufficient financial coverage in respect of their potential liability under Directive 85/374/EEC, without prejudice to more protective measures under national law. Such a provision, however, seems too broad and generic, and it does not guarantee any protection in the case of bankruptcy of the manufacturer, or of its representative in the EU. Moreover, maintaining the Member States' competence causes an uneven regulation in the internal market. For example, in France, under Art. L. 1142-2 of the *Code de la santé publique*, it is mandatory for manufacturers and distributors of medical devices to provide adequate insurance coverage; on the contrary, in other Member States, similar obligations do not exist.

A compulsory European-wide insurance coverage would have been a more effective solution. Such a requirement, however, has been introduced solely in relation to NBs. As mentioned above, NBs must take out appropriate liability insurance for their conformity assessment activities; the MDR states that the scope and overall financial value of the liability insurance must correspond to the level and geographic scope of activities of the NB and be commensurate with the risk profile of the devices certified [78] .

As observed by Hancher and Földes , the new EU legal framework is likely to bring about new legal challenges for supervisory authorities because it establishes directly applicable rules, whereas the previous rules, provided for by directives, required national implementation measures [79].

In fact, it might be observed that the MDR strengthens the multilevel protection of public health. The Member States have the duty to take all appropriate interim measures to withdraw medical devices from the market, or to prohibit or restrict their being placed on the market or put into service, if they present risks to the health or safety of patients or users or other persons. In such circumstances, the Commission shall, in turn, examine whether those interim measures are justified and, where that is found to be indeed the case, it must immediately inform the Member State which initiated the measures, as well as the other Member States.

13.3.4 Final remarks on the legal framework

The MDR is likely to be a stable discipline for endorobots for the next few years.

As mentioned above, the European Parliament recently adopted a Resolution with recommendations to the Commission on a civil liability regime for AI. The Resolution does not clarify the interplay with the MDR, but seems to permit additional claims for damages resulting from regulations on product liability; the definition of the scope of application, however, appears rather ambiguous (see Art. 2 par. 3). It is worth noting that the Parliament proposes to introduce an obligation of the operators of "high-risk AI systems" to take out an insurance, fitting the maximum amounts of compensation provided for by its Art. 5 (two million euro in the event of death or harm of one person). Such compulsory insurance regime may prevent the drawbacks mentioned in the case *RB c. TUV Reinland et a.* [77].

Concerning the above mentioned proposal entitled "Artificial Intelligence Act" [19]. The proposal is part of the Commission's overall digital strategy, closely linked to the Data Governance Act [80], the Open Data Directive [81] and other initiatives under the EU strategy for data [82], which aim to establish trusted mechanisms and services for the re-use, sharing and pooling of data that are essential for the development of data-driven AI models of high quality. The Commission highlighted the need for consistency with specific EU legislation (including the MDR), applicable to sectors where high-risk AI systems are already used or likely to be used in the near future. In fact, the proposal uses a risk-based approach, similar to that of the MDR. It provides that high-risk AI systems, including AI systems that constitute medical devices, may access the European market if they comply with certain mandatory requirements, and on condition that they have carried out a prior conformity assessment [83].

Indeed, the development of new technologies does not appear to require regulatory changes relating to civil liability; moreover, the safety rules of the MDR seem to provide adequate guarantees for patients.

In any case, it seems that both the MDR and the (forthcoming) Artificial Intelligence Act will strengthen the regulatory framework applicable to surgical robots, reconciling the internal market objectives with the fundamental rights, including health and protection of personal data.

13.4 Conclusions

For centuries, the arts of medicine and surgery have been scientific disciplines. Their tools have significantly improved since the "bronze lancet", allowing today operations from the inside of patient's body and presenting new ethical and legal dilemmas. In this sense, endorobotics is perhaps the most innovative field in healthcare, with a potential that can bring extraordinary benefits to patients, likewise posing many risks that cannot be underestimated. Technological assessment, in such an important sector, should be left neither to a pure pragmatical balance of costs nor to the imposition of a set of undiscussed moral values.

In this chapter, we observed how the technological evolution has been followed by the development of the ethical and legal framework: as we have witnessed over time, in the transition from the simple transposition of generic industrial robots to the creation of specialized devices, theoretical approaches have changed from the general application of preexisting tenets and methods in this field to the development of specific criteria and tools.

We can argue that the role of developers will be crucial in the future for designing devices that are not only effective but also suitable for a fair use and transparent toward both the individual patient and, in general, the healthcare system. In this sense, at least in the near future, it could be useful to foster the development of an ethical code of conduct for endorobotics designers, even if it would be left to the spontaneous adoption by stakeholders.

Likewise, the challenge raised by the difficulties in assigning liability for damages and malpractice could be reduced with the deployment of "black box" features that could allow the storage of relevant information—even in cloud archives—not only for maintenance of mechanical or electronic components, or for training human operators, but also—and especially—for forensic purposes in order to facilitate the collection of forensic data. An effort with this aim, globally coordinated, could contribute to increasing the trust of these devices and to spreading their adoption worldwide.

Author contributions

This chapter is the result of a joint research of co-authors; the individual contribution can be attributed as follows: Federico Costantini, paragraphs 1, 2, and 4; Fabio Balducci Romano, paragraph 3.

References

[1] Rositani A. Work and wages in the Code of Hammurabi. Egitto Vicino Oriente 2017;40:47–72.

[2] Hockstein NG, Gourin CG, Faust RA, Terris DJ. A history of robots: from science fiction to surgical robotics. J Robot Surg 2007;1:113–8.

[3] Gomes P. Surgical robotics: reviewing the past, analysing the present, imagining the future. Robot Comput Integr Manuf 2011;27:261–6.

[4] Ewing DR, Pigazzi A, Wang Y, Ballantyne GH. Robots in the operating room—the history. Semin Laparosc Surg 2016;11(2):63–71. Available from: https://journals.sagepub.com/, https://doi.org/10.1177/107155170401100202.
[5] Leal Ghezzi T, Campos CO. 30 years of robotic surgery. World J Surg 2016;40:2550–7. https://doi.org/10.1007/s00268-016-3543-9.
[6] Cole AP, Trinh Q-D, Sood A, Menon M. The rise of robotic surgery in the new millennium. J Urol 2017;197:S213–5. Available from: https://www.auajournals.org/, https://doi.org/10.1016/j.juro.2016.11.030.
[7] Lane T. A short history of robotic surgery. Ann R Coll Surg Engl 2018;100:5–7. Available from: https://publishing.rcseng.ac.uk, https://doi.org/10.1308/rcsann.supp1.5.
[8] Marino MV, Shabat G, Gulotta G, Komorowski AL. From illusion to reality: a brief history of robotic surgery. Surg Innov 2018;25:291–6. https://doi.org/10.1177/1553350618771417.
[9] Fenn J, Blosch M. Understanding Gartner's hype cycles. Gartner; 2018. Available from: https://www.gartner.com/en/documents/3887767/understanding-gartner-s-hype-cycles.
[10] Lin P, Hughes JJ, Abney K, Bekey GA, Allen C, Beavers A, et al. Robot ethics: the ethical and social implications of robotics. Cambridge, MA: MIT Press; 2011.
[11] Kassim I, Phee L, Ng WS, Gong F, Dario P, Mosse CA. Locomotion techniques for robotic colonoscopy. IEEE Eng Med Biol Mag 2006;25:49–56. http://ieeexplore.ieee.org/document/1636351/.
[12] SAE Mobilus. Safety-relevant guidance for on-road testing of SAE level 3, 4, and 5 prototype automated driving system (ADS)-operated vehicles (J3018 ground vehicle standard). Available from: https://saemobilus.sae.org/content/j3018_201909; 2019.
[13] Yang G-Z, Cambias J, Cleary K, Daimler E, Drake J, Dupont PE, et al. Medical robotics—regulatory, ethical, and legal considerations for increasing levels of autonomy. Sci Robot 2017;2:eaam8638. Available from: https://robotics.sciencemag.org, https://doi.org/10.1126/scirobotics.aam8638.
[14] Inagaki T, Sheridan TB. A critique of the SAE conditional driving automation definition, and analyses of options for improvement. Cogn Tech Work 2019;21:569–78. https://doi.org/10.1007/s10111-018-0471-5.
[15] Kim W, Kim H-S, Lee S-J, Kim J, Yoon D. Transitions from autopilot to manual control in highly automated driving: cognitive simulations. In: 2018 IEEE 88th vehicular technology conference (VTC-Fall), Chicago, IL, USA; 2018. p. 1–5. https://doi.org/10.1109/VTCFall.2018.8690796.
[16] Beer JM, Fisk AD, Rogers WA. Toward a framework for levels of robot autonomy in human-robot interaction. J Hum-Robot Interact 2014;3:74–99. https://doi.org/10.5898/JHRI.3.2.Beer.
[17] Onnasch L, Roesler E. A taxonomy to structure and analyze human–robot interaction. Int J Soc Robot 2020. https://doi.org/10.1007/s12369-020-00666-5.
[18] European Parliament, Resolution 20 October. Civil liability regime for artificial intelligence, https://www.europarl.europa.eu/doceo/document/TA-9-2020-0276_EN.html; 2020.
[19] Proposal for a Regulation of the European Parliament and of the Council laying down harmonised rules on Artificial Intelligence (Artificial Intelligence Act) and amending certain Union legislative acts, COM/2021/206 final, of 21 April 2021, https://eur-lex.europa.eu/legal-content/EN/TXT/?uri=CELEX:52021PC0206.
[20] Morozov E. To save everything, click here. The folly of technological solutionism. New York: Public Affairs; 2013.
[21] Ellul J. The technological society. New York: Vintage Books; 1964.
[22] Lyotard J-F. The postmodern condition: a report on knowledge. Manchester: Manchester University Press; 1984.

[23] Coeckelbergh M. AI ethics. Cambridge: MIT Press; 2020.
[24] Wiener N. The human use of human beings: cybernetics and society. Boston: Houghton Mifflin; 1950.
[25] Fortunati L, Cavallo F, Sarrica M. Multiple communication roles in human–robot interactions in public space. Int J Soc Robot 2020;12:931–44. https://doi.org/10.1007/s12369-018-0509-0.
[26] Floridi L, editor. The Cambridge handbook of information and computer ethics. Cambridge: Cambridge University Press; 2010.
[27] Veruggio G, Operto F. Roboethics: social and ethical implications of robotics. In: Siciliano B, Khatib O, editors. Springer handbook of robotics. Berlin, Heidelberg: Springer; 2008. p. 1499–524. https://doi.org/10.1007/978-3-540-30301-5_65.
[28] Wallach W, Allen C. Moral machines: teaching robots right from wrong. Oxford: Oxford University Press; 2008.
[29] Pagallo U. The laws of robots: crimes, contracts, and torts. Netherlands: Springer; 2013. Available from: https://www.springer.com/gp/book/9789400765634.
[30] Moor JH. The nature, importance, and difficulty of machine ethics. IEEE Intell Syst 2006;21:18–21. Available from: http://ieeexplore.ieee.org/document/1667948/.
[31] Dodig Crnkovic G, Çürüklü B. Robots: ethical by design. Ethics Inf Technol 2012;14:61–71. https://doi.org/10.1007/s10676-011-9278-2.
[32] The IEEE Global Initiative on Ethics of Autonomous and Intelligent Systems. Ethically aligned design. A vision for prioritizing human well-being with autonomous and intelligent systems. IEEE; 2019. Available from: https://standards.ieee.org/content/dam/ieee-standards/standards/web/documents/other/ead1e.pdf.
[33] Bartneck C, Lütge C, Wagner A, Welsh S. Ethics in AI and robotics: a strategic challenge. In: Bartneck C, Lütge C, Wagner A, Welsh S, editors. An introduction to ethics in robotics and AI. Cham: Springer International Publishing; 2021. p. 101–4. https://doi.org/10.1007/978-3-030-51110-4_12.
[34] Sparrow R. Virtue and vice in our relationships with robots: is there an asymmetry and how might it be explained? Int J Soc Robot 2020;13(1):23–9. https://doi.org/10.1007/s12369-020-00631-2.
[35] Coeckelbergh M. Does kindness towards robots lead to virtue? A reply to Sparrow's asymmetry argument. Ethics Inf Technol 2021. https://doi.org/10.1007/s10676-021-09604-z.
[36] van den Hoven J, Vermaas PE, van de Poel I. In: Handbook of ethics, values, and technological design. Dordrecht: Springer; 2015.
[37] Floridi L. The ethics of information. Oxford, New York: Oxford University Press; 2013.
[38] Morley J, Machado CCV, Burr C, Cowls J, Joshi I, Taddeo M, et al. The ethics of AI in health care: a mapping review. Soc Sci Med 2020;113172. Available from: http://www.sciencedirect.com/science/article/pii/S0277953620303919.
[39] Morley J, Floridi L. Enabling digital health companionship is better than empowerment. Lancet Digit Health 2019;1:e155–6. Available from: http://www.sciencedirect.com/science/article/pii/S2589750019300792.
[40] Morley J, Floridi L. An ethically mindful approach to AI for health care. Lancet 2020;395:254–5. Available from: http://www.sciencedirect.com/science/article/pii/S0140673619329757.
[41] Friedman B. Value-sensitive design. Interactions 1996;3:16–23. https://doi.org/10.1145/242485.242493.
[42] Kheirandish S, Funk M, Wensveen S, Verkerk M, Rauterberg M. A comprehensive value framework for design. Technol Soc 2020;62:101302. Available from: http://www.sciencedirect.com/science/article/pii/S0160791X19303951.

[43] Dignum V. Responsible artificial intelligence. How to develop and use AI in a responsible way. Cham: Springer; 2019. https://link.springer.com/book/10.1007%2F978-3-030-30371-6.

[44] Regulation (EU) 2016/679 of the European Parliament and of the Council of 27 April 2016 on the protection of natural persons with regard to the processing of personal data and on the free movement of such data, and repealing Directive 95/46/EC (General Data Protection Regulation), in OJ L, 119; 4.5.2016. p. 1–88.

[45] van Wynsberghe A. Healthcare robots: ethics, design and implementation. London: Routledge; 2016.

[46] Umbrello S, Capasso M, Balistreri M, Pirni A, Merenda F. Value sensitive design to achieve the UN SDGs with AI: a case of elderly care robots. Minds Mach 2021;31(3):395–419. https://doi.org/10.1007/s11023-021-09561-y.

[47] Stahl BC, Coeckelbergh M. Ethics of healthcare robotics: towards responsible research and innovation. Robot Auton Syst 2016;86:152–61. https://doi.org/10.1016/j.robot.2016.08.018.

[48] Burget M, Bardone E, Pedaste M. Definitions and conceptual dimensions of responsible research and innovation: a literature review. Sci Eng Ethics 2017;23:1–19. https://doi.org/10.1007/s11948-016-9782-1.

[49] Von Schomberg R. Prospects for technology assessment in a framework of responsible research and innovation. Technikfolgen abschätzen lehren: Bildungspotenziale transdisziplinärer Methoden. Wiesbaden: Springer VS; 2011. p. 39–61.

[50] Stilgoe J, Owen R, Macnaghten P. Developing a framework for responsible innovation. Res Policy 2013;42(13):1568–80. https://doi.org/10.1016/j.respol.2013.05.008.

[51] Olaronke I, Oluwaseun O, Rhoda I. State of the art: a study of human-robot interaction in healthcare. Int J Inf Eng Electron Bus 2017;9:43. https://doi.org/10.5815/ijieeb.2017.03.06.

[52] Andolfi C, Umanskiy K. Mastering robotic surgery: where does the learning curve lead us? J Laparoendosc Adv Surg Tech A 2017;27:470–4. Available from: https://www.liebertpub.com/, https://doi.org/10.1089/lap.2016.0641.

[53] Kassite I, Bejan-Angoulvant T, Lardy H, Binet A. A systematic review of the learning curve in robotic surgery: range and heterogeneity. Surg Endosc 2019;33:353–65. https://doi.org/10.1007/s00464-018-6473-9.

[54] Collins JW, Dell'Oglio P, Hung AJ, Brook NR. The importance of technical and non-technical skills in robotic surgery training. Eur Urol Focus 2018;4:674–6. https://doi.org/10.1016/j.euf.2018.08.018.

[55] Expert Group on Liability and New Technologies. Liability for artificial intelligence and other emerging digital technologies. Liability for artificial intelligence and other emerging digital technologies. Luxembourg: Publications Office of the European Union; 2019, doi:10.2838/573689.

[56] High-Level Expert Group on AI. Ethics guidelines for trustworthy AI. Brussels: European Union; 2019 August. Available from https://ec.europa.eu/newsroom/dae/document.cfm?doc_id=60419.

[57] High-Level Expert Group on AI. Sectoral considerations on the policy and investment recommendations for trustworthy artificial intelligence. 16; 2020. Available from: https://futurium.ec.europa.eu/sites/default/files/2020-07/Sectoral%20Considerations%20On%20The%20Policy%20And%20Investment%20Recommendations%20For%20Trustworthy%20Artificial%20Intelligence_0.pdf.

[58] High-Level Expert Group on AI. Assessment list for trustworthy artificial intelligence (ALTAI) for self-assessment. Available from: https://futurium.ec.europa.eu/en/european-ai-alliance/pages/altai-assessment-list-trustworthy-artificial-intelligence; 2020.

[59] European Parliament. Civil law rules on robotics. Thursday, 16 February 2017 – Strasbourg. Available from: https://www.europarl.europa.eu/doceo/document/TA-8-2017-0051_EN.html; 2017.

[60] Council Directive 93/42/EEC of 14 June 1993 concerning medical devices, OJ L 169; 12.07.1993. p. 1–43.

[61] Art. 2 Regulation (EU) 2017/745 of the European Parliament and of the Council of 5 April 2017 on medical devices, amending Directive 2001/83/EC, Regulation (EC) No 178/2002 and Regulation (EC) No 1223/2009 and repealing Council Directives 90/385/EEC and 93/42/EEC, OJ L 117; 5.5.2017. p. 1–175.

[62] www.intuitivesurgical.com.

[63] Holder C, Khurana V, Harrison F, Jacobs L. Robotics and law: key legal and regulatory implications of the robotics age (part I of II). Comput Law Secur Rev 2016;32:383–402. https://doi.org/10.1016/j.clsr.2016.03.001.

[64] European Commission proposal for a Regulation of the European Parliament and of the Council laying down harmonised rules on artificial intelligence (Artificial Intelligence Act) and amending certain Union legislative acts, COM/2021/206 final, available at: https://eur-lex.europa.eu/legal-content/EN/TXT/?qid=1623335154975&uri=CELEX%3A52021PC0206.

[65] Art. 114 of the Treaty on the Functioning of the European Union (TFEU). Available from: https://eur-lex.europa.eu/collection/eu-law/treaties/treaties-force.html;. Previously: Art. 95 of the Treaty establishing the European Community, Available from: https://eur-lex.europa.eu/collection/eu-law/treaties/treaties-other.html.

[66] Art. 35 of the Charter of Fundamental Rights of the European Union. Available from: https://eur-lex.europa.eu/collection/eu-law/treaties/treaties-force.html.

[67] Lobmayr B. An assessment of the EU approach to medical device regulation against the backdrop of the US system. Eur J Risk Regul 2010;1:137–49. https://doi.org/10.1017/S1867299X00000222.

[68] Chai JY. Medical device regulation in the United States and the European Union: a comparative study. Food Drug Law J 2000;55:57–80. 12296349.

[69] Frank S. An assessment of the regulations on medical devices in the European Union. Food Drug Law J 2001;56:99–122. Available from: http://www.jstor.org/stable/26660247.

[70] Holder C, Khurana V, Hook J, Bacon G, Day R. Robotics and law: key legal and regulatory implications of the robotics age (part II of II). Comput Law Secur Rev 2016;32:557–76. https://doi.org/10.1016/j.clsr.2016.05.011.

[71] Horton LR. Medical device regulation in the European Union. 34; 1995. Available from: http://aei.pitt.edu/7464/.

[72] See: ECJ, judgment of 7 December 2017, Syndicat national de l'industrie des technologies médicales (Snitem), case C-329/16, ECLI:EU:C:2017:947, p. 34–35.

[73] See: MDR, recital 19.

[74] Council Directive 85/374/EEC of 25 July 1985 on the approximation of the laws, regulations and administrative provisions of the Member States concerning liability for defective products. OJ L 210; 7.8.1985. p. 29–33.

[75] See: ECJ, judgment of 21 June 2017, N. W. and others v Sanofi Pasteur MSD SNC and others, case C-621/15, ECLI:EU:C:2017:484, p. 25–27.

[76] Bouquet V, Fouassier E. Dispositifs médicaux: responsabilité civile du mandataire après l'entrée en vigueur des règlements 2017/745et 2017/746. Méd Droit 2019;2019:71–6. https://doi.org/10.1016/j.meddro.2018.09.001.

[77] See: ECJ, judgment of 11 Jun 2020, RB v TÜV Rheinland, case C-581/18, ECLI:EU:C:2020:453.

[78] See: MDR, Annex VII, par. 1.4.

[79] Hancher L, Földes ME. Revision of the regulatory framework for medical devices in the European Union: the legal challenges. Eur J Risk Regul 2013;4:429–35. https://doi.org/10.1017/S1867299X0000307X.

[80] See: Proposal for a Regulation of the European Parliament and of the Council on European data governance (Data Governance Act), COM/2020/767 final, available at https://eur-lex.europa.eu/legal-content/EN/TXT/?uri=CELEX:52020PC0767.

[81] See: Dir. (EU) 2019/1024 of the European Parliament and of the Council of 20 June 2019 on open data and the re-use of public sector information (Open Data Directive), ELI: http://data.europa.eu/eli/dir/2019/1024/oj.

[82] See: Proposal for a Regulation of the European Parliament and of the Council on a Single Market For Digital Services (Digital Services Act) and amending Directive 2000/31/EC, COM/2020/825 final, available at https://eur-lex.europa.eu/legal-content/en/TXT/?qid=1608117147218&uri=COM%3A2020%3A825%3AFIN.

[83] See: Artificial Intelligence Act, cit. note n. 55, Explanatory memorandum, par. 5.2.3.

Chapter 14

Healthcare data governance in the EU: Main challenges in personal data protection

Federico Costantini and Giada Soncini
University of Udine, Udine, Italy

14.1 Introduction: Healthcare data and endorobots

Information Communication Technologies (ICTs) in healthcare offer an extraordinary range of options in terms of infrastructures and devices. Thus, data workflow can reach an extreme level of complexity. Indeed, information can be collected by many devices (ordinary diagnostic tools, endorobots, or even miniaturized capsules) in many different ways and digital formats (text, parameters, images, video recordings, sonic, or electronic tracks). Information can also be exchanged among artificial agents and human operators, who can be on-site or remotely connected. Information can then be processed, shared, communicated, or transferred abroad for different purposes, for example, medical diagnosis, device monitoring, and academic or industrial research. Furthermore, it can be transferred to different environments—private clinics, public authorities, and homes, among others. Finally, it can be archived in datasets located anywhere worldwide that can be retrieved in the future. Given their relevance, it has been claimed that healthcare data have become the "new blood" [1], and in fact, some could argue that in the relatively near future, this digital fluid will become even more lucrative than that which flows through human veins.

The widespread increasing adoption of robotics escalates such complexity, thus posing many risks. In general, it is known that a robot can collect huge amounts of information (maps of buildings, personal daily routines, sounds, images, videos) even unbeknownst to the people involved or without valid legal grounds (with the flawed consent of minors, elders, the mentally impaired, or people with other personal vulnerabilities). Furthermore, such data can be exploited for illegal purposes (mainly through cyberattacks), causing damages that can be even severe. This can also be predicated for endorobots, since they interact directly with patients, posing concerns not only for their personal

Endorobotics. https://doi.org/10.1016/B978-0-12-821750-4.00014-1

safety—they could be physically harmed, in fact—but also for their privacy, especially if robots are connected to open networks. Endorobotics, however, representing a very specific field of healthcare robotics, is not a frequent topic in the literature regarding healthcare data protection. Indeed, scholars tend to focus on kinds of "social robots" (see the definition in Chapter 13) that raise more immediate concerns, such as care robots [2] or robotic surgeons [3].

In this chapter, we will raise the most relevant issues concerning personal data protection as regards the field of endorobotics, providing an overview of their context as well. To begin with, we provide a preliminary introduction to the European Union legal framework on e-healthcare, offering an overview of information security, on the one hand, and "open data" and "non-personal data" provisions, on the other. Then, we will focus on personal data, analyzing the most significant concerns related to endorobots. For the sake of simplicity, we will model an ideal workflow process into two phases: "before" the collection of data, where we will put emphasis on the Data Protection Impact Assessment (DPIA) and data subject's consent, and "after" data processing, where we will consider privacy by design related to data sharing and transfer. Finally, we will offer a thorough overview and some final evaluations.

14.2 An overview of healthcare data governance in the European Union

Of the many policies implemented by the EU to foster the creation of a digital single market, some are remarkable for their indirect impact on healthcare data governance. On the one hand, the European High Performance Computing Strategy [4] aims at creating a shared infrastructure suitable for allowing the deployment of super-computing technologies and fast connectivity. On the other hand, the European Cloud Initiative [5] intends to widen the use of computing, storage, and analysis capacity to the private sector while maintaining high levels of security. In a recent communication, the Commission has envisioned a comprehensive "European Strategy for data" that aspires for the EU to become a world-leading data economy in the near future [6]. In this strategy, the intention to create a "Common European health data space" is specifically mentioned, i.e., an ecosystem in which to build a shared infrastructure among Member States and promote the digitalization of national healthcare systems.

As regards specifically the digital transformation of the healthcare sector, the European Commission has declared the intention to pursue the ambitious goal of a "triple win" that "benefits people, health, and market" [7]. Indeed, digitalization offers many opportunities to increase the well-being of citizens, promote health, and prevent diseases, especially through the adoption of new care models that allow the conversion from the current hospital-centered to a community-based system. However, many challenges have to be tackled, such as the lack of optimization of processes, the fragmentation of the EU market, and the absence of interoperability among different platforms, sometimes even within the same country. Hence, strong cooperation among Member States is required in order to integrate technologies, provide sustainable public services, and allow citizens to access high-quality diagnostic tools, surgical procedures, and personal assistance. In order to foster a "meaningful use of digital innovation" in healthcare, the EU has decided to prioritize specific actions, such as the definition of EU-wide standards, especially in electronic health records, and the creation of open-exchange formats. Such initiatives will be incorporated within the current EU legal framework provided by the Directive 2011/24/EU on the application of patients' rights in cross-border healthcare [8], which facilitates cooperation among Member States within the pre-existing Regulation (EC) 883/2004 on social security systems [9].

In our view, EU efforts toward digital transformation in healthcare can be better understood by highlighting the underlying necessity to balance between two opposite needs: confidentiality and openness. On the one hand, it is crucial to protect the strategic value incorporated in data, defending it from unlawful appropriations or from abusive exploitations. On the other hand, enabling direct access to data is essential to spread transparency, especially in public decisions and procedures, and to promote fair competition, research, and innovation, particularly in the industrial sector. Both aspects have been addressed by EU legislation in recent years, as summarized in the next two subsections. With the COVID19 pandemic, the EU has reached a higher awareness of the importance of digitalization in healthcare. Many initiatives have been adopted by EU institutions in order to tackle the spread of virus, coordinating national efforts, fostering pharmaceutical research, financing political and economical resilience with "NextGenerationEU" funding [10] and proposing common or interoperable technological infrastructures for data exchange [11]. Borrowing a quote from a recent contribution, it is confirmed by recent events that in healthcare "sharing is caring" [12].

14.2.1 Confidentiality and information security

Regarding information security, scholars have argued that sensible risks are raised by internal lack of competencies and digital skills by staff members, or scarce awareness by management [13]. Yet, as recent events have shown, the main danger is posed by external threats. Indeed, the number and scale of cyberattacks against hospitals and healthcare institutions has dramatically increased recently, as confirmed by the fact that in the past 35 years academic literature tackling this topic has exponentially grown [14–17]. In this regard, the international call endorsed in May 2020 by the International Committee of the Red Cross against cyberattacks targeting specifically healthcare institutions during the COVID-19 pandemic [18] and the declaration of the President of the European Commission in June 2020 condemning such worrisome events [19] are remarkable.

Concerning cyberattacks, the adoption of the Directive (EU) 2016/1148 (henceforth "Network Information Security Directive," or "NIS Directive") [20] is noteworthy. The main purpose of this piece of legislation is to directly incorporate a mechanism suitable for improving the overall "resilience" of the European economy into what will be the backbone of the digital single market, strengthening the security level of "network and information systems" (Article 4(1)(1) NIS Directive). More specifically, this Directive requires Member States to adopt a national strategy on cybersecurity (Article 7 NIS Directive) designating competent authorities and points of contact to liaise with other institutions (Article 8 NIS Directive). Moreover, the duties of assisting and coordinating national "Computer Security Incident Response Teams" (CSIRTs) and cooperating with other authorities at an international level are assigned to the "European Network and Information Security Agency" (ENISA), established as an independent Agency in 2004 [21] and recently reinforced by means of Regulation (EU) 2019/881 [22].

The most relevant aspect refers to the definition of "operator of essential services" (Article 4(1)(4) NIS Directive). In order to fall into that definition, a public or private entity has to comply with the three criteria set forth by Article 5(2). To be precise, such an entity (a) should be "essential for the maintenance of critical societal and/or economic activities," (b) should be required for the provision of "network and information systems," and (c), if afflicted by an incident, "significant disruptive effects" should be addressed (Article 6). Furthermore, such an entity should be listed in Annex II where, among others crucial sectors (for example, energy and transport), "healthcare providers" are included. According to the definition of point (g) of Article 3 of Directive 2011/24/EU, a healthcare provider is "any natural or legal person or any other entity legally providing healthcare."

The definition of "operator of essential services" is crucial for designing medical devices. If an institution or a company can be qualified as such, it is due to its ability to enhance the level of security by taking "appropriate and proportionate technical and organizational measures to manage the risks" (Article 14(1) NIS Directive) and "appropriate measures to prevent and minimize the impact of incidents" (Article 14(2) NIS Directive), promptly notifying CSIRTs in the case of incidents (Article 14(3) NIS Directive). In other words, according to the NIS Directive, any kind of hospital or medical clinic that falls into the definition of "healthcare provider"—which is extremely wide—has to devote resources, develop personal competencies, and deploy procedures in order to set a higher level of control of information. To sum up, when it comes to cybersecurity, everyone has to do their part, especially those who could become easy targets.

Therefore, the need to comply with the NIS Directive is relevant for endorobot engineering and procurement. Indeed, in designing devices that have to be placed in ecosystems that are expected to abide by rigid security protocols, the achievement of such higher standards is inevitably mandatory. Consequently, since the level of security embedded in such devices has to be included in the assessment of the measures adopted by the institutions, a more protected equipment should be preferred to others in the procurement processes. For these reasons, cybersecurity features and standards, such as ISO/IEC 27000:2018 [23] and ISO 27799:2016 [24], should at least be considered—if not entirely fulfilled—in producing and marketing medical devices.

14.2.2 Openness and transparency: "Open data" and "non-personal data"

Considering the aspect related to openness, it is useful to observe, as a preliminary remark, that the free flow of information is essential in modern democratic societies under many perspectives, both between governments and citizens and among private entities. With special regard to the public sector, and at least in Western countries, freedom of information is guaranteed by international treaties and constitutions (e.g., Article 19 of the Universal Declaration of Human Rights of the United Nations). More recently, it has been enforced by means of the widespread attribution of the right to information by individual citizens (e.g., Article 42 of the European Charter of Fundamental Rights in the European Union) even at the national level. In other terms, rephrasing the glorious motto coined by the Washington Post, we could agree that "without transparency, democracy dies" [25].

Digital technologies have added new ways whereby such circulation of information can be declined in the public sphere. Indeed, not only has freedom of expression expanded in a new realm, such as the Internet, but citizens have also obtained the right to access information owned by authorities in order to scrutinize their exercise of power, especially in public spending. In this sense, it is noteworthy that in Western countries, specific regulations have been adopted in order to increase transparency that are similar to the original Freedom of Information Act enacted in the USA in 1966. Regarding the EU, the revision of the Regulation (EC) No 1049/2001 [26], currently in force, has recently been restarted after a long interruption due to a misalignment formed among EU institutions regarding the balance between confidentiality and transparency [27].

In addition, the advancement of ICTs has made possible a new relationship between public institutions and information. Indeed, data owned by the public sector represent a resource that can be exploited if publicly shared, fostering social progress and economic growth. In other words, any kind of information archived by institutions can be given new life by allowing its re-use by private citizens and businesses. In this regard, the legislation of the European Union has significantly evolved, following technological innovation aimed at creating fair, proportionate, and nondiscriminatory conditions for the re-use of information and allowing for the provision of data by public bodies upon the request of private citizens and companies with minimal costs. The original Directive

2003/98/CE, or "Public Sector Information" Directive (henceforth "PSI Directive") [28], was amended by Directive 2013/37/EU [29], providing a wider definition of "document" and introducing the obligation of Member States to make such information re-usable, of course, if they neither contain confidential information nor their access is restricted or excluded by national rules. Furthermore, since 2013, the necessary adoption of machine-readable formats and the incorporation of meta-data to allow interoperability and cross-border circulation [30] have been established. More recently, the discipline was repealed by Directive (EU) 2019/1024 (also named the "Open Data Directive") [31], which puts forward the ambitious goal of establishing real-time access to dynamic data—even through the adoption of APIs (Application Programming Interfaces)—in order to make them suitable for processing even by artificial agents, extending the duty to release open information to public undertakings and expanding the adoption of an open approach to research data stemming from public funding.

Therefore, it can be said that these last-mentioned provisions, which Member States are expected to implement by July 17, 2021, will likely produce a significant benefit in the field of healthcare, since more data will be available to be collected, processed, and deployed for the designing of services and products. Moreover, and with specific regard to medical devices, it can be argued that in order to facilitate their clients to be compliant with the "Open Access" Directive, manufacturers should not only design tools that permit the release of data in order for them to be re-used but also automate such processes, for example, including built-in anonymization and conversion of information in the formats required. Lastly, the fact that Member States are entitled to include computer programs in the definition of "document," as stated in Recital (30) of the "Open Access Directive," can be considered an opportunity for the development of new software applications and even the deployment of new business models.

One last aspect needs to be addressed as regards the legal framework of EU data governance. Because there is increasing circulation of data, regardless of the public or private nature of the source, and the rising adoption of cloud computing technologies, which makes it possible, the EU established a regulation in which the free flow of information has become a legal cornerstone and the demand for services is fulfilled by operators in fair competition. This is the *rationale* behind Regulation (EU) 2018/1807 [32], and we will point out a few remarkable aspects.

First of all, the Regulation applies only to "non-personal data" (Article 3 Regulation (EU) 2018/1807). The fact that such a legal provision specifically excludes "personal data" from this concept without offering a more detailed explanation does not affect its interpretation. Indeed, in this definition, not only can data concerning individuals, namely, "personal data," be included after it has been anonymized but also information generated by devices while in operation and related to their working conditions (for example, internal processes, event logging, and so on). Therefore, we can observe that this piece of legislation is crucial in building the EU-wide "single data market", addressing a strategic resource not only in the design, maintenance, and monitoring of machines but also for improving their effectiveness through machine-learning applications.

A second noteworthy aspect is the prohibition of data localization requirements (Article 4 par. 1 Regulation (EU) 2018/1807). With this provision, the EU established the general principle of free movement of data, binding Member States to impose limitations only if "justified on grounds of public security in compliance with the principle of proportionality." In order to implement such obligations, Member States have to immediately communicate to the Commission every draft act that introduces a new data localization requirement or changes an existing one (Article 4 par. 2), and they have to repeal any conflicting provision by May 30, 2021 (Article 4 par. 3). Conversely, public authorities benefit directly from this discipline, since they are allowed to access data regardless of the fact that information is stored in another Member State (Article 5). Moreover, Article 5 (2) introduces a specific case of assistance between competent authorities. Indeed, it is stated that if a competent authority does not obtain access to a requested user's data, they can request assistance from a competent authority in another Member State by following the procedure described in Article 7.

The third remarkable provision aims at avoiding vendor lock-in business practices (Recital (2), (3), (4), (5), (6), Regulation (EU) 2018/1807), facilitating the porting of data for professional users in order to increase competitive advantages and create a competitive data economy based on the principles of transparency and interoperability (Article 6). In this regard, in accordance with the encouragement of the EU, a Code of Conduct has recently been released by stakeholders (Switching Cloud Providers and Porting Data) [33]. Here, we can observe another case where the involvement of operators becomes essential in developing flexible regulations.

The regulation on "non-personal data" has a significant impact on healthcare. The EU Commission in COM(2019) 250 [34], which offers comprehensive guidance on provisions addressed especially to small-medium enterprises, warns operators that healthcare data are frequently contained in "mixed databases," including both non-personal and personal data. The division between such categories has become blurred mainly in mobile devices that are used, for example, to measure vital parameters, to remind people about medications, or to track fitness progress. We can argue that, in order to exploit the benefit of free circulation of "non-personal data" while preserving confidentiality on personal data, manufacturers should implement workflows that compartmentalize by default non-personal and personal data to avoid the creation of mixed databases where possible.

14.3 Personal data

Personal data protection is ensured both at an international and at the European level. Indeed, aside from the general safeguards on personal intimacy against arbitrary interferences (Article 12 of the Universal Declaration of Human Rights of the United Nations, Article 8 of the European Convention on Human Rights, and Article 7 of the Charter of Fundamental Rights of the European Union), it is specifically incorporated into several treaties and declarations (Convention n. 108 of 1981 for the Protection of Individuals with regard to Automatic Processing of Personal Data of the Council, Article 8 of the Charter of Fundamental Rights of the European Union). Consequently, respecting individual privacy is mandatory as an expression of fundamental rights and freedoms of the individual to whom such data pertain. Those principles, of course, are not only addressed to the sovereign country that adheres to the conventions but are also enforced on each handler of personal data, regardless of whether the handler is a single professional, a company, or a public institution. Apart from a "vertical" application of data protection from States to private citizens, there is a "horizontal" dimension whose impact is consistent, especially in the relationship between customers and providers of goods and services.

In the European Union, the legal framework based on Directive 95/46/EC [35] was recently repealed by Regulation (EU) 2016/679, namely, the "General Data Protection Regulation" (hereinafter, GDPR) [36], enacted together with Directive (EU) 680/2016, namely, "Law Enforcement Directive" [37] and Directive (EU) 681/2016, namely, the "Passenger Name Record Directive" [38] in the attempt to heighten personal data protection[a] and, at the same time, support the growth of the digital market. Such balance is maintained due to a complex regulation, which, on the one side, endorses the rights of the "data subject"—the individual to whom the data refers, enabling her/him to ultimately retain some control of her/his information; on the other hand, it overrules her/his consent, allowing the sharing of information according to contractual obligations and legal provisions, but also recognizing the legitimate interest of the "data controller" or third parties (Article 6 GDPR) [39]. Since in our digital society the flow of data cannot be stopped, it is crucial to balance the principles (Article 5(1) GDPR)[b] that the "data controller" (Article 4(7) GDPR) or the "data processor" (Article 4(8) GDPR) have to fulfill, the security measures to be adopted (Article 32 GDPR) and, more importantly, the procedures that have to be enforced in order to comply with the requirement of "accountability"

a. According to art. 4(1), "personal data" means "any information relating to an identified or identifiable natural person ('data subject'); an identifiable natural person is one who can be identified, directly or indirectly, in particular by reference to an identifier such as a name, an identification number, location data, an online identifier or to one or more factors specific to the physical, physiological, genetic, mental, economic, cultural, or social identity of that natural person."

b. Lawfulness, fairness and transparency, purpose limitation, data minimization, accuracy, storage limitation, integrity, and confidentiality are the principles in Article 5(1) GDPR.

(Article 5(2) GDPR). This latter feature[c] is the core and the real challenge of the GDPR for companies and institutions.

The balance between individual privacy and data sharing is obviously more difficult in healthcare, given the peculiar nature of the data processed, which has important legal implications. Indeed, "data concerning health" are included in a discipline (Article 9 GDPR) concerning special categories of data whose processing is restricted to a few purposes and reduced conditions, while more severe security measures are required for processing them. In this regard, it is useful to underline that in order to increase the protection of the "data subject," the legislation provides many tools such as, *inter alia*, "privacy by design" and "by default" (Article 25 GDPR), with the aim of integrating the necessary safeguards into the technologies adopted in the earliest stage of data processing.

The difficulty in balancing between the conflicting interests of the "data controller" and "data processor," on the one side, and the "data subject," on the other, can be appreciated also by considering that the application of the principle of "purpose limitation" (Article 5(1)(b) GDPR) is derogated by the fact that the GDPR allows further processing. Indeed, data can be used outside the original scope of their collection if their employment is "compatible" with the original purpose. The assessment concerning "compatibility" has to include, among other elements, the link between purposes, the context, the nature of personal data, the consequences, and the adoption of safeguards (Article 6(4) GDPR). Regarding endorobots, therefore, it can be said that the exploitation of personal data requires that it should be duly justified, if not specifically aimed at academic research (Article 89 GDPR), in which case the compatibility is presumed (Article 5(1)(b) GDPR). This latter case expresses, evidently, the interference of the protection of the freedom to do research, which in this case prevails over other values.

14.3.1 Personal data protection issues before data collection

It is important to observe that GDPR requires data protection to be deployed before starting any data processing. This principle has two main practical consequences. The first regards the security measures that need to be implemented, and the second pertains to the transparency regarding the "data subject" who has to be informed on relevant aspects even if her/his consent is not required. In this section, we will outline both perspectives.

Concerning the DPIA, it can be observed that in order to determine the most suitable security measure, GDPR provides a tool, which is the DPIA (Article 35 GDPR). It is prescribed as mandatory whenever data processing is most likely to affect the rights and freedoms of individuals. As shown by the following

c. "Accountability" means to "*put in place internal mechanisms and control systems that ensure compliance and provide evidence to demonstrate it*," according to the European Data Protection Supervisor Glossary (https://edps.europa.eu/data-protection/data-protection/glossary)

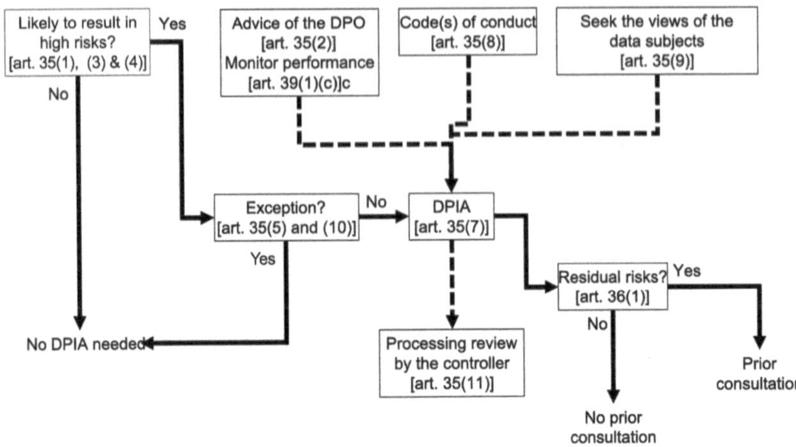

FIG. 14.1 Guidelines on data protection impact assessment by Article 29 Working Group, April 4, 2017 (page 7).

scheme (Fig. 14.1), a DPIA entails an evaluation that follows a risk-based approach, so it allows the identification of possible threats, assesses their probable impact, and estimates likely damages. Once such assessment is done, technical and organizational measures can be implemented in the data strategy.

It is important to point out that the DPIA is the result of a very complex procedure. The WG29, now the European Data Protection Board (EDPB), has clearly stated that such an evaluation is embodied in a document that, once released, has to be continuously reviewed and regularly re-assessed, as explained in the scheme below (Fig. 14.2).

Hence, in designing an endoscopic device, it is important to consider that it will very probably undergo a DPIA, so data processing features should be implemented in order to allow or facilitate it.

As regards the second aspect, transparency and the right to information, which represents its paramount expression, it can be observed that the GDPR requires the "data subject" to be informed "at the time when personal data are obtained" (Article 13 and Article 14 GDPR). This requirement is aimed at empowering individuals, allowing them to understand the details of the processing and to appreciate how to make a responsible choice to defend their rights and freedoms as well.

While designing endorobots, therefore, there should be consideration of a way to ensure the information of the patient is protected before beginning the endoscopic medical procedure, or before acquiring any data. In this sense, the appropriate instrument to provide any information related to processing is the privacy policy, and the correct moment of its submission is during the first contact with the patient (e.g., before booking an appointment or before the start of a clinical trial), even if it can be delivered "within a reasonable period after obtaining

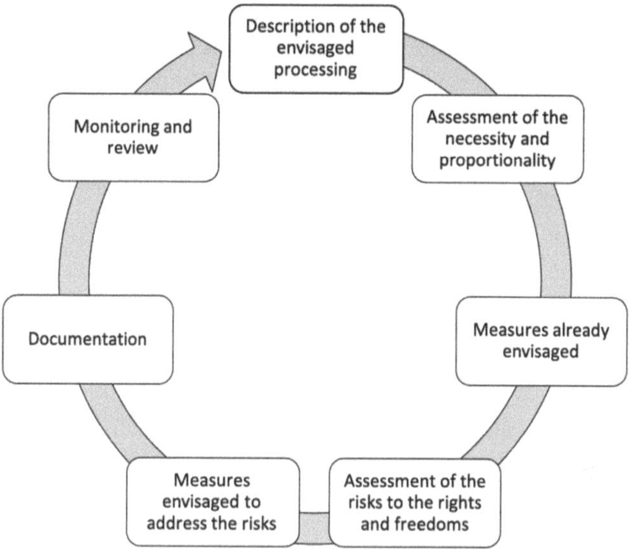

FIG. 14.2 Guidelines on data protection impact assessment by Article 29 Working Group, April 4, 2017 (page 16).

personal information" (Article 14(3)(a) GDPR). Such a document must be transparent, containing information about data processing as well as communication and modalities for the exercise of the rights of data subjects. It must use clear and plain language, especially if minors are involved (Article 12(1) GDPR). Should it be difficult, or even impossible, to give the required information (e.g., unconscious patients) before data processing, the GDPR requires health operators to fulfill the obligation of transparency in other ways, such as informing the data subject's next-of-kin.

An important observation needs to be made in order to clarify the intersection between data protection and bioethical legislation, since they often cross in the field of endorobots. In both cases, the right to information and the consent of the patient are relevant, and yet the balance of rights is set differently. Regarding the GDPR, in particular, we have already observed that while the fulfillment of the obligation of transparency is always required, the consent of the "data subject" is not always needed to legitimize data processing, since an alternative legal basis can become relevant. The legal grounds that allow the processing of data can be different and may relate to preventive or occupational medicine, medical diagnosis, provision of health or social care or treatment, or the management of health or social care systems and services, reasons of public interest in the area of public health, scientific research, etc. (Article 9(2) GDPR). Conversely, the consent of the patient is a stricter requirement under the Declaration of Helsinki of the World Medical Association, released in 1964 and revised in 2013, and the Oviedo Convention of the Council of Europe—open to signatures since 1997—as underlined by the European Data Protection Board [40], since the aim of the informed consent in this regard is to ensure the protection of human dignity and of integrity, which are also included in the Charter of Fundamental Rights of the European Union, thus binding EU Member States. Consequently, informed consent is required both for data processing and for clinical trials or medical procedures, since the one does not include the other, and vice versa.

Finally, it has to be considered that consent is the weakest legal basis for processing personal data according to the GDPR. Indeed, as stated before, the GDPR requires strict conditions to legitimize data processing (Article 4(11) GDPR and WP259) [41], since a "clear affirmative act" has to be put in place. Yet, the "data subject" is entitled to withdraw her/his consent at any time. Consequently, if consent is withdrawn, all data processing remains lawful but the controller shall stop the processing, and if there is no other legal ground that can justify the retention of data for further processing, it shall be deleted (Article 17(1)(b) GDPR). Hence, an efficient privacy policy concerning healthcare devices should aim at finding, where possible, other legal grounds to justify the processing aside from consent.

14.3.2 Personal data protection issues after collection

"Data controllers" and "data processors" are required to comply with data protection regulations not only with respect to their fundamental principles (Article 5 GDPR, in particular) but also before adopting and constantly updating adequate security measures (Article 32 GPDR) in order to prevent the risks of unauthorized access to data, data loss, data destruction, or data breach in general. In healthcare, such precautions—data encryption, pseudonymization (Article 4(5) GDPR), two-factor authentication, firewall, antivirus, and so on—become crucial with the wide-spread adoption of cloud computing, since it exposes patients' data remarkably. Even endorobots, if connected to cloud infrastructure, could be impacted by such vulnerabilities and, consequently, special precautions have to be taken.

One of the most valuable resources for enhancing security measures is the adoption of "privacy by design" precautions and the "privacy by default" option, which are included in Article 25 GDPR. Apart from the latter, which is essentially a standard choice of the most favorable setting for the "data subject," to whom the ultimate decision of raising the level of interaction has to be left, the concept of "privacy by design" deserves to be explored, since it can be of interest for engineers and developers of endoscopic devices.

As many know, the expression "privacy by design" was first coined by Ann Cavoukian at the beginning of the Millennium to label a set of seven Foundational Principles: (1) proactive not reactive, which focuses on prevention; (2) privacy as the default setting; (3) privacy embedded into design; (4) full functionality (not zero-sum!); (5) end-to-end security; (6) visibility and transparency; and (7) respect for user privacy [42]. As the same inventor remarked, the incorporation of such principles into technological devices inhibits their infringement but, likewise, introduces a higher level of complexity into the creative process, elevating costs of production. While some critics have pointed out that such principles are vaguely and generically formulated [43], other experts have stressed that such a paradigm is characterized by a high degree of intrinsic complexity [44]. The fact is that "privacy by design" has been adopted in different ways [45], with remarkable results in informed consent [46] and the sharing of medical records [47], showing that such an approach is quite promising despite the conflicting attitudes of those who are directly involved in it [48], especially if considering the adoption of blockchain technologies [49–51].

Regardless of the scenario depicted by the "privacy by design" approach, we can observe that there are other issues, such as the data transfer, which can hardly be solved definitively by the mere design of devices and processes. Indeed, after their collection, personal data can be transferred to or be shared with institutions, companies, or organizations in other countries. For example, a medical diagnosis could be uploaded into a dataset owned by a third party and substantially moved into another jurisdiction with a lower level of protection, thus avoiding legal liabilities for unlawful processing. On the contrary, such an elusive strategy for data protection is not allowed by the GDPR, which provides the "data subject" with a strong level of protection, since it requires that whenever data are transferred abroad, an "adequate level of data protection" has to be ensured, i.e., protection similar to that provided within the EU. In particular, the "adequacy" is assessed by the European Commission (Article 45 GDPR) either country by country or regarding a specific international organization, and it is affirmed by appropriate safeguards provided by the controller and processor (Article 46 GDPR), such as binding corporate rules (Article 47 GDPR). If none of these conditions apply, a transfer of personal data to third countries or international organizations can take place only if one of the conditions of Article 49 of the GDPR is met. Consequently, before transferring or sharing data, health operators must verify if the level of data protection granted to individuals is adequate in the target country and act responsibly and accordingly.

14.4 Conclusions

Data governance represents one of the most important current challenges in healthcare. Indeed, not only does digital transformation require integrating traditional resources (physical infrastructures, medical devices, personnel) with new ones (for example, artificial agents) but so does the exchange of information even at an international level and the combination of the needs of the public sector (especially in medical care) with those of the market (specifically, the protection of innovation). As we have seen, the EU is putting a remarkable effort not only into the improvement of quality in healthcare assistance in ordinary conditions but also in facing emergencies such as the one we are experiencing in these times as the COVID-19 pandemic is challenging the resilience of national systems [52].

Information security has become a fundamental requirement for every resource or device deployed in healthcare. Indeed, healthcare is a strategic sector but data also represents an extremely valuable asset distributed in a very wide ecosystem whose absence of defined boundaries makes it difficult to protect because data have to be considered together with all other components that inevitably depend on it: digital services, technological infrastructures, buildings and facilities, workers and, of course, the lives of patients. Every operator is required to safeguard the information under her or his control as well as she or he safeguards medical tools or adheres to workplace hygienic measures [53]. In other words, everyone needs to understand precisely the relevant rules in data ownership and data processing, as well as the roles of the data controller, data processor, mere recipient, or third party, since different kinds of responsibility are assigned and decision-making power is assigned at each stage of the data workflow.

Data protection has become—and will be even more—a matter to be taken into serious consideration in the design of medical devices and even of endorobots. The current legislation offers, in this regard, many tools to improve the value of personal data and to exploit their potential benefit, such as the principles of accountability (Article 5(2) GDPR), privacy by design and by default (Article 25 GDPR), and the possibility of setting up and adopting certification schemas (Article 46 GDPR). Compliance with GDPR provisions is certainly a challenge for engineers and developers of medical devices, since it has to be implemented in every action, instrument, and tier of any procedure [54]. However, these requirements can also be considered opportunities for exploiting the real potential of data while safeguarding the fundamental rights and freedoms of patients.

As regards the future, it is noteworthy that, among the numerous initiatives aimed by the European Union at "shaping a digital future" [55], the proposal of Regulation called "Artificial Intelligence Act" approved on 21 April 2021 [56], which is the outcome of many years of discussions and consultations at different levels, a special consideration is devoted to privacy. Indeed, according to Article 10(5), the processing of special categories of personal data is allowed by "high risk" AI systems only under strict requirements.Therefore, for designers, engineers, and developers, it becomes crucial to keep the pace with the discussion on this topic in order to be prepared with the many significant changes that will be introduced in the next few years.

Author contributions

This chapter is the result of joint research of the co-authors. Individual contributions can be attributed as follows: Federico Costantini, paragraphs 1, 2 and 4; Giada Soncini, paragraph 3.

References

[1] Perakslis E, Coravos A. Is health-care data the new blood? Lancet Digit Health 2019;1:e8–9. Available from https://doi.org/10.1016/S2589-7500(19)30001-9.

[2] Lutz C, Tamò A. Privacy and healthcare robots—an ANT analysis. We robot, http://robots.law.miami.edu/2016/wp-content/uploads/2015/07/We_Robot_Lutz_Tamo_ANT_Healthcare_Robots_final.pdf; 2016.

[3] Rueben M, Aroyo AM, Lutz C, Schmölz J, Van Cleynenbreugel P, Corti A, et al. Themes and research directions in privacy-sensitive robotics. In: 2018 IEEE workshop on advanced robotics and its social impacts (ARSO) Genoa (IT). IEEE; 2018. p. 77–84. Available from https://doi.org/10.1109/ARSO.2018.8625758.

[4] Council Regulation (EU) 2018/1488 of 28 September 2018 establishing the European High Performance Computing Joint Undertaking. OJ L 252, 8.10.2018, p. 1–34. Available from: http://data.europa.eu/eli/reg/2018/1488/oj/.

[5] COM(2016) 178 final. European Cloud Initiative – building a competitive data and knowledge economy in Europe. Available from: https://eur lex.europa.eu/legal-content/EN/TXT/?uri=CELEX%3A52016DC0178.

[6] COM/2020/66 final. A European strategy for data COM/2020/66 final. Available from: https://eur-lex.europa.eu/legal-content/EN/TXT/?qid=1593073685620&uri=CELEX%3A52020DC0066.

[7] COM/2018/233 final. On enabling the digital transformation of health and care in the digital single market; empowering citizens and building a healthier society. Available from: https://eur-lex.europa.eu/legal-content/EN/TXT/?uri=COM%3A2018%3A233%3AFIN.

[8] Directive 2011/24/EU of the European Parliament and of the Council of 9 March 2011 on the application of patients' rights in cross-border healthcare. OJ L 88, 4.4.2011, p. 45–65. Available from: http://data.europa.eu/eli/dir/2011/24/oj.

[9] Regulation (EC) No 883/2004 of the European Parliament and of the Council of 29 April 2004 on the coordination of social security systems. OJ L 166, 30.4.2004, p. 1–123. Available from: http://data.europa.eu/eli/reg/2004/883/oj.

[10] Council Regulation (EU, Euratom) 2020/2093 of 17 December 2020 laying down the multiannual financial framework for the years 2021 to 2027 OJ L 433I , 22.12.2020, p. 11–22, ELI: http://data.europa.eu/eli/reg/2020/2093/oj.

[11] COM/2021/118 final of 9 March 2021, 2030 Digital Compass: the European way for the Digital Decade, https://eur-lex.europa.eu/legal-content/EN/TXT/?uri=CELEX:52021DC0118.

[12] Hulsen T. Sharing is caring—data sharing initiatives in healthcare. Int J Environ Res Public Health 2020;17(9):3046. https://doi.org/10.3390/ijerph17093046.

[13] Sardi A, Rizzi A, Sorano E, Guerrieri A. Cyber risk in health facilities: a systematic literature review. Sustainability (Switzerland) 2020;12. https://doi.org/10.3390/su12177002.

[14] Kruse CS, Frederick B, Jacobson T, Monticone DK. Cybersecurity in healthcare: a systematic review of modern threats and trends. Technol Health Care 2017;25:1–10. https://doi.org/10.3233/thc-161263.

[15] Jalali MS, Razak S, Gordon W, Perakslis E, Madnick S. Health care and cybersecurity: bibliometric analysis of the literature. J Med Internet Res 2019;21. https://doi.org/10.2196/12644.

[16] Pandey AK, Khan AI, Abushark YB, Alam MM, Agrawal A, Kumar R, et al. Key issues in healthcare data integrity: analysis and recommendations. IEEE Access 2020;8:40612–28. https://doi.org/10.1109/ACCESS.2020.2976687.

[17] Tully J, Selzer J, Phillips JP, O'Connor P, Dameff C. Healthcare challenges in the era of cybersecurity. Health Secur 2020;18:228–31. https://doi.org/10.1089/hs.2019.0123.

[18] CyberPeace Institute. A call to all governments: work together now to stop cyberattacks on the healthcare sector. Available from: https://cyberpeaceinstitute.org/campaign/call-for-government; 26 May 2020.

[19] European Commission. President von der Leyen on the EU-China Summit. Available from: https://ec.europa.eu/commission/presscorner/detail/en/statement_20_1162; 22 June 2020.

[20] Directive (EU) 2016/1148 of the European Parliament and of the Council of 6 July 2016 concerning measures for a high common level of security of network and information systems across the Union. OJ L 194; 2016. p. 1–30. Available from https://eur-lex.europa.eu/eli/dir/2016/1148/oj.

[21] Regulation (EC) No 460/2004 of the European Parliament and of the Council of 10 March 2004 establishing the European Network and Information Security Agency. OJ L 77, 13.3.2004, p. 1–11. Available from: https://eur-lex.europa.eu/eli/reg/2004/460/oj.

[22] Regulation (EU) 2019/881 of the European Parliament and of the Council of 17 April 2019 on ENISA (the European Union Agency for Cybersecurity) and on information and communications technology cybersecurity certification and repealing Regulation (EU) No 526/2013 (Cybersecurity Act). OJ L 151, 7.6.2019, p. 15–69. Available from: https://eur-lex.europa.eu/eli/reg/2019/881/oj.

[23] ISO/IEC 27000. Available from https://www.iso.org/cms/render/live/en/sites/isoorg/contents/data/standard/07/39/73906.html; 2018.

[24] ISO 27799. Available from https://www.iso.org/cms/render/live/en/sites/isoorg/contents/data/standard/06/27/62777.html; 2016.

[25] Leetaru K. Without transparency, democracy dies in the darkness of social media. Forbes; 2018. Available from: https://www.forbes.com/sites/kalevleetaru/2018/01/25/without-transparency-democracy-dies-in-the-darkness-of-social-media/.

[26] Regulation (EC) No 1049/2001 of the European Parliament and of the Council of 30 May 2001 regarding public access to European Parliament, Council and Commission documents. OJ L 145, 31.5.2001, p. 43–48. Available from: https://eur-lex.europa.eu/eli/reg/2001/1049/oj.

[27] Resolution on the deadlock on the revision of Regulation (EC) No 1049/2001. Procedure File: 2013/2637(RSP). Available from: https://oeil.secure.europarl.europa.eu/oeil/popups/ficheprocedure.do?lang=en&reference=2013/2637(RSP).

[28] Directive 2003/98/EC of the European Parliament and of the Council of 17 November 2003 on the re-use of public sector information. OJ L 345, 31.12.2003, p. 90–96. Available from: http://data.europa.eu/eli/dir/2003/98/oj.

[29] Directive 2013/37/EU of the European Parliament and of the Council of 26 June 2013 amending Directive 2003/98/EC on the re-use of public sector information. OJ L 175, 27.6.2013, p. 1–8 Available from: http://data.europa.eu/eli/dir/2013/37/oj.

[30] Lassinantti J. Re-use of public sector open data - characterising the phenomena. Int J Public Inf Syst 2019;13. Available from: http://www.ijpis.net/ojs/index.php/IJPIS/article/view/158.

[31] Directive (EU) 2019/1024 of the European Parliament and of the Council of 20 June 2019 on open data and the re-use of public sector information. OJ L 172, 26.6.2019, p. 56–83. Available from: http://data.europa.eu/eli/dir/2019/1024/oj.

[32] Regulation (EU) 2018/1807 of the European Parliament and of the Council of 14 November 2018 on a framework for the free flow of non-personal data in the European Union. OJ L 303, 28.11.2018, p. 59–68. Available from: https://eur-lex.europa.eu/eli/reg/2018/1807/oj.

[33] European Commission. Presentation of Codes of Conduct on cloud switching and data portability. 26/11/2019. Available from: https://ec.europa.eu/digital-single-market/en/news/presentation-codes-conduct-cloud-switching-and-data-portability.

[34] COM/2019/250 final. Guidance on the regulation on a framework for the free flow of non-personal data in the European Union. Available from: https://eur-lex.europa.eu/legal-content/EN/TXT/?uri=COM:2019:250:FIN.

[35] Directive 95/46/EC of the European Parliament and of the Council of 24 October 1995 on the protection of individuals with regard to the processing of personal data and on the free movement of such data. OJ L 281, 23.11.1995, p. 31–50. Available from: http://data.europa.eu/eli/dir/1995/46/oj/.

[36] Regulation (EU) 2016/679 of the European Parliament and of the Council of 27 April 2016 on the protection of natural persons with regard to the processing of personal data and on the free movement of such data, and repealing Directive 95/46/EC (General Data Protection Regulation). OJ L 119, 4.5.2016, p. 1–88. Available from: http://data.europa.eu/eli/reg/2016/679/oj.

[37] Directive (EU) 2016/680 of the European Parliament and of the Council of 27 April 2016 on the protection of natural persons with regard to the processing of personal data by competent authorities for the purposes of the prevention, investigation, detection or prosecution of criminal offences or the execution of criminal penalties, and on the free movement of such data, and repealing Council Framework Decision 2008/977/JHA, OJ L 119, 4.5.2016, p. 89–131. Available from: https://eur-lex.europa.eu/eli/dir/2016/680/oj.

[38] Directive (EU) 2016/681 of the European Parliament and of the Council of 27 April 2016 on the use of passenger name record (PNR) data for the prevention, detection, investigation and prosecution of terrorist offences and serious crime OJ L 119, 4.5.2016, p. 132–49. Available from: https://eur-lex.europa.eu/eli/dir/2016/681/oj.

[39] Voigt P. The EU general data protection regulation (GDPR): a practical guide. Cham: Springer; 2017.

[40] European Data Protection Board, Opinion 3/2019 concerning the Questions and Answers on the interplay between the Clinical Trials Regulation (CTR) and the General Data Protection regulation (GDPR), adopted on 23 January 2019, https://edpb.europa.eu/our-work-tools/our-documents/dictamen-art-70/opinion-32019-concerning-questions-and-answers_en.

[41] European Data Protection Board, Guidelines 05/2020 on consent under Regulation 2016/679, Version 1.1, adopted on 4 May 2020, https://edpb.europa.eu/our-work-tools/our-documents/guidelines/guidelines-052020-consent-under-regulation-2016679_en.

[42] Cavoukian A. Understanding how to implement privacy by design, one step at a time. IEEE Consum Electron Mag 2020;9:78–82. Available from https://doi.org/10.1109/MCE.2019.2953739.

[43] Fosch-Villaronga E, Felzmann H, Ramos-Montero M, Mahler T. Cloud services for robotic nurses? Assessing legal and ethical issues in the use of cloud services for healthcare robots. In: 2018 IEEE/RSJ international conference on intelligent robots and systems (IROS). Madrid, Spain: IEEE; 2018. p. 290–6. Available from https://doi.org/10.1109/IROS.2018.8593591.

[44] Pagallo U. On the principle of privacy by design and its limits: technology, ethics and the rule of law. In: Gutwirth S, Leenes R, De Hert P, Poullet Y, editors. European data protection: in good health? Dordrecht: Springer Netherlands; 2012. p. 331–46. Available from https://doi.org/10.1007/978-94-007-2903-2_16.

[45] Semantha FH, Azam S, Yeo KC, Shanmugam B. A systematic literature review on privacy by design in the healthcare sector. Electronics 2020;9:452. Available from https://doi.org/10.3390/electronics9030452.

[46] O'Connor Y, Rowan W, Lynch L, Heavin C. Privacy by design: informed consent and Internet of Things for smart health. Procedia Comput Sci 2017;113:653–8. https://doi.org/10.1016/j.procs.2017.08.329 [cited 2020 Sep 22].

[47] Stevovic J, Bassi E, Giori A, Casati F, Armellin G. Enabling privacy by design in medical records sharing. In: Gutwirth S, Leenes R, de Hert P, editors. Reforming European data protection law. Dordrecht: Springer Netherlands; 2015. p. 385–406. https://doi.org/10.1007/978-94-017-9385-8_16.

[48] Bednar K, Spiekermann S, Langheinrich M. Engineering privacy by design: are engineers ready to live up to the challenge? Inf Soc 2019;35:122–42. https://doi.org/10.1080/01972243.2019.1583296.

[49] Dimitrov DV. Blockchain applications for healthcare data management. Healthc Inform Res 2019;25:51–6. https://doi.org/10.4258/hir.2019.25.1.51.

[50] Albanese G, Calbimonte J-P, Schumacher M, Calvaresi D. Dynamic consent management for clinical trials via private blockchain technology. J Ambient Intell Humaniz Comput 2020. https://doi.org/10.1007/s12652-020-01761-1.

[51] Shahnaz A, Qamar U, Khalid A. Using blockchain for electronic health records. IEEE Access 2019;7:147782–95. https://doi.org/10.1109/ACCESS.2019.2946373.

[52] Becher S, Gerl A, Meier B, Bölz F. Big picture on privacy enhancing technologies in e-health: a holistic personal privacy workflow. Information (Switzerland) 2020;11. https://doi.org/10.3390/info11070356.

[53] Lea NC, de Meyer F. A primer on cybersecurity for physicians. Acta Medica Port 2019;32:563–4. Available from: https://pesquisa.bvsalud.org/portal/resource/pt/mdl-31493357.

[54] Larrucea X, Moffie M, Asaf S, Santamaria I. Towards a GDPR compliant way to secure European cross border healthcare industry 4.0. Comput Stand Interfaces 2020;69. https://doi.org/10.1016/j.csi.2019.103408.

[55] European Commission. Shaping europe's digital future; 19 February 2020. https://doi.org/10.2759/091014.

[56] Proposal for a Regulation of 21 April 2021 laying down harmonised rules on artificial intelligence (Artificial Intelligence Act) and amending certain union legislative acts, https://eur-lex.europa.eu/legal-content/EN/TXT/?uri=CELEX:52021PC0206, Com/2021/206 final.

Chapter 15

Project management

Pietro Raffaini[a] and Luigi Manfredi[b]
[a]*Project Control Management, London, United Kingdom,* [b]*Division of Imaging Science and Technology, School of Medicine, University of Dundee, Dundee, United Kingdom*

15.1 Introduction

Project management (PM) appeared as a business and social process after the Second World War, aiming to enable the successful completion and development of technology and infrastructure. However, the PM discipline and its application emerged in the early 1940s in the Manhattan Project, whereby a combination of existing methodologies and innovative methods were developed in the context of an extraordinary and unprecedented endeavor. Mainly concentrated on the use of techniques and tools as planning, measuring, controlling, and organizing teams [1], it is extensively used by the U.S. Department of Defence and Nasa, construction, and the oil and gas and civil engineer sectors. Those businesses have made a major contribution toward the economic development of various countries in several areas (e.g., engineering, technological, scientific, and social) [2]. Therefore, the financial appeal and associated intrinsic riskiness of those sectors have raised the discipline of PM to the top of the agenda of several multinational companies. Such industries mainly involve a combination of different engineering, procurement and construction operations. Thus, a combination of systemized processes and aptitudes to transfer human ideas into reality is required [3].

Many scholars have attempted to define PM as the capability to complete the project about given specifications and within the cost and time constraints. The criteria for measuring success have been defined as the "iron triangle". Indeed, it is the most widely used criteria to measure performance, mainly focused around the use of schedule, cost, and quality compliances techniques. It relies upon planning, measuring, controlling all aspects of a project and motivating all those engaged in it (i.e., organizing teams) to fulfil project requirements on time and under specific cost and quality constraints. The discipline includes many management theories and "best practices," which have been deployed over the years to create the status of PM as a theory [4]. Those theories and best practices applied in PM are rooted in a mix of scholarly research and business

Endorobotics. https://doi.org/10.1016/B978-0-12-821750-4.00015-3

practices applied in different industries, aiming to deliver a specific project [5]. Hundreds of people, members of the PM institute, worked to collect those PM theories and best practices in the Guide of Project Management: PMBOK (Project Management Body of Knowledge) [6]. The PMBOK covers a standardized collection of recognized PM practices, knowledge areas.

Over the last few decades, the necessity to standardize those procedures under uncertainty, over project planning, execution and control, enhanced the development of theories seeking to identify the appropriate method to secure project success. Nonetheless, projects have witnessed the importance of strategy, external factors such as business-driven changes, socioeconomic factors, and team culture. Those factors conform with the necessity conveyed in scholars' research, to reduce the gap between theoretical approaches and empirical realities, as shown in Morris works [7].

This chapter, using theory and practical examples, aims to bridge the above-mentioned gap by perfecting and crafting a comprehensive body of knowledge that has the capacity to support PM in decision-making in different backgrounds.

15.2 Literature review and project management (PM) technique

15.2.1 Iron triangle and project performance

The key literature on PM acknowledges the critical role played by the coordination of resources to deliver a project under time, cost, and quality constraints (i.e., iron triangle) (Fig. 1).

The concept of the iron triangle is described as a triangle with specific criteria on the vertices (time, quality, and cost). The iron triangle evaluates

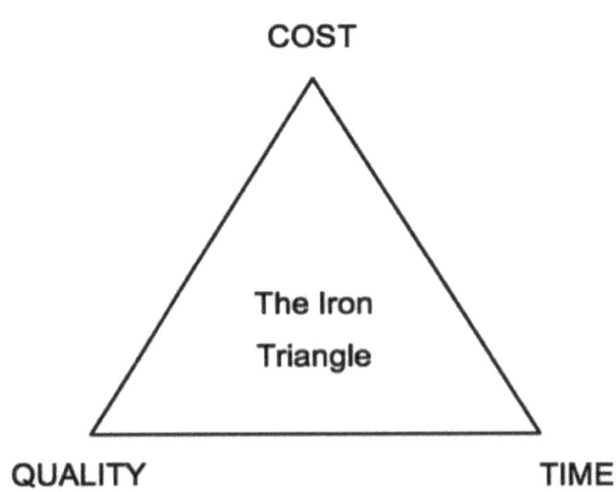

FIG. 1 The iron triangle: a PM tool to describe criteria to assess project success, such as time, quality, and cost.

objective success criteria, which are the most widely cited measures of project success. According to de Wit, project phases define the importance of time, cost, and quality through the lifetime of the project [8]. The time and the schedule of the works in the early phase of any project plays a primary role. But later in the project, cost gains momentum as the key factor and becomes the main concern [9]. Consequently, the Triangle vertices move according to the phase in which the project is at, at any specific point.

Despite its well-known reputation among scholars, the quintessence of project performance has profoundly changed over time [10]. The complexity of each project context has demonstrated that when a project is delivered misreading and misinterpreting the iron triangle, that can lead to project failure [9]. Complexity itself involves a number of different interrelated factors that affect the project objectives of time, cost, and quality through the lifetime of the project [11]. With regard to the iron triangle and its application to megaprojects, the experience across the world demonstrates that the "pathological" use of the criteria leads to strategic misrepresentations and planning fallacies [12]. Once the project is set, the time and schedule are no longer the only dimensions of the project. Cost embodies the other factor—as part of the iron triangle—that gathers momentum in the subsequent stage of the project.

There is a project control management technique that provides quantitative methods of schedule performance. In the applicable terminology, this is known as the *earned value method* (EVM) [13]. This criterion of PM is used to check and observe schedule performance with cost performance indicators, which benchmark the budgeted cost of the works performed against what was originally scheduled and the actual costs of the works performed. Therefore, cost and planning are interdependent dimensions. After the project is completed, the schedule and the costs are no longer core issues and quality becomes the most important factor.

The commentator Atkinson has criticized the technique of evaluating the criteria costs and time against the schedule, as "only guesses" and quality as a "phenomenon," in light of the reality that 50%–70% of the projects end in a budget or schedule overrun [14]. Therefore, original estimates are often inadequate. Quality, as one element of the iron triangle "vertices", has been treated as a subjective criterion by the commentators Morris, Hough, and Heisler [15]. And the third dimension, the quality, according to Turner, is nothing more than an ambiguous method [15]. However, he is one of the few academics who has attempted to more clearly describe project quality comprising two dimensions such as product quality and process quality.

Turner's view of quality—product quality and process quality—forms part of the total quality management (TQM) approach. It is widely embraced in the manufacturing and service industries to deliver products and services with no defects, matching customers' requirements.

The concept of "quality", despite its non-"objective" definition, has been an essential element in the success of projects in several industries and its trade-off between cost criteria has been included in the latest research (e.g., cost of non-quality) [16]. Further, common management procedures employed on a day-to-day basis have been lifted to the top of the agenda, aiming to create the right attitude in the management and teams involved in various projects. However, according to Wilkinson et al. [17], the quality dimension has not yet been accurately addressed within the existing literature. The various theories supporting the establishment of a wider approach toward PM "quality" have been on the increase in the past few decades: theories have split short-term and long-term success criteria, as part of teams and stakeholders' involvement in the strategy of a project [18]. Short-term success criteria are predominantly oriented around delivering a project on time, in budget, and within scope. Conversely, long-term success criteria involve not only projects but organizations. Therefore, the criteria become wider and include integration, communications, human resources, procurement-supply chains, and risks. Long-term success criteria can be included in "program-portfolio" management (i.e., different projects, either independent or managed and coordinated as a group and forming part of the same organizational objective), as a broader view over PM and bring to attention of the audience, raising the role of organizations as active representatives in establishing TQM culture. Flyvbjerg et al. have retraced the external role played by organizations in the projects to set long-term success criterion, which goes beyond the concept of the iron triangle.

In the late 1980s, as illustrated in De Wit's research, another methodology brought the role of stakeholders to the fore, through the project life cycle in measuring project success. Stakeholder satisfaction and team member relationships play a pivotal role. An excessive focus on the iron triangle, as shown by Maylor et al.'s studies, can limit organization effectiveness and stakeholder satisfaction [19].

Since the development of the PM discipline, academics and professionals have attempted to encapsulate in the iron triangle theory the core elements of project success and project failure. The measurement of performance is undoubtedly an essential part of project control, but this is not the only degree of measurement of success. As noted above, there are different factors affecting the project and the most appropriate criteria for success change during the life cycle of the project. Academics and professionals have adopted different theories and guidelines, as they have identified a sequence of factors, either from experience or from studies, that are essential to success.

Nowadays, PM has been applied as a practical discipline to enable organizations operating in different industries to standardize and allocate human, financial, and technical capabilities to reach an objective. Hence, the introduction of a global instrument as PMBOK as part of the standardization process enables knowledge sharing to take place.

Conceptually, a project presents a difficult know-how problem: "How do we execute a project?"

All guidelines have the execution discipline approach: delivering the project on time, on budget, and under quality constraints, with the attention concentrated on the stakeholders' side (i.e., internally and externally). The PMBOK approach introduced a role for stakeholders with a widely shared understanding of what a project manager should do regarding the stakeholder's objectives. However, the stakeholder and human resources areas in the PMBOK are based on theories stemming from the 1970s. As such, contemporary stakeholder theories might offer new insights into our understanding of project stakeholder management. The objectives of the stakeholders involved in a project are unlikely to be compatible, as the aim of the party might differ.

Further in the application part, this chapter will go into various details demonstrating the variety of objectives in different industries and among stakeholders (internal-external factors).

In addition, the role of project managers will be analyzed in greater detail in light of their crucial role at understanding how micro (the project) and macro expectations (the organization and the stakeholder) are evaluated.

Different techniques support the PM discipline portrayed in the literature. The most used include the Gantt chart, the critical path method (CPM), and the project evaluation and review technique (PERT).

15.2.2 Gantt chart

This method was developed by Henry Gantt in 1910 [20] with the aim of representing the establishment of repetitive routine processes through the scientific management of work. The main focus of the method is on the efficient and effective use of resources and productivity [21].

According to Wren, a Gantt chart is a developed visualization tool that enables foremen and workmen to check the efficiency of employees, thus reducing idle times [22]. The first version of a Gantt chart managed the schedule based on quantities. But the charts were mainly focused on the past and missed the future dimension. Consequently, the second version of the Gantt chart, which is more like what is used today, emphasizes not only the present, but also the comparison between the planned and actual results. In addition, it provides a visualization of the issues limiting efficiency (i.e., critical path). An updated version of the Gantt charts was not restricted to the coordination of production, but also broadened its scope to recognize the patterns of failures, enabling counterchecks to be activated by managers. The current version applied in PM software uses the Gantt chart as the key factor to plan and manage projects (i.e., Microsoft Project, Primavera P6, Fig. 2). In addition, PM software and Gantt charts have been used for planning and cost purposes.

A Gantt chart shows different aspect of the project schedule, such as the *critical path* [23]. The *critical path* is a sequence of *work package* (WP) with its duration, from the start of the project up till its end. Each WP includes tasks and subtasks. WPs are small units of work of a project. Tasks are grouped into WPs based on the geographical area, engineering discipline, technology, or time to accomplish them. WPs can be set at the beginning in parallel if the natural

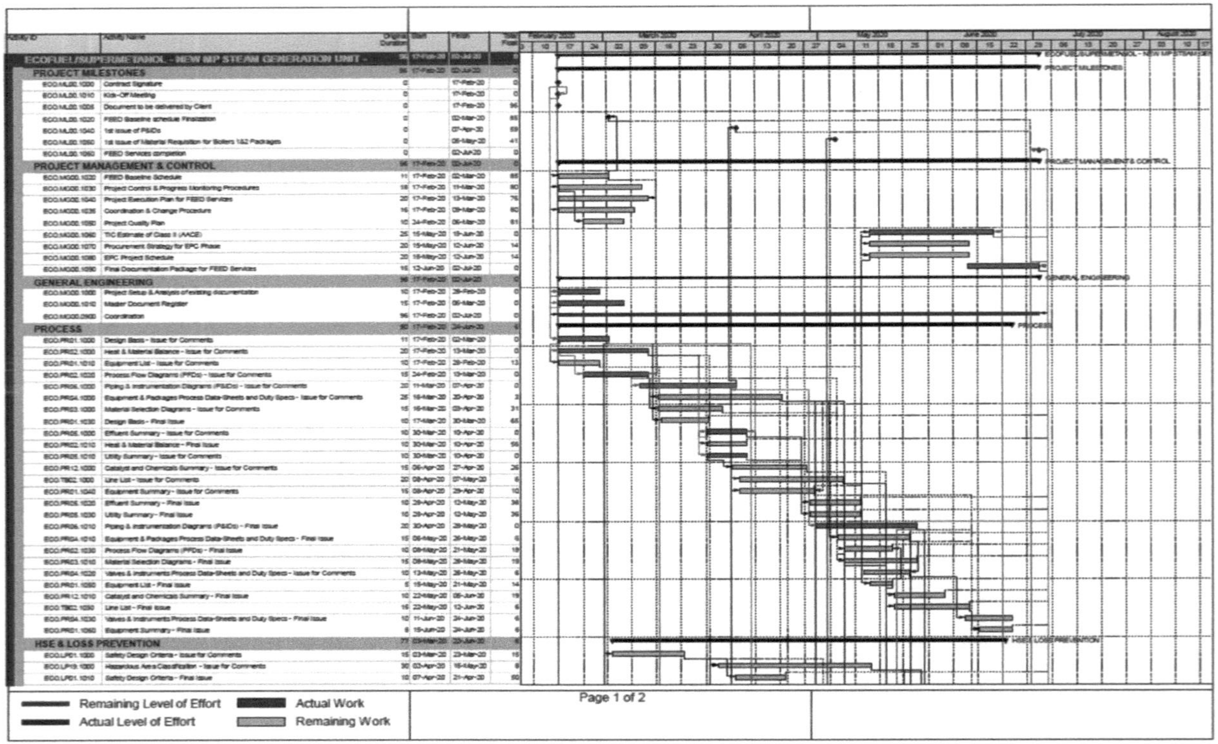

FIG. 2 Use of Gantt chart and CPM in Primavera P6 software. The red rectangle (*gray color in print version*) indicates that the activity is in a critical path, green rectangle (*light gray color in print version*) shows the remaining work, and blue rectangle (*dark gray color in print version*) indicates the activities performed.

work logic and availability of resource allow this process. Otherwise, these can be set by changing the *logical dependency* between work packages. There are four types of dependency [23]:

- *End-to-Start*: WP B cannot start until WP A is finalized.
- *End-to-End*: WP B cannot be finalized until WP A is finalized.
- *Start-to-Start*: WP B cannot start until WP A starts.
- *Start-to-End*: WP B cannot be finished until WP A starts.

A logical approach to start a Gantt chart is by putting all the WP with no dependency. After carefully checking each WP, the dependencies can be assigned. Creation of too many WPs can be difficult and can bring to lose the control at the large scale. The estimation of the duration of the WP should be done on the basis of the planned lead time and not on the basis of resource expenditure. Only the most important dependencies of the WPs should be indicated in a Gantt chart; otherwise, it quickly becomes difficult to read.

The results and outputs of a project are the *deliverables*. They are expected by the stakeholders, such as a product, service, as well as paperwork and note. They are needed to evaluate the progress or completion of the project. Example of paperwork includes but is not limited to signed contracts, expense reports, or prototypes or components of a product. Deliverables are reported by project managers and are presented to stakeholders. Different stakeholders have different needs, and this changes the way to report.

Estimation on the duration of WP and tasks requires important skills. These estimations are vital to the success of a project. Frequently, team members experience time of idleness; in different circumstances, team members can have too much pressure to accomplish the expected output. This estimation can involve other team member for feedbacks and suggestions. Each task has assigned available resources. This helps to avoid problem such as scheduling conflicts or over-allocating work. Each work carried by a resource can be tracked and recorded by many available software packages to help to utilize the resource with maximum efficiency.

15.2.3 Critical path method (CPM)

CPM or critical path analysis is a project modeling technique to schedule a set of project activities, commonly used in conjunction with a PERT (project evaluation and review technique) chart. It includes a list of activities required to complete a project, the time for each activity to be completed, the dependencies between the activities, and the end-points such as milestones or deliverables.

15.2.4 The project evaluation and review technique (PERT) chart

The PERT is a statistical technical tool to analyze the tasks involved in a project, particularly the time needed for each task and the minimum time required to complete the total project. It relies on arrow and node diagrams of activities and events. The arrows represent the activity or work needed to achieve the events or nodes that indicate each completed phase of a project. PERT was invented in 1958, in order to examine the uncertainty of research projects, which had not been performed earlier. The PERT technique was introduced in the Polaris nuclear submarine program by the U.S Navy [24]. The purpose of the PERT theory is to help organizations undertake new projects and to determine the anticipated time for the completion of a project. As time and costs are part of the project constraints, PERT analysis supports budgeting, costs, and resources control.

The use of PERT involves the breakdown of the project into different tasks at the initial stage, which enables project managers and project directors to take the necessary action to accomplish the requisite objective on time and under costs control. PERT is based on the definition of a project as a well-defined system of activities and milestones combined with mathematical and statistical variables to calculate the "critical path" [25]. A PERT mesh is a graphical representation to plan and control the development of a project. This process starts by defining the activities that must be executed to carry out a project and the defined requirements. Each activity has a predecessor and the expected duration, estimated from people with experience in a similar task. The duration of the activity is estimated by using the following formula:

$$te = \frac{t_a + 4t_m + t_b}{6}$$

whose variance is given by:

$$\sigma^2 = \frac{(t_b - t_a)^2}{6}$$

And a standard deviation:

$$\sigma = \frac{t_b - t_a}{6}$$

where t_a is the shorter time that an activity can last, t_m is the most probable time, and t_b is the most pessimistic time that an activity can last.

The drawing of a PERT includes nodes and arcs. Nodes represent instances in time, including the start time of one or more activities and the end-time of several other activities. The arches represent the activity, with an initial node and a final node.

The implementation of a PERT starts by determining the tasks that a project requires and those that must be completed. There are two types of network diagram: activity on arrow (OAO) and activity on node (AON). AON is the simpler to create and interpret and it is recommended, but not necessary, to start with a node named start, with duration zero (0). Each activity that does not have a predecessor activity is then drawn, and they are connected with an arrow from the start to each node (Fig. 3).

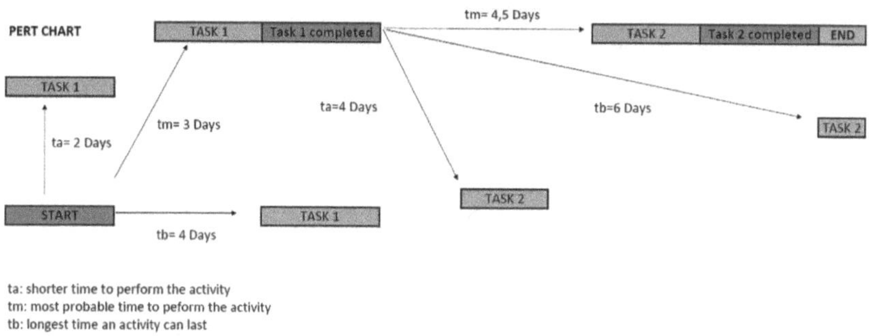

FIG. 3 PERT chart, where t_a, shorter time to perform the activity, t_m, most probable time to perform the activity, t_b, longest time an activity can last.

This process continues until the final activity connected to the last node, which is labeled *end* or *finish*. A project manager can use it to develop a realistic timeframe for the project. This can be done by analyzing the longest path based on the estimation entered in the diagram. This analysis is done by including tasks that take longer to be completed.

15.2.5 Comparison between PERT, Gantt, and CPM

In terms of PERT theory, the use of CPM is pivotal at the initial stage of the project in order to plan and control the progress of the activities, considering the interdependency of the tasks on each other (i.e., critical task), schedule tasks, track the progress, and find the dependence: CPM associated with Gantt chart [26]. Although a network diagram does not give much more information than a Gantt chart, it can be expanded to display more information and a different visual inspection. A PERT is easy to organize in complex projects with a variety of moving parts by visualizing the dependencies between each step of the process.

A PERT chart is a useful tool to determine the critical path in a project in order to guarantee that all the deadlines are met; display the various interdependencies between the tasks; estimate the amount of time that is needed to complete a project; understand the workflow; and select the most efficient and beneficial path. The analysis of a PERT chart requires data from various departments within an organization. The combination of all this information helps to identify each responsible team within the company. It also facilitates each department to take responsibility for their work. With this visual analysis, deadlines can be more predictable, interdependencies clarified, and the appropriate order identified in respect of the completion of each of the relevant tasks.

However, there are a variety of disadvantages that can be laid at the door of a PERT chart. The full financial position can be lost because of a strict focus on deadlines. If any calculation is inaccurate, this can cause delay, bottlenecks, and negative impact in the final schedule time. In the absence of experience, the project manager can produce unreliable data or expectations in respect of the cost and time frame of the project. The creation of a PERT chart is also a labor-intensive task, requiring additional time and resources, with ongoing reviews to keep the chart consistent.

In the same case, a PERT chart is preferred over a Gantt chart because it clearly illustrates task dependencies. However, on the other hand, a PERT chart can be much more difficult to interpret, especially on complex projects. A project manager often uses both techniques, and software can easily allow both visualizations. Both charts help to simplify the PM process by using a breakdown structure that divides the project into smaller tasks and identifies any constraints. These tools allow managers to increase efficiencies and view the project as a whole. They both help in identifying dependencies to improve time management.

A PERT chart is the best tool to use before a project begins, and also to plan and estimate the length of each task. This can be compared with the Gantt chart, which with CPM is the best option to use while a project is running because of how it enables visualization by date and the amount of completed work.

15.3 Internal and external factors influencing PM, organizations management, and technology readiness level

15.3.1 Stakeholders

Stakeholders are individuals, groups, or an entire organization that have specific interests, who may be impacted, or could have a business interest in the outcome of the project. They can be the members of the team such as the project managers, executives, sponsors, customers, users, suppliers, vendors, or contractors. They could have invested and be affected by the project at any point. Their input could have an impact upon the outcome. The management of stakeholders is an essential consideration, and it is good practice to keep communication and collaboration flowing throughout the project [23].

Key project stakeholders are those who have the influence and authority to judge whether and when a project is a success or not. Their objectives must be satisfied with the project outcomes. In the same vein, even though deliverables are fulfilled and budget requirements are met, if key project stakeholders are dissatisfied, the project will be considered a failure.

Examples of key project stakeholders are as follows:

- *Customers* are the users or a product or a service. They can be internal or external to the organization that is executing a project.
- *Project manager*, the leader of the project.
- *Project team members*, the group that executes the project.
- *Project sponsor*, who finances the project.

- *Steering committee*, which are the advisory group providing guidance on key decisions. They can include the sponsors.
- *Executive*, the top management that executes the project and leads the strategy of the organization.
- *Resource managers*, who control resources that are needed for the execution of the project.

Many different key stakeholders can be part of a project, including sellers and suppliers, contractors, owners, government agencies, and media outlets.

Stakeholders in PM can be *internal* or *external* [27]. *Internal stakeholders* are those within the organization, including the top management, the project and resource manager, project team members, and internal customers. *External stakeholders* are not part of the organization. They can include external customers, government entities, contractors, subcontractors, and suppliers. The execution of a project plan would benefit from the identification of the different internal and external stakeholders. This identification process is needed to guarantee that each stakeholder is informed and involved in the project plan and successfully satisfied by the outcome.

A *stakeholder management plan* is a written document that outlines the team plans to manage goals and expectations of key stakeholders and is used to guarantee this process and usually kept by the project manager. The project manager can periodically review and update this plan by involving the project team and gathering input and feedback from the stakeholders as well. The intention is that this will better serve the needs of the project and keep stakeholders informed.

15.3.2 Product development and management association (PDMA)

PDMA is a global community of professional members whose skills, expertise, and experience drive the most recognized and respected innovative companies in the world. It is a nonprofit professional society that communicates and organizes information about the development of new products, process, strategy innovation, market research, tools and metrics, organizational issues, and portfolio management. It was founded in 1976 and has currently about 3500 members worldwide. They provide a training program leading to New Product Development Professional (NPDP) certification, and a not-profit internationally recognized body of knowledge.

15.3.3 Workgroup, team, functional, and cross-functional team

A *workgroup* consists of member that can be from one unit (functional area) or several units (functional areas). *Dependent-level workgroups* are work units or department groups with a supervisor who plays a strong role as the line manager. Each member has a specific job and works in close collaboration with the supervisor. Most of the problem solving, work assignments, and other decisions affecting the group are directly made by the supervisor. This strategy can work well in the short term. *Independent-level workgroups* members are responsible for their own main area. The supervisor is not controlling but instead assigns general direction with minimal supervision. Examples are sales representatives, research scientists, lawyers, and teachers. *Interdependent-level workgroups* rely on each other to achieve the same goals. Members can either have their own roles or share responsibilities. They coordinate with each other to produce an overall product or a set of outcomes.

A *team* is a group of people sharing a common purpose, goals, or objectives, and holds its members mutually accountable [28]. A team can have members with complementary skills working together through a coordinate effort to maximize strength and minimize the weaknesses of each member. The challenge for a team is the process in which each member learns how to help the others in order to maximize their potential, overcoming personal limitations [29]. Communication between the members is a critical aspect of the success of the team relationship. Team members depend on each other and an effective team organization can lead to greater productivity and more effective implementation of resources to improve the overall decision-making process, problem solving, and the outcomes generated by a product or service, including a higher degree of innovation and originality. Managing a team is often challenging, especially when members come from different backgrounds and cultural settings. The soft skills required include emotional well-being and interactions as members need to trust each other, possess a sense of group identity and efficiency.

Functional teams can have different definitions. PDMA defines the functional team as a type of team. They are composed of members from several vertical levels of the organizational hierarchy who perform specific functions. They typically have several subordinates and a manager to control internal operations and the external relationships of a specific department or division of the organization. For example, accounting, marketing, finance, and human resources are functional teams [30].

Cross-functional teams are defined by PDMA as "a team consisting of representatives from the various functions involved in product development, usually including members from all key functions required to deliver a successful product." They are from a different part of the organization and have different functional expertise, but are working toward a common goal [30]. The members can be either from inside or from outside the organization. Interaction between cross-functional teams can be cooperative or competitive. The organization, in some cases, forces the teams to compete and cooperate simultaneously: this complex relationship may affect the performance of an organization.

15.3.4 Management strategies in medical devices

In specific industries, there are particular quality system regulations and standards. The introduction of a new medical device in particular requires effective

PM policies and practices. Moreover, it is critical that there is strict compliance with specific quality system regulations and international requirements in standards, such as the ISO 13487 quality management system (QMS). These principles apply to almost any business. The manufacture of medical device is particularly affected by these regulations as such projects are complex and require expertise and input from multiple areas within the organization. The project must comply with rigorous regulatory requirements to develop, produce, and commercialize a safe and effective product. Owing to the level of complexity involved, the project scope requires multiple communication channels across multiple functional areas. A medical device company will be required to establish an internal quality system that ensures compliance with design control and applies risk management principles. This complex task involves corporate administrators, directors, and those responsible for the implementation and control of the project.

15.3.4.1 Top-down versus bottom-up management

The top-down management style ensures that goals, projects, and tasks are determined by the senior leader of an organization independently from their team's members. These tasks and goals are then communicated to the rest of the organization. This is the most common approach to organizational management. As a particular strategy, it ensure that team objectives are aligned with the mission and vision of the company. However, the team members may be disconnected from the top management team since they are not involved in aligning goals to the business. This approach relies on the strength, focus, and thoughtfulness of the senior management [31]. To make it successful, the top-level goals need to be communicated with clarity to the rest of the team.

Bottom-up management ensures that goals, projects, and tasks are determined largely from feedback obtained from employees. Employees are involved in the assignation of the goal settings, in the same way by simply providing feedback, and also having a stake in the decision. The outcome of the decision-making is then communicated by each team to senior leadership. With this strategy, team members across all levels of the organization will feel engaged and involved with their team leader. But one limitation of this approach is that team members often do not have full access to the top-level insights across the organization. This limits their capacity to identify and set appropriate goals that are aligned with the mission and long-term vision of the organization.

The decision on what is the most appropriate strategy to pursue in an organization is often governed by its particular culture—which is unique—and incorporates various levels of risks. Before deciding on the strategy, it is important to consider several aspects, such as the extent to which team members have bought into the vision and mission of the business. Likewise, the organization could be suited to adopting a mix of both approaches. In the case of PM, the top leaders can determine goals. Once these goals are defined and communicated to the team members, they can decide how to implement their personal goals with the management team in a bottom-up style.

15.3.4.2 Matrix management

In the case of matrix management, an organization member of a team can report to more than one team leader. In this structure, there is no strict vertical business control. This form of management is commonly used to share employees and resources across functions. A team member owes his primary duties to his team leader or line manager, while also working with other people in management. For example, this approach is adopted in the case of a marketing team who has to report to both the engineering and management teams. It is an ideal strategy to apply in order to share skills across departments and/or divisions and in order to develop new products. In particular, it enables staff with different skills within the organization to work under a project manager to create new technologies that are specifically tailored to the production, e.g., of a medical device. Team members with diverse skill sets from multiple disciplines can collaborate to strengthen the project team, with more efficiency than many functional teams [32].

There are some circumstances where this style of management approach is not ideal. In the case of a long-term project, a dedicated team with a permanent assignment may be optimal. When a particular member function is critical, sharing this individual may reduce the effectiveness of that function.

The matrix management can be *weak matrix*, *balanced matrix*, and *strong matrix* [23]. In the *week matrix*, the project manager communicates with the functional manager of each team member. Each member reports to a functional manager for any task on functions and project assigned. In the *balanced matrix*, the team members report to the project manager and the functional manager. In this case, members are working on the project and their function to keep both managers informed. The *strong matrix* is when the team members have a clear line of reporting to the project manager. In this case, the project manager is the functional manager until the project ends.

The matrix management approach has many benefits, but also involves certain challenges. Members may endure conflicts between various managers and priorities. Communication can become confused between and across projects and functions and those responsible for the evaluation of performance might lose control.

There are a variety of explanations for this: the responsibility for mentoring, coaching, and professional development may lack a certain degree of clarity. And the efficiency of individuals can become compromised as resources often end up being shared across too many initiatives.

To improve transparency and avoid any sort of conflict between functions and projects, responsibilities should be defined in the project charter and agreed at the manager level. The members of the team assigned to a project should be evaluated to ensure that they can handle any increase in workload. Different matrix styles should be investigated to ensure that the proper option is chosen in relation to the capabilities of the team members and the organization.

All the actions and strategies should be documented throughout the project, thus providing clarity and relevant information to the team members.

The adoption of the matrix management model can also create stress among the team members where the amount of work ends up exceeding the time and resources available. On the other hand, it offers access to specialized expertise on a temporary basis in a more cost-effective manner as opposed to having a dedicated team.

15.4 Technology readiness level (TRL)

The technology readiness level (TRL) approach, following planning experience in the PERT and CPM methods, gives the project director and project managers an opportunity to plan the development of a particular technology and take specific decisions regarding projects [33].

TRL is a method used to assess the maturity of a technology. NASA developed it during 1974 in respect of planning the Jupiter Orbiter design team. The original definition of TRL involved seven levels of development stages, but the current one, which has been adopted by NASA and European Union, now has nine levels [34]:

TRL 1—Basic principles observed.

TRL 2—Technology concept formulated.

TRL 3—Experimental proof of concept.

TRL 4—Technology validated in laboratory.

TRL 5—Technology validated in relevant environment (industrially relevant environment in the case of key enabling technologies).

TRL 6—Technology demonstrated in relevant environment (industrially relevant environment in the case of key enabling technologies).

TRL 7—System prototype demonstration in operational environment.

TRL 8—System complete and approved.

TRL 9—Actual system proven in its operational environment (competitive manufacturing in the case of key enabling technologies; or in space).

The TRL starts at phase one, where the technology is at an embryonic stage and the more mature stage is level nine, where the technology has been tested and launched. This tool provides a quick view of the maturity of the technology and helps management in taking decisions on the development and transition of a technology. It is not a stand-alone tool, but one among several others that are needed to manage the progress of research, development, and testing activities within an organization. Investors and national funding agencies use this tool to identify the type of project that best fits with their objectives.

TRL phases one to three are considered to be the *ideas levels*. The basic research has been done, the fundamental concept investigated, and a proof of

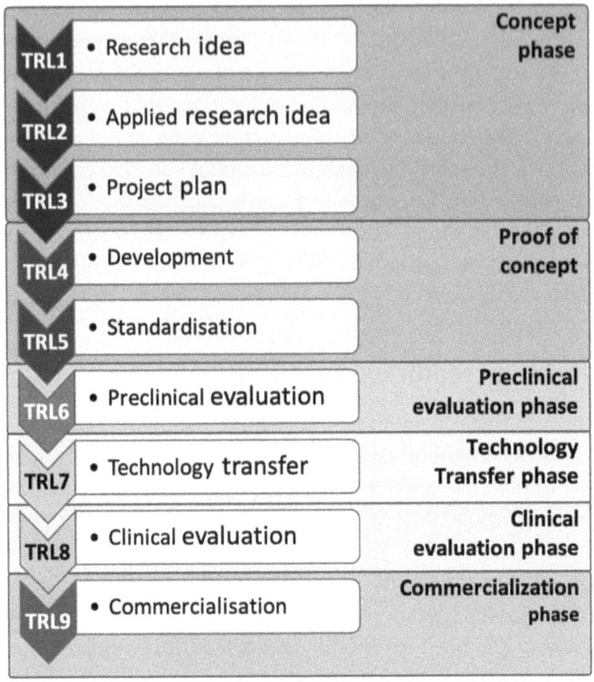

FIG. 4 Technology readiness levels for medical products.

concept developed. In TRL phases four and five at the *prototype level*, the technology has been developed and approved in a laboratory or in some other relevant environment. In TRL phases six and seven, the technology finds itself at the *validation level*, with demonstrations of its capacity occurring at a relevant operational environment. The final TRL phases eight and nine involve the production level. The technology is complete and approved and proved to work in its operational environment.

For medical devices, the method has the same number of levels, but with more specific references to medical development (Fig. 4).

In TRL phases one to three, the technology is in at the *concept phase*, starting with an idea at TRL phase one, and a hypothesis tested at the initial proof of concept (PoC) phase two TRL, and a more consolidated proof of concept carried out at TRL phase three. At TRL phases four and five, referred to as the *proof of concept phase*, the PoC has been demonstrated *in vitro/ex vivo/in vivo* conditions, and further improvements are done to assess the prototype's performance. TRL phase six includes a *preclinical evaluation phase* with a prototype board tested and a preclinical evaluation completed. In TRL phase seven, which is the *technology transfer phase*, the technology is tested within the appropriate operational system. The clinical evaluation phase begins at TRL phase eight, and at this stage, the device production and certifications are submitted and approved. TRL phase nine is the *commercialization phase*, where distributors and marketing strategies are assessed, and the technology is made fully available.

15.5 Applications

There are many examples of PM case studies in the military industry, aerospatial projects, civil engineering, and the oil and gas sector, where mega projects are often involved. However, the use of PM is transitioning from the traditional engineering-based projects over to IT (information technology) projects [35]. Despite the variety of projects and businesses that rely on PM guidelines, the PM discipline regulates the processes to plan, execute, control, and deliver the objective in one simple code. Past examples have shown how the homogeneous techniques of scheduling create unrealistic expectations. Cost practices have revealed similar outcomes. Any project, in fact, does not take place in a vacuum, but in a specific environment where external and internal conditions affect the course of the project. According to Bounds, the PM industry in 1998 was valued at US$850 billion, with a forecasted growth rate of 20% each year. However, the aggregate value of PM business value is now around US$8.500 billion [36]. The use of PM techniques, which started with the Second World War, has since been integrated and implemented in several civilian projects. The Military was the first industry that developed PM models to deliver a project. The aerospace industry, as part of the military business, has witnessed the extensive use of PM theory in many programs: Apollo project, Eurofighter project. [37] The Apollo program costs more than US$100 billion in today's money and involved hundreds of people and 2000 different organizations in its execution. The complexity of the project required the project managers to separate the "land on the moon" component of the project into its own distinct phases and milestones [38]. The scope of the project was clear: "put a man on the moon and land him back safely". At the first phase, project managers set the plan for preparing and controlling the relevant activities by selecting the specific team and the equipment required (i.e., material, capsules, rocket). The initiating process determines the scope of the project and its requirements. If we consider the nature of this project, the initial planning and scheduling process was necessary to break down the various activities into separate divisions using a top-down list. In the first phase, attention focused on procuring the equipment and the technologies ready for the scope. At this stage, the training of the team is prioritized. Three full missions, before the first landing in 1969, involved a passage around the moon to test the equipment, the docking of the lunar module, and landing procedure. On the planning dimension, the project required three years to prepare the moon approach and almost a decade to test the equipment and the ability of the astronauts to fly a long-duration mission. A detailed and meticulous schedule enhanced the chances of success. The cost-budget control was less critical than what it would be today as the federal budget in 1960 was in surplus and economists were pushing for government spending increases. As such, budget performance as reported on "Lunar Landing and Lunar Exploration Program Cost Summary" was not the main objective of this project. The magnitude of the project overshadowed the cost and schedule dimension to focus on the success of the mission. Measuring the success of a project should consider a wider range of factors, rather than internal factors only, such as the cost and duration. The criterion of long-term success involves not only the project but the organizations involved (i.e., NASA at the US nation). Therefore, a unique standard to define success or failure is excessively narrow. The Apollo project demanded a forward-looking approach to evaluate the outcome. Many actors involved in the project have shared the same interest (i.e., a team culture): the achievement of the goal. Less visible factors have been relegated to the lower level as the main objective was to accomplish what the internal and external stakeholders expected: "to see a man on the moon."

A better understanding of the different "dimensions" of success and failure is illustrated by the Eurofighter Typhoon project. The objective of the project was the gradual reduction of European defense expenditure and the enhancement of technological content for modern systems development. The project's scope was to provide Europe with an independent combat aircraft production capacity that was not dependent on the US. Despite the solid planning activities started in 1987, among the UK, Germany, Italy, and Spain, the project has suffered a series of difficulties. These range from errors and malfunctions in the system, costs increases and over-budgeting, and the quality of the project itself (i.e., initial performance and capabilities of the aircraft). The malfunction affected the reliability of the project, as the malfunctioning of an inherently unstable airframe led to crashes and loss of life. The costs overruns incurred as an effect of the malfunctioning, and the required additional investment in R&D reached an increase of 75% of the program total cost. Most of the mistakes were made on the cost estimation, which tends to be over-optimistic and based on inefficient commercial and technical arrangements. Commercial and technical inefficiency has been constantly part of the Mega Project schedule, budget, and quality misalignment. Failing to deliver the project on time, due to malfunctioning and technical shortcomings, created budget misalignment and additional costs to the project. However, the project has delivered substantial benefits to stakeholders. It has increased cross-border cooperation and technological content, creating employment and delivering an air superiority fighter. While internal project constraints, as time and cost, seem to be utopian to achieve, quality and stakeholders' benefits appear more likely.

Besides the military and aerospace industries, other businesses are involved in the application of the PM discipline. The oil and gas and civil engineering sectors are well-known examples of PM application. However, in certain cases, despite following PM guidelines, some businesses have been reported as behind schedule in different types of project. The delay factor plays a substantial role in the delivery of a project under budget and quality constraints. According to Bourn, 70% of UK construction projects carried out by public agencies are delayed [39]. As much as the military and aerospatial sectors, the construction and oil and gas industries depend on several phases that occur from various sources, involving the performance of contractors, contractual relationships among stakeholders, and procurement and logistic analysis. Consequently, delays in construction projects are likely to happen, and the magnitude of these delays can impact on project costs and quality. Those projects are often characterized by large investments and complex engineering efforts. Managing energy projects involves engineering, technologies, and human resource management [40]. Poor design in the engineering phase, inadequate understanding of the technologies utilized during the tendering stage, and incompetence of teams are often the main reasons for the partial failure of projects or incomplete success. The projects delays are almost operational in the sector. However, project managers might reduce delays by coordinating the team in the execution phase of the process, thus reducing idle times, inefficiency, and cost savings. Technical and operational challenges in delivering oil and gas plants and offshore platforms involve planning, scheduling, and controlling. In addition, cooperation between different disciplines to achieve the desired corporate objective is essential [40]. On the one hand, project managers have to deliver the project under objective project constraints, while protecting stakeholders' investments from poor performance. The PM model could adopt different strategies: labeled as "quality first," "limited resources," and "get it done no matter what" scenarios [41]. And in fact, improving critical thinking to deal with the various scenarios is the countermeasure that project managers and teams embrace against complexity [42].

The project complexity is defined with soft skills; robust leadership combined with a soft skills culture of the team increases the chances of succeeding with the project. According to Edum-Fotwe and McCaffer, there are skills outside the traditional role of project manager and construction management professionals, which demand a wider range of knowledge and skills [43]. Those skills can be specific skills and general skills. The former refers to specific knowledge of the sector, while the latter provides a foundation for the development of skills such as leadership, communication, and team motivation. These skills can assist project managers in contributing more efficiently to projects. Leadership improves project managers' vision and develops long-term strategies. Team cooperation and alignment to long-term strategies inspire subordinates to overcome complexity in the project [43]. However, as the PM discipline is adopted in different contexts, project final intention and project success are dependent on one's perception and view.

15.6 Conclusions

The PM framework emerged in the twentieth century as a business procedure to deliver complex projects based on the time involved to complete, the costs, and quality constraints. The use of the discipline as an objective formula to successfully complete projects has been applied in several contexts.

The advent of complex projects in new emerging sectors such as the military, energy, and civil engineering has expedited the need for tailored theories and best practices to support future challenges.

Academics and practitioners have developed dynamic theories starting with the iron triangle, which aim to supply project managers with adequate tools to direct and control projects during different phases.

Internal and external demands from stakeholders and complexity within environments have required project managers to deploy several techniques: the Gantt chart, the critical path method, and the Pert method.

In fact, projects are not situated in a vacuum but in a dynamic environment, in which different teams cooperate with different activities and tasks to deliver the project under the guidance of project managers and senior managers.

The advent of the Project Management Institute and PMBOK as standardized collections of theories and best PM practices has eased the spread of the procedures and standards among different industries worldwide. Theories and best practices are precious forms of guidance that enable directors to keep track of progress and drive the project in the right direction. Notwithstanding this context, it should be stressed that stakeholders and organizations differ in their objectives, and the definition of "project success" does not rely only on following one exclusive standard procedure. Team culture, organizational culture, and adaptive management structure set the additional probability of project managers being able to adapt to different scenarios.

Projects have different design goals, effects on stakeholders, and benefits to organizations. The application section above explored different projects and established that one's perception of project success or failure could well differ among industries and organization.

A design goals failure, which is not meeting the budget, allocated time, or quality specifications, in such a context might not be translated into a failure, as long as benefit to the organization and stakeholders are achieved, and vice versa.

Regardless of the context, project complexity is challenged with soft skills as team culture and leadership. Ranging between civil engineering complex projects and smaller-scale endorobots, the outcome is strictly linked to team philosophy, managers' guidance, and organizational support.

The next challenge for the PM discipline ranges from its application to medical devices, to nano technologies, with the main purpose of promoting the approach seen in mega projects in smaller-scale developments.

References

[1] Morris PW. Updating the project management bodies of knowledge. Proj Manag J 2001;21–30. https://doi.org/10.1177/875697280103200304.

[2] Sabri HR, Rahim AA, Yew WK, Ismail S. Innovation and sustainable economic competitive advantage: from regional development to global growth. In: The 26th International Business Information Management Association Conference; 2017.

[3] Safra EB, Antelo SB. Integrated project management applied in world-class gas-field development projects: from theory to practice. In: SPE Latin American & Caribbean Petroleum Engineering Conference. Lima: Society of Petroleum Engineers; 2010.

[4] Oisen R. Can project management be defined? Proj Manage Q 1971;12–4.

[5] Cicmil S, Williams T, Thomas J, Hodgson D. Rethinking project management: researching the actuality of projects. Int J Proj Manag 2006;675–86. https://doi.org/10.1016/j.ijproman.2006.08.006.

[6] Hodgson D, Cicmil S. Making projects critical. New York: Palgrave MacMillan; 2006.

[7] Morris PW. Project Management: a profession with a hole in its head or, why a change in the culture of academic support is needed for the profession. Eng Proj Org J 2014;147–51. https://doi.org/10.1080/21573727.2013.873717.

[8] De Wit A. Measurement of project management success. Int J Proj Manag 1998;164–70. https://doi.org/10.1016/0263-7863(88)90043-9.

[9] Pollack J, Helm J, Adler D. What is the Iron Triangle, and has it changed. Int J Manag Proj Bus 2018;11:527–47. https://doi.org/10.1108/IJMPB-09-2017-0107.

[10] Kloppenborg T, Opfer W. The current state of project management: trends, interpretations and predictions. Proj Manag J 2002;5–18. https://doi.org/10.1177/875697280203300203.

[11] Baccarini D. The concept of project complexity—a review. Int J Proj Manag 1996;201–4. https://doi.org/10.1016/0263-7863(95)00093-3.

[12] Lehtonen M. Evaluating megaprojects: from the 'Iron Triangle' to network mapping. Sage 2014;278–95. https://doi.org/10.1177/1356389014539868.

[13] Kim Y-W, Ballard G. Management thinking in the earned value method system and the last planner system. J Manag Eng 2010;223–8. https://doi.org/10.1061/(ASCE)ME.1943-5479.0000026.

[14] Atkinson R. Project management: cost, time and quality, two best guesses and a phenomenon, its time to accept other success criteria. Int J Proj Manag 1999;337–42. https://doi.org/10.1016/S0263-7863(98)00069-6.

[15] Basu R. Managing quality in projects. Farnham: Gower; 2012.

[16] Panuwatwanich K, Nguyen TT. Influence of organisational culture on total quality management implementation and firm performance: evidence from the Vietnamese Construction Industry. Manage Prod Eng Rev 2017;5–15. https://doi.org/10.1515/MPER-2017-0001.

[17] Wilkinson A, Redman T, Snape E, Marchington M. Managing with total quality management. London: Palgrave Macmillan; 1998.

[18] Pinto M, Pinto J. Determinants of cross-functional cooperation in the project implementation process. Proj Manag J 1991;15–20.

[19] Maylor H, Brady T, Cooke-Davies T, Hodgson D. From projectification to programmification. Int J Proj Manag 2006;663–74. https://doi.org/10.1016/j.ijproman.2006.09.014.

[20] Petersen PB. The evolution of the Gantt chart and its relevance today. J Manage Issues 1991;131–55.

[21] Geraldi J, Lechter T. Gantt charts revisited: A critical analysis of its roots and implications to the management of projects today. Int J Manag Proj Bus 2012;578–94. https://doi.org/10.1108/17538371211268889.

[22] Wren DA, Bedeian AG. The evolution of management thought. Hoboken: John Wiley & Sons; 1976.

[23] PMI. A guide to the project management body of knowledge PMBOK guide. Project Management Institute; 2017.

[24] Trojanowska J, Dostatni E. Application of the theory of constraints for project management. Manage Prod Eng Rev 2017;87–95. https://doi.org/10.1515/MPER-2017-0031.

[25] Engwall M. PERT, Polaris, and the realities of project execution. Int J Manag Proj Bus 2012;595–616. https://doi.org/10.1108/17538371211268898.

[26] Punmia BC, Khandelwal KK. Project planning and control with PERT and CPM. New Delhi: Laxmi Publications LTD; 2006.

[27] Karlsen JT. Project stakeholder management. Eng Manag J 2002;19–24. https://doi.org/10.1080/10429247.2002.11415180.

[28] Katzenbach JR, Smith DK. The rules for managing cross-functional reengineering teams. Plan Rev 1993;2:12–3. https://doi.org/10.1108/eb054404.

[29] Naresh J. Build teams to run marathons, not sprints. Retrieved from O'Reilly; 2014. https://www.oreilly.com/library/view/97-things-every/9780596805425/ch48.html.

[30] Kahn KB. The PDMA handbook of new product development. John Wiley & Sons; 2004.

[31] Lester DH. Critical success factors for new product development. Res Technol Manag 1998;36–43. https://doi.org/10.1080/08956308.1998.11671182.

[32] Larson EW, Gobeli DH. Matrix management: contradictions and insights. Calif Manag Rev 1987;126–38. doi:10.2307%2F41162135.

[33] Altunok T, Cakmak T. A technology readiness levels (TRLs) calculator software for systems engineering and technology management tool. Adv Eng Softw 2010;769–78. https://doi.org/10.1016/j.advengsoft.2009.12.018.

[34] NASA. Technology Readiness Nasa website level. Retrieved from Nasa website https://www.nasa.gov/directorates/heo/scan/engineering/technology/txt_accordion1.html; 2021.

[35] Van Donk DP, Molloy E. From organising as projects to projects as organisations. Int J Proj Manag 2008;129–37. https://doi.org/10.1016/j.ijproman.2007.05.006.

[36] Bounds G. The last word on project management. IIE Solut 1998;41–3.

[37] Heinrich MN. The Eurofighter Typhoon programme: economic and industrial implications of collaborative defence manufacturing. Def Stud 2015;340–59.

[38] NASA. Nasa Apollo Mission. Retrieved from Nasa Apollo Mission https://www.nasa.gov/mission_pages/apollo/missions/index.html; 2019.

[39] Bourn J. Modernising construction. National Audit Office; 2003.

[40] El-Reedy M. Project management in the oil and gas industry. Beverly: Scrivener Publishing; 2016.

[41] Chatzoglou P, Macaulay L. Requirements capture and analysis: the project manager's dilemma. Int J Comput Appl Technol 1995;190–202. https://doi.org/10.1504/IJCAT.1995.062411.

[42] Ramazani J, Jergeas G. Project managers and the journey from good to great: the benefits of investment in project management training and education. Int J Proj Manag 2015;41–52. https://doi.org/10.1016/J.IJPROMAN.2014.03.012.

[43] Edum-Fotwe F, McCaffer R. Developing project management competency: perspectives. Int J Proj Manag 2000;111–24. https://doi.org/10.1016/S0263-7863(98)90075-8.

Chapter 16

Future trends

Luigi Manfredi
Division of Imaging Science and Technology, School of Medicine, University of Dundee, Dundee, United Kingdom

16.1 Introduction

We live in an era where technology is part of our lives, and in some cases, it is difficult to move forward with our daily routines without using smart devices. The way we use technology has also changed. We have devices that are getting smaller and smarter, are connected together, and are easy to use. These changes have affected several industrial sectors, and medicine is one of them, bringing advancements in many applications. Medical procedures have moved from laparotomy, or open surgery, at the beginning of the century [1], when the access to the internal organs was allowed by having big incisions and by using the surgeon's hands, to laparoscopy at the end of the century, where surgeons use instruments to access internal organs through small incisions. This has requested advancements in miniaturization of the mechanical components, instruments, as well as miniaturization of cameras, at the beginning by using optical fiber and then using semiconductors. Thanks to CMOS (complementary metal-oxide-semiconductor) cameras, surgeons have been able to see and have high-quality videos inside the human body in two and three dimensions (2D and 3D) [2]. Laparoscopy, also known as minimal access surgery (MAS), or minimal invasive surgery (MIS), or keyhole surgery [3], has brought several benefits compared to open surgery. Fast recovery, reduced infection, and improved cosmetic outcome are some of the advantages of this disruptive technology when introduced. However, the adoption of this technology has faced several issues related to the required training [4] and learning a "new way" to control and maneuver the instruments [5]. In MIS, instruments are an interface between the surgeons' hands and the manipulated organs. This has reduced the haptic feedback [6] and thus complicated the performance of surgical tasks because of the unnatural sensory-motor coordination [7]. Other issues are the amplification of the tremor because of the long distance of the shaft which connects the surgeon hand to the end-effector [8] as well as because the hands are being suspended over the patient. Handling instruments for MIS requires more trained

Endorobotics. https://doi.org/10.1016/B978-0-12-821750-4.00016-5

skills to perform tasks in a limited space inside the human body with a limited degree of movements. The access through a pivot point limits the dexterity and mirrors the X-Y plane movement. Most of the medical instruments have limited degrees of freedom compared to human hands, and this requires a long training to perform complex tasks. Due to these limitations, robotic devices have provided great advantages in improving the precision of the control and simplifying the performance of complex tasks. This has been possible, at the beginning, by modifying industrial robots with medical instruments as end-effectors, and later by providing robotics specifically designed for medical applications. This generation of medical robots has higher degrees of mobility to the end-effector, improving the controllability of surgeons' hands inside the human body. An external console provides comfort to the surgeons by reducing fatigue and tremor, and by improving precision and shortening the learning curve of complex procedures [9]. Many of the surgical robots still lack a haptic feedback, partially compensated by having an advanced 3D vision system [10]. This type of devices represents the first generation of medical robotics. They have rigid instruments limiting the access to the most remote parts of the body. A more advanced generation of medical robots is designed by using flexible and articulated instruments that enable the access to remote areas. Natural orifice transluminal endoscopic surgery (NOTES) is an emerging field where natural lumina are used to access and to perform diagnostic and therapeutic procedures [11]. These new medical devices [12], many of them coming from research institutes [13], have recently brought to the market new products: just to mention some, the flexible robotic bronchoscope by Auris Health Robotic Monarch, recently acquired by Johnson and Johnson [14]; the Ion robotic catheter system (Intuitive Surgical Inc., US), which has recently been approved for the diagnosis of lung diseases; or the Hominis Surgical System developed by Tel Aviv, Israel-based Memic Innovative Surgery [15], a transvaginal robotic surgery. Great advancements on the miniaturization of electronics and cameras components have allowed the development of endoscopy capsules that can be swallowed and allow the inspection of the entire gastrointestinal (GI) tract in a minimally invasive way. Wireless capsules endoscopy (WCEs) has recently started to show how a patient can swallow a medical device to inspect internal organs. They have on-board battery, camera, and a wireless communication with an external apparatus to record pictures of the internal GI tract and to communicate with the patient, all in a small volume of few cubic centimeters [16]. The video recorded is then analyzed and approved by a clinician [17]. Recent advantages in AI have shown promising results of convolutional neural network for the process of WCE videos to identify various lesions such as erosion/ulcers, angioectasias, polyps, and bleeding lesions [18]. This automatic process has saved time and costs for health care centers [19]. Limitations of the current WCEs are the transition time, which relies on a natural peristalsis locomotion with no precise control by the user, as well as the lack of onboard instrumentation for treatment or biopsy samples collection.

16.2 Design approach: Tethered vs. untethered

Endorobots can be classified into two major categories: *tethered* and *wireless or untethered*. They have different challenges and roadmap for the development of the technology.

Tethered endorobots have a physical connection to an external console that can provide energy, high bandwidth to exchange data, and high-quality video, if needed. A catheter can be used for the passage of instruments or to remove tissue samples, to supply air or CO_2 as well as for water irrigation. A physical connection can be used to pull out the device in case of malfunction. The control can be done by using an external pushing force or by using an internal self-propelled mechanism. The design of a small and tethered endorobot with a self-propelled locomotion is of paramount importance for some medical applications [16] and requires facing four major challenges: (1) to produce enough force to overcome the tether drag; (2) to provide a speed to perform the procedure in a time comparable to the current procedure [20]; (3) to reduce the force applied to the internal organs to reduce pain and discomfort as well as avoiding any adverse events during the procedure [21]; (4) to allow a precise control of the device to perform surgical tasks. The light weight of a small device reduces the use of friction-gravity locomotion ($F_t = P \cdot \mu$) due to the low friction coefficient (μ) (in colonoscopy, this relates to the mucosa [22]) together with a light weight (P). This will produce a very limited traction force (F_t). This approach design limits the size of an endorobot since it cannot go down to millimeters due to the limited locomotion force that is able to produce. A tether endorobot cannot be used for an implantable device, and many of the internal locations of the body are inaccessible or difficult to reach by a tether endorobot.

Wireless or untethered endorobots need to include all the required functional components onboard. This type of endorobots can be very agile; however, they have many challenges ahead related to miniaturization. Limited space requires a compromise in the design of the components, including the energy supplier, instrumentation, and data connection to an external console. When the size becomes very small, the physics interaction requires a different approach to design a locomotion solution [23]. Including instruments is very challenging when the size becomes too small. At the current stage, endorobots at the micro-scale do not have therapeutic or diagnostic medical functions yet [24].

16.3 Anatomy

Defining the medical procedures is the starting point to identify the clinical needs that will address the required functionalities as well as the overall size of the endorobot and the choice of including or not a tether connection. The challenges in the design and the choice of materials, sensors, and actuation solutions relate to the size of the lumen to access specific organs. The next section is intended to provide a brief overview of the human anatomy with consideration of sizes to gain access and to operate through natural lumina or with very limited invasiveness. In this regard, it is worth to consider in the design of a medical device the fact that some organs or canals in the human body present a virtual lumen. In the case of the esophagus, the lumen is normally collapsed, and it distends approximately 2 cm to the content when introduced. For its part, other anatomical structures like the trachea have a real lumen provided by a significantly rigid structure that confers a well-defined shape. Hence, the anatomy of human organs can help to identify the design challenges linked to the miniaturization of the device to access specific organs.

16.3.1 Gastrointestinal (GI) tract

The GI or digestive tract starts from the mouth, goes through the pharynx, esophagus, stomach, the small (duodenum, jejunum, and ileum), and large bowel. The GI tract ingests and digests the food to extract and absorb the nutrients and expels the waste as feces. The small intestine is the most difficult part to access due to its total length of about 6 m. The diameter of the small intestine is about 2.5 cm [25]. The large bowel has a total length of up to 1.6 m and an internal diameter ranging from 3 to 8 cm [26], with internal folds named haustra.

WCE (wireless capsule endoscopy) has had a huge impact and has revolutionized the way to screen the GI tract in the last decade. WCE can visualize, localize, and measure the lesion in the GI tract. It moves by peristalsis within a constrained path and takes about 8 to 10 h to be evacuated from the anus. Retention in the large bowel can occur [27,28] because of the rigid WCE shape and its overall size, with a diameter of 11.6 mm and a total length of 32.3 mm (PillCam Colon 2, Medtronic). Having a controllable device can provide advantages in analyzing specific sections of the bowel wall with more accuracy. When inspecting the large bowel, WCE finds some limitations related to the much bigger diameter, which produces the twist of the capsule during its journey. This has been solved by having two cameras on both sides to improve the quality of the screening. Bowel preparation is still an issue for the screening and plays an important role in this concern. A clean environment, however, requires the use of heavy laxative medicine that increases the discomfort of the procedure. In the case of WCE, this process is more intense than traditional colonoscopy [29]. Current limitations relate to the passive and noncontrollable locomotion and the lack of on-board instruments requiring a further inspection by colonoscopy if any abnormal lesion is detected.

Navigation inside the small intestine requires a maximum diameter of the device to be in the range of 1–2 cm, and it is likely to use the wireless approach design if all the tract needs to be investigated. In the large bowel, 2 cm can be considered a suitable size. If a device is designed to be swallowed, the locomotion can take advantage of the peristaltic movement; however, this takes more than 10 h for the current WCE to inspect the entire GI tract with, in the same case, battery issues when recording videos of the final tract of the large bowel [30].

Increasing the locomotion speed is an important aspect to consider since it could reduce the time spent for the procedure and for the analysis of the recorded videos [31]. The screening of the colon often uses CO_2 for inflation to increase the quality of the inspection. This requires a device to have a gas supplier; thereby, a tether design approach is likely to be needed. A device to inspect the large bowel is inserted through the anus. The internal slippery surface requires an effective locomotion solution to move the device forward and overtake the tether drag without damaging the colonic mucosa. Low forces applied to the colonic wall can reduce the pain and discomfort of the current colonoscopy procedure as well as avoid any adverse events during the procedure [21].

16.3.2 The respiratory system

The respiratory system consists of a series of organs responsible for taking up oxygen and expelling carbon dioxide. The gas exchange takes place in millions of small air sacs located in the lungs and named alveoli. The communication with the external environment is through a system of airways, or hollow tubes. The air goes through the nasal or oral cavities to the larynx and trachea (diameter of 13–25 mm in men, 10–23 mm in women [32]) that branches to the middle of the chest in the two main bronchi (15 mm [32]). Each of them enters into the lung branching into progressively narrowed subbranches and then into numerous smaller tubes, named bronchioles (less than 2 mm in diameter [33]), that are connected to the alveoli, the terminal section.

Respiratory endoscopy is used for early detection, diagnostic procedures, endoscopic ultrasound, and endobronchial interventions. Navigation in the respiratory system can be done by inserting an instrument through the nose or mouth [34] by using an endoscope. Its diameter can be very small, about 4 mm, and with a length of 100 cm to inspect the lung. Inspection through the nose is named transnasal endoscopy. To access the smaller bronchioles, a flexible bronchoscope has a diameter that can be 3.5 mm, with a 1.2-mm suction instrument port, or even smaller. Instruments with a diameter of 1.8 mm are now commonly used and are provided with a suction channel of 0.7 mm. The movement of the tip ranges from 180 degrees to 220 degrees [35], and optical fiber is used to reduce the size of the instrument. Complications during a bronchoscopy can occur [36], and it is a procedure that takes time to achieve competency [37].

A feasibility study has shown a robotic bronchoscope as an alternative solution to address the limitations of current manual bronchoscopy techniques for biopsy of suspected peripheral lung lesions [38]. The device is called Monarch, developed by Auris surgical Robotics, Redwood City, CA, USA. The outer sheath has an outer diameter of 5.7 mm, with an inner telescoping endoscope measuring 4.4 mm in diameter [39]. Electromagnetic navigation is included in the system to navigate for guidance. It relies upon an external electromagnetic field generator. The physician uses an external hand-held controller to guide the robotic scope into the biopsy position. A working channel of 2.1 mm is included.

16.3.3 The circulatory system

The cardiocirculatory system is a network comprising the heart and vessels that allow blood to circulate and transport nutrients, oxygen, carbon dioxide, hormones, and blood cells. It also plays a fundamental role in helping fighting diseases, stabilizing the temperature and the pH. This system includes the lymphatic system, with vessels with a diameter ranging from 0.2mm up to 2mm [40]. In the circulatory system, two different pathways come from the heart, i.e., the pulmonary circulation and the systemic circulation. The pulmonary circulation consists of a short loop between the heart and the lungs for the gas exchange. The systemic circulation connects the heart to the rest of the body to distribute oxygenated blood through the arteries and drop off the carbon dioxide through the veins.

The sizes of the internal vessels are arteries with a typical diameter (D) of 4mm and wall thickness (WT) of 1mm [41]; coronary arteries, $D=2$–5mm [42]; veins $D=5$mm, WT$=0.5$mm; portal vein, $D=7$-15mm [43]; heart, coronary vessels, $D=2$-4mm [44]; arterioles, $D=30$mm, WT$=20$mm; aorta, $D=25$mm, WT$=2$mm, capillaries, $D=8\,\mu$m, WT$=1\,\mu$m.

Access to this system is relevant since nearly every site of the body can be accessed by blood. This makes this system the most important access area for endorobots. Applications can range from targeted drug delivery [45], atherectomy (removing plaques from a blood vessel), thrombolysis (destroying blood clots), maintaining blood flow by using stents, removing occlusions that starve a region of nutrition, and administering therapy for aneurysms. The flow of the blood may require the ability to move against it with swimming capability.

Many disorders can affect this system, including the cardiovascular diseases affecting the cardiovascular system and lymphatic diseases affecting the lymphatic system.

16.3.4 The central nervous system

The central nervous system includes the brain and the spinal cord. It integrates the received information, coordinates and influences the activity of all parts of the body. A clear and colorless fluid with similar properties to water, named cerebrospinal fluid, circulates around this system as well as in the cerebral ventricles and in the central canal of the spinal cord. The navigation through this fluid requires an endorobot with a size of 2.5mm in diameter to fit approximately 50% of the population, and below 1.5mm in diameter to fit 85% of the population [23]. Access to the ventricles of the brain can be gained from the spine through the cerebral aqueduct, with a diameter within 1–2.2mm and a length up to 18mm [46]. Craniotomy, the removal of part of the bone from the skull to expose the brain, is now a conventional process to access the brain. Endorobots can access the cerebrospinal fluid through lumbar puncture between the L3 and L4, in a similar way to the one currently used to collect a fluid sample. Then, they could potentially navigate through the brain, leaving the skull intact and performing an intervention. Nowadays, the percutaneous intraspinal navigation is similarly performed with the use of catheters [47].

16.3.5 The urinary system and the prostate

The urinary system includes two kidneys that produce urine. This waste product exits the kidney via minor calyces, followed by the major calyces that ultimately form the renal pelvis. From here, urine continues its flow into the ureter. The two ureters enter the bladder where the urine is stored to be subsequently expelled from the body through the urethra. The kidney filters the blood and produces urine, with density and viscosity similar to the water, to eliminate waste from the body. It also regulates the blood volume and blood pressure, controls the levels of electrolytes and metabolites, and regulates the blood pH. The length of each ureter is about 250 mm with a typical internal diameter of 3 mm. The bladder is a hollow muscular sac in the pelvis that stores urine, with an average capacity between 400 and 600 mL. To prevent backflow of urine to the kidney, the ureter connects to the bladder's wall with an oblique angle to close off the ureteral orifice as the bladder expands. The typical urethral length is 22 cm and diameter of 9 mm in male [48,49], while in female it presents an approximate length of 4 cm [50] and a diameter of 7 mm [51]. The access to the urethra in men uses a catheter size from 14Fr (OD = 4.67 mm, ID = 0.184) to 16Fr (OD = 5.33 mm, ID = 0.210), where 1 "French," or "Fr," is equivalent to 0.33 mm.

Ureteroscopy is a procedure to address kidney stones through the urethra and bladder up to the ureter where the stone is located. It uses an ureteroscope, which has a camera with illumination and an instrument channel to insert a snare with a basked retrieval device to pull out the stone. If the diameter of the stone is large, a laser is used to previously fragment the stone into small pieces and then remove them. In males, this procedure must be done gently to avoid injuries to the fragile urethral mucosa that would create secondary risks, such as urethral stenosis [52]. The current ureteroscopes have a diameter between 7.5 and 9 Fr (3 mm), with a working channel up to 3.5 Fr (1.17 mm) [53].

16.3.6 The visual system

The visual system includes the eye (the sensory organ) and part of the central nervous system, giving the organisms the sense of sight. The eye is a very delicate organ heavily supplied by nerves. Vitrectomy is a surgical procedure to remove the vitreous humor gel that fills the eye cavity [54]. This provides a better access to the retina for a variety of interventions, including the removal of scar tissue, laser repair, or repair and treatment of macular holes. Laser can be used to treat nonrefractive conditions, such as laser corneal surgery, to reshape the surface of the eye to correct myopia, hypermetropia, and astigmatism.

The surgical procedure to remove the eye's crystalline lens is named cataract surgery, and this requires steady hands since it is performed by using an ultrasonic handpiece and aspiration from the eye. Then, irrigation of balanced salt solution is used to replace the removed fluid. Considering how delicate this area is, endorobots could improve the intervention by precisely controlling the miniaturized tools [55]. These systems can help to overcome human physiologic tremor and to improve the outcome of the surgical performance, such as scaling the motion and reducing the tremor [56]. Microelectromechanical systems (MEMS) have been used to develop a minimally invasive wireless optical sensor to measure the oxygen level inside the eye. Access is granted through the vitreous cavity via a small scleral incision [57]. This concept can be extended and applied for a localized delivery of drugs into the vitreous cavity. Untethered endorobots can provide an alternative way to retinal procedures. Their small size and low force can facilitate the interaction within the visual system. In the pars plana vitrectomy, the access is gained to the posterior segment with the instrument introduced into the eye through the pars plana, using 20-gauge (0.9 mm) instruments. The durable pars plana can be an appropriate point to insert and remove the device. The small size would reduce the invasiveness and could potentially improve the recovery of the retina after the procedure compared to the use of traditional larger tools. Magnetic control is an approach investigated to propose microrobots for intraocular procedures. Visual servoing has been used to track the movement through the pupil [58].

16.3.7 The auditory system

The auditory system provides the sense of hearing and includes the ears (sensory organs) and the auditory parts [59]. The sound waves are propagated through the auditory canal up to the tympanic membrane, a part of the external ear that separates the external ear from the middle ear. The eardrum has a diameter of 8–10 mm and is stretched by means of small muscles. The sound waves make the eardrum vibrate. These vibrations are transmitted to the inner ear via three bones contained in the middle ear (tympanic cavity) (typical diameter of 18 mm [60]): the hammer (malleus), the anvil (incus), and the stirrup (stapes). The sound waves are then transmitted to the oval windows 20 times higher. This is due to the difference in size between the larger surface of the eardrum and the smaller surface of the oval windows. The inner ear includes the cochlea, where the vibration of the sound causes the stereocilia to move and to convert these vibrations into nerve impulses, which are taken up to the brain to be interpreted. The cochlea is a spiral tube that is coiled two and a half turns around a hollow central pillar, the modiolus. The diameter is approximately 9 mm and the base about 5 mm in height. The total length of the tube, when stretched out, is approximately 30 mm. The smaller dimension of the tympani and the vestibuli of the cochlea varies linearly from 1 mm at the beginning down to 0.3 mm at the end [61].

16.3.8 The fetus

Fetal surgery covers surgical techniques to unborn babies, hence in a growing fetus. This surgery can prevent malformations and, in some cases, save precious life. Access to the fetus by using minimal access or robotics surgery is challenging. It is noteworthy that the instruments need to go through different organs, such as the abdominal wall of the mother and the wall of the uterus. This long path restrains the dexterity of the tools. Preterm labor is a risk that can be minimized when the instruments are thin in diameter, $D = 2$ mm or 3 mm [62]. Minimally invasive fetoscopic surgery can be useful for some fetal conditions, which include twin-twin transfusion syndrome (TTTS) and spina bifida [63]. The TTTS procedure uses a small endoscope ($D = 2$ mm [64]) with a laser that is used to coagulate the vessel to stop the blood flow between the two twins [65]. Small endorobots offer promising perspectives to address new procedures entering through the cervix (2–3 cm in length and $D = 25$–30 mm [66]) to the uterus.

16.4 Challenges ahead: AI & ML, miniaturization of smart materials, ethics, and regulations

The next generation of endorobots needs to be cost-effective, dexterous to provide a precise control of the instruments, and helpful in the most powerful tool of clinicians: the clinical judgment. Five are the main challenges for the improvement of the current generation of endorobots: (1) AI (artificial intelligence) and ML (machine learning) algorithms to improve the controllability and the autonomy level; (2) fast, stable, and safe connectivity to access and share data as well as to allow remote control; (3) miniaturization of smart materials for actuation and sensing; (4) ethics; and (5) medical regulation.

16.4.1 AI & ML

AI has become more available in health care centers with software that can assist and help clinicians in making a decision and that can potentially reduce or avoid human errors [67]. ML algorithms can improve the controllability of a medical device and increase the level of autonomy. Semiautonomous and fully autonomous devices [68,69] will be able to perform surgical tasks in a future where a clinician will train and supervise smart endorobots performing complex procedures. A smart generation of endorobots will be able to improve the autonomy and not just being limited to help making a clinical decision. AI will allow endorobots to take action by themselves [70]. They will learn from clinicians how to perform a specific task and they will be able to reproduce it by themselves, or, even to improve what they have learned. This level of autonomy will face several concerns, related to ethics and legislations. At the current stage, there is no robot available to the market that can perform a surgical task autonomously. This is not only related to technical challenges but also to medical regulations. However, things are likely to change in the near future. The technology will take advantages of the speed of the network communication to have smart and advanced endorobots connected and capable of sharing data and experiences. They will have access to an enormous database and being capable to learn from each other. The learning process is something that has always differentiated humans from machines [71]. We can think on how fast an infant can learn by using the five senses and learn how to perform complex tasks just by getting information around, with a fast trial-and-error process. This mechanism in humans is impressive, however, constrained by the data that an infant can access. If this process could be replicated by using ML, the result would be exponential if we consider the massive quantity of information that is already available and accessible to all of us through the web. The management of all this information will be the main challenge and defining how much we can or we are willing to control in this learning process will be crucial. Having too much control will reduce the speed on the learning and will reduce the efficacy of this process. Having too few control will cause unexpected results and will lose the control of the final expectation. These complex learning processes will require a powerful combination of software and hardware. In this direction, an important role will be played by the advancement achieved in mobile technology, with processors able to find a compromise between high performance and energy efficiency [72]. Analog chip systems can replicate a biological neuron and are even faster. The energy efficiency of our neurons is very high [73]. However, recent studies on neural network chips have largely improved this efficiency becoming closer to a human neural network [74]. High efficiency is an essential requirement for moving forward technologies on autonomous endorobots that can last for a very long time with no external power or by using wireless power [75,76]. WCEs are strongly limited in size and performance because of the volume occupied by the on-board battery. If the efficiency could be improved, the size of the battery could become smaller, and consequently, the performance would be improved. Performance and efficiency are the main challenges that the constructors of microcontrollers face every year. Research centers try to overcome this challenge by making the die size of a chip smaller [77] to reduce the consumption, by increasing the performance and reducing the price per surface unit [78]. This process has been defined in 1965 by Gordon Earle Moore, when he predicted the technology trend till 1975 [79]. An alternative way to design the next generation of computers involves the quantum technology [80,81]. This technology can be used to have a distributed intelligence, and an endorobot can include minimal essential functionalities that are later processed and elaborated outside by having a wireless or wired connection. The design of a smart system is accomplished by having a smart sensory motor system, which includes sensors, actuators, and a mechanical connection. These three elements connected together with a smart controller will define a new era of embody intelligence.

The telecommunication network has increased the bandwidth and reduced the delay of the transmission. This has allowed the implementation of faster closed-loop controllers with more stability, which allow telepresence technology to be available for remote control [82].

16.4.2 Smart materials

The current generation of medical devices has greatly helped to reduce the trauma to access and operate inside the human body by using natural orifices. They are reducing size and improving flexibility to go around narrow spaces and to access remote areas with limited or null damage to the surrounding organs. These devices can change mechanical properties, becoming soft or stiff [83]. Richard Feynman, an American theoretical physicist, in his talk given at an American Physical Society meeting in California said to the audience in 1959 *"Although it is a very wild idea, it would be interesting in surgery if you could swallow the surgeon. You put the mechanical surgeon inside the blood vessel, and it goes into the heart and 'looks' around. It finds out which valve is the faulty one and takes a little knife and slices it out."* The movie "Fantastic Voyager" in 1966 envisaged a journey into a human body to repair a damage of the brain of an injured scientist. The crew and its submarine were shrunk to micro scale and injected into the vein. The statement made by Feynman and the movie "Fantastic Voyager" addressed many challenges at that time related to the miniaturization of medical devices that even today remain unsolved. The size of an endorobot is related to the space inside the organ where it needs to operate. In the GI tract, this space is of a magnitude of a couple of centimeters, in veins goes down to a few millimeters, in the bronchioles below 1 mm, and so down to nanoscales at the cell level.

16.4.3 Ethics and regulations

Regulatory bodies have a very delicate and balanced role to play to protect people when new technologies are brought to the market. However, they try to do it without slowing down the innovation since this is what improves healthcare. Innovation is needed but in a way that it is safe and effective in its development. Regulatory bodies need to provide an effective way to regulate development that protects people while it does not restrain innovation.

16.4.4 Usability

Another important aspect of a new technology is the learning curve when using a new device. For this concern, the user interface defines the way to interact with the device, to manipulate instruments, and to interact with the internal organs. This interface aims to extend the surgeons' sense by reducing the physical and physiological stress [84,85]. The physical stress can be addressed by having a "comfortable" interface where the user can control the device by reducing the forces applied to the instrument, hence the muscles fatigue. Current robotic-assisted laparoscopy (RALS) has shown a beneficial impact on physiological indicators of stress [86]. This benefit should be extended to other procedures, for example, colonoscopy, where technical skills to use pure mechanical instruments are still a common practice [87]. The posture and fatigue suffered by clinicians during a colonoscopy procedure can cause varying degrees of injuries [88–90]. The years of experience proportionally correlates with better outcomes [91,92], affecting the pain and discomfort for the patient as well as the quantity of sedation used during the procedure [93,94].

16.4.5 Commercial challenges

What is important to bear in mind in medical devices is the long process that takes to bring a new technology to the market. This process involves several aspects, including funding and compliance with medical regulations. When the product fulfills all the requirements to be used in the clinical practice, the commercial and marketing strategy will decide its success.

Including all these technological functionalities needs to be cost-effective to speed the use of the technology in health centers. Having a cost-effective smart endorobot is a requirement that should be addressed since the beginning of the development, starting from the early R&D. In many cases, this requirement is difficult to accomplish when a new device is just commercialized, since the spread of the technology can eventually push down the price. However, costs should be taken into account when a technology is at early stage, possibly after the validation of a proof of concept.

At the end of this long process, the last barrier is represented by the acceptance of a new medical technology, which is influenced by various factors. These are related to the organization of the management control, or in some cases, the clinical staff welcomed a new technology because this can provide more opportunities to interact with patients and their careers [95]. Acceptability can be increased if all the stakeholders are involved in the development process.

In this regard, Fig. 16.1 shows the diffusion of the innovation theory (DIT) proposed by Rogers in 1962 [96]. He defined different categories of adopters.

Innovators are the people that are keen to be the first to try the innovation. These users are enthusiastic and are very willing to take risks. These are often the first to develop a new idea. Very little effort needs to be done to appeal these users.

Early adopters are people that represent opinion leaders. They play leadership roles and embrace change opportunities. These users are aware of the need to change and are comfortable in using and adopting new ideas. There is no need of information to convince them to change.

Early majority are rarely leaders. They adopt new ideas before the average users. They need to see evidence that the innovation works before adopting it.

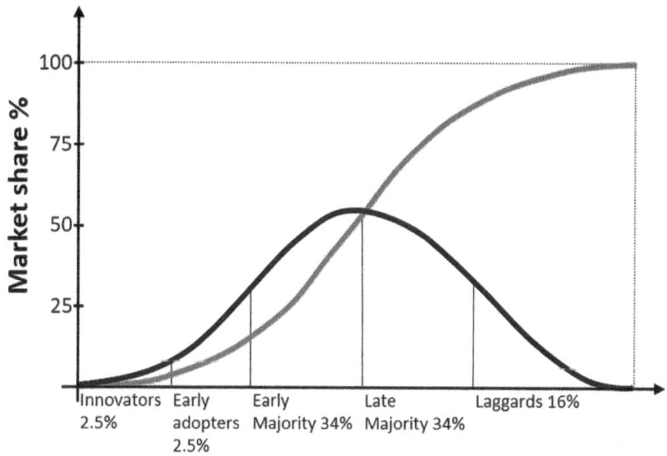

FIG. 16.1 Diffusion of innovation theory proposed by Rogers in 1962 [96], where the blue line (*dark gray color in print version*) represents the adoption of a new technology, broken into sections of adopters. The yellow line (*gray color line in print version*) represents its market share that eventually reaches the saturation level.

Success stories and evidence of the innovations are strategies that appeal these adopters.

Late majority are skeptical to change. They will adopt an innovation only after it has been tried by the majority. Strategies to appeal these adopters include information on how other people have tried and successfully adopted an innovation.

Laggards are traditional and very conservative adopters. They are skeptical to change and are the most difficult group to convince. Statistical evidence and pressure from the other adopter groups are strategies to convince these people.

16.5 Conclusions

Endorobots are becoming smaller and smarter with the potential to perform medical tasks that are difficult to envisage at the current stage. They will lead to the development of therapies that are not yet conceived. They will open a new era of precision medicine, or precision robotics, where micro- and nano-endorobots will be specifically programmed based on individual needs to target specific pathologies. Endorobots can improve current medical procedures by proposing minimally invasive procedure and, in some cases, cost-effective solutions. In the last decades, many medtech companies have embraced the challenge and have brought innovative solutions to the market. Overcoming technical challenges is just one of many other important aspects to make a technology widely used in health care. Health outcomes are another important aspect that, according to Porter, can be defined as quality divided by cost [97].

Optimizing this process can be achieved either by increasing quality without increasing cost or by reducing the cost without compromising the quality. Ideally, this should be achieved by simultaneously improving quality while reducing costs.

The design and introduction of a new technology have many challenges that are not only limited to technical aspects. Many other factors need to be faced in a big effort that involves substantial changes in an increasing complex healthcare environment.

References

[1] Mehasseb M. Laparoscopic surgery: when to convert to laparotomy? In: Gynecologic and obstetric surgery; 2016.

[2] Dirie NI, Wang Q, Wang S. Two-dimensional versus three-dimensional laparoscopic systems in urology: a systematic review and meta-analysis. J Endourol 2018;32(9). https://doi.org/10.1089/end.2018.0411.

[3] Ochsner JL. Minimally invasive surgical procedures. Ochsner J 2000;2(3). https://doi.org/10.1007/s10039-011-1814-0.

[4] Wolfe BM, Szabo Z, Moran ME, Chan P, Hunter JG. Training for minimally invasive surgery—need for surgical skills. Surg Endosc 1993;7(2). https://doi.org/10.1007/BF00704386.

[5] King CR, Lum D. Techniques in minimally invasive surgery for advanced endometriosis. Curr Opin Obstet Gynecol 2016;28(4). https://doi.org/10.1097/GCO.0000000000000291.

[6] van der Putten EPW, Goossens RHM, Jakimowicz JJ, Dankelman J. Haptics in minimally invasive surgery—a review. Minim Invasive Ther Allied Technol 2008;17(1). https://doi.org/10.1080/13645700701820242.

[7] White AD, et al. Does monitor position influence visual-motor performance during minimally invasive surgery? J Surg Simul 2016;3. https://doi.org/10.1102/2051-7726.2016.0001.

[8] Kumar A, Kumar S, Kaushik A, Kumar A, Saini JS. Real time estimation and suppression of hand tremor for surgical robotic applications. Microsyst Technol 2020. https://doi.org/10.1007/s00542-019-04736-1.

[9] Marengo F, Larraín D, Babilonti L, Spinillo A. Learning experience using the double-console da Vinci surgical system in gynecology: a prospective cohort study in a University hospital. Arch Gynecol Obstet 2012;285(2). https://doi.org/10.1007/s00404-011-2005-8.

[10] Zwart MJW, et al. Added value of 3D-vision during robotic pancreatoduodenectomy anastomoses in biotissue (LAEBOT 3D2D): a randomized controlled cross-over trial. Surg Endosc 2021;35(6). https://doi.org/10.1007/s00464-020-07732-z.

[11] Zou X, et al. Natural orifice transluminal endoscopic surgery in urology: the Chinese experience. Asian J Urol 2020;7(1). https://doi.org/10.1016/j.ajur.2019.07.001.

[12] Zhao J, Feng B, Zheng MH, Xu K. Surgical robots for SPL and NOTES: a review. Minim Invasive Ther Allied Technol 2015;24(1). https://doi.org/10.3109/13645706.2014.999687.

[13] Azizi Koutenaei B, Wilson E, Monfaredi R, Peters C, Kronreif G, Cleary K. Robotic natural orifice transluminal endoscopic surgery (R-NOTES): literature review and prototype system. Minim Invasive Ther Allied Technol 2015;24(1). https://doi.org/10.3109/13645706.2014.992907.

[14] Anon. Johnson & Johnson announces agreement to acquire Auris Health, Inc, https://www.jnj.com/johnson-johnson-announces-agreement-to-acquire-auris-health-inc; 2021. (Accessed 28 April 2021).

[15] Anon. Hominis surgical system, https://www.mddionline.com/robotics/fda-authorizes-first-surgical-robot-help-transvaginal-hysterectomies; 2021. (Accessed 28 April 2021).
[16] Manfredi L. Endorobots for colonoscopy: design challenges and available technologies. vol. 8; 2021. p. 1–10. https://doi.org/10.3389/frobt.2021.705454.
[17] Flemming J, Cameron S. Small bowel capsule endoscopy: indications, results, and clinical benefit in a University environment. Medicine 2018;97(14).
[18] Yang YJ. The future of capsule endoscopy: the role of artificial intelligence and other technical advancements. Clin Endosc 2020;53(4). https://doi.org/10.5946/ce.2020.133.
[19] Piccirelli S, Milluzzo SM, Bizzotto A, Cesaro P, Pecere S, Spada C. Small bowel capsule endoscopy and artificial intelligence: first or second reader? Best Pract Res Clin Gastroenterol 2021;52–53. https://doi.org/10.1016/j.bpg.2021.101742.
[20] Jain D, Goyal A, Zavala S. Predicting colonoscopy time: a quality improvement initiative. Clin Endosc 2016;49(6):555–9. https://doi.org/10.5946/ce.2015.110.
[21] Ben-Menachem T, et al. Adverse events of upper GI endoscopy. Gastrointest Endosc 2012;76 (4). https://doi.org/10.1016/j.gie.2012.03.252.
[22] Ortega Alcaide J, et al. Tether-colon interaction model and tribological characterization for front-wheel driven colonoscopic devices. Tribol Int 2021;156:106814. https://doi.org/10.1016/j.triboint.2020.106814.
[23] Nelson BJ, Kaliakatsos IK, Abbott JJ. Microrobots for minimally invasive medicine. Annu Rev Biomed Eng 2010;12(August):55–85. https://doi.org/10.1146/annurev-bioeng-010510-103409.
[24] Sitti M. Miniature soft robots—road to the clinic. Nat Rev Mater 2018;3(6):74–5. https://doi.org/10.1038/s41578-018-0001-3.
[25] Helander HF, Fändriks L. Surface area of the digestive tract-revisited. Scand J Gastroenterol 2014;49(6). https://doi.org/10.3109/00365521.2014.898326.
[26] Alazmani A, Hood A, Jayne D, Neville A, Culmer P. Quantitative assessment of colorectal morphology: implications for robotic colonoscopy. Med Eng Phys 2016;38(2):148–54. https://doi.org/10.1016/j.medengphy.2015.11.018.
[27] Rondonotti E. Capsule retention: prevention, diagnosis and management. Ann Transl Med 2017;5(9):1–8. https://doi.org/10.21037/atm.2017.03.15.
[28] Karagiannis S, Faiss S, Mavrogiannis C. Capsule retention: a feared complication of wireless capsule endoscopy. Scand J Gastroenterol 2009;44(10):1158–65. https://doi.org/10.1080/00365520903132039.
[29] Spada C, et al. Accuracy and safety of second-generation PillCam COLON capsule for colorectal polyp detection. Therap Adv Gastroenterol 2012;5:173–8. https://doi.org/10.1177/1756283X12438054.
[30] Friedel D, Modayil R, Stavropoulos S. Colon capsule endoscopy: review and perspectives. Gastroenterol Res Pract 2016;2016. https://doi.org/10.1155/2016/9643162.
[31] Ashour AS, et al. Colored video analysis in wireless capsule endoscopy: a survey of state-of-the-art. Curr Med Imaging 2020;16(9). https://doi.org/10.2174/1573405616666200124140915.
[32] Patwa A, Shah A. Anatomy and physiology of respiratory system relevant to anaesthesia. Indian J Anaesth 2015;59(9). https://doi.org/10.4103/0019-5049.165849.
[33] Seneterre E, Paganin F, Bruel JM, Michel FB, Bousquet J. Measurement of the internal size of bronchi using high resolution computed tomography (HRCT). Eur Respir J 1994;7(3):596–600. https://doi.org/10.1183/09031936.94.07030596.
[34] Stierschneider M, Franz S, Baumgartner W. Endoscopic examination of the upper respiratory tract and oesophagus in small ruminants: technique and normal appearance. Vet J 2007;173 (1):101–8. https://doi.org/10.1016/j.tvjl.2005.09.002.

[35] Wheeler M, Ovassapian A. Fiberoptic endoscopy—aided techniques. In: Benumof's airway management: principles and practice. Mosby; 2007. p. 399–438.
[36] Stahl D, Richard K, Papadimos T. Complications of bronchoscopy: a concise synopsis. Int J Crit Illn Inj Sci 2015;5(3):189. https://doi.org/10.4103/2229-5151.164995.
[37] Fielding DI, Maldonado F, Murgu S. Achieving competency in bronchoscopy: challenges and opportunities. Respirology 2014;19(4):472–82. https://doi.org/10.1111/resp.12279.
[38] Rojas-Solano JR, Ugalde-Gamboa L, MacHuzak M. Robotic bronchoscopy for diagnosis of suspected lung cancer: a feasibility study. J Bronchol Interv Pulmonol 2018;25(3): 168–75. https://doi.org/10.1097/LBR.0000000000000499.
[39] Krimsky WS, Pritchett MA, Lau KKW. Towards an optimization of bronchoscopic approaches to the diagnosis and treatment of the pulmonary nodules: a review. J Thorac Dis 2018;10(Suppl 14):S1637–44. https://doi.org/10.21037/jtd.2018.04.38.
[40] Margaris KN, Black RA. Modelling the lymphatic system: challenges and opportunities. J R Soc Interface 2012;9(69):601–12. https://doi.org/10.1098/rsif.2011.0751.
[41] Costa-Hong V, Katayama KY, Consolim-Colombo FM. Methods to investigate endothelial function in humans. Endothel Cardiovasc Dis Vasc Biol Clin Syndr 2018;217–28. https://doi.org/10.1016/B978-0-12-812348-5.00016-7.
[42] Dodge JT, Brown BG, Bolson EL, Dodge HT. Lumen diameter of normal human coronary arteries: influence of age, sex, anatomic variation, and left ventricular hypertrophy or dilation. Circulation 1992;86(1):232–46. https://doi.org/10.1161/01.CIR.86.1.232.
[43] Geleto G, Getnet W, Tewelde T. Mean normal portal vein diameter using sonography among clients coming to radiology department of Jimma University Hospital, Southwest Ethiopia. Ethiop J Health Sci 2016;26(3):237–42. https://doi.org/10.4314/ejhs.v26i3.6.
[44] Zhou FF, et al. Coronary artery diameter is inversely associated with the severity of coronary lesions in patients undergoing coronary angiography. Cell Physiol Biochem 2017;43(3):1247–57. https://doi.org/10.1159/000481765.
[45] Hu M, Ge X, Chen X, Mao W, Qian X, Yuan WE. Micro/nanorobot: a promising targeted drug delivery system. Pharmaceutics 2020;12(7). https://doi.org/10.3390/pharmaceutics12070665.
[46] Duffner F, Schiffbauer H, Glemser D, Skalej M, Freudenstein D. Anatomy of the cerebral ventricular system for endoscopic neurosurgery: a magnetic resonance study. Acta Neurochir 2003;145(5):359–68. https://doi.org/10.1007/s00701-003-0021-6.
[47] Purdy PD, Fujimoto T, Replogle RE, Giles BP, Fujimoto H, Miller SL. Percutaneous intraspinal navigation for access to the subarachnoid space: use of another natural conduit for neurosurgical procedures. Neurosurg Focus 2005;19(1):17–21. https://doi.org/10.3171/foc.2005.19.1.12.
[48] Kohler TS, Yadven M, Manvar A, Liu N, Monga M. The length of the male urethra. Int Braz J Urol 2008;34(4):451–4. https://doi.org/10.1590/s1677-55382008000400007.
[49] Abelson B, et al. Sex differences in lower urinary tract biology and physiology. Biol Sex Differ 2018;9(1):1–13. https://doi.org/10.1186/s13293-018-0204-8.
[50] Pomian A, et al. Demographic features of female urethra length. Neurourol Urodyn 2018;37(5):1751–6. https://doi.org/10.1002/nau.23509.
[51] Chang YL, Lin ATL, Chen KK, Chang LS. The caliber and length of the urethra in asymptomatic Taiwanese adult females. Urol Sci 2010;21(2):93–5. https://doi.org/10.1016/S1879-5226(10)60019-X.
[52] Gkougkousis EG, Jain S, Mellon JK. Urologic issues for the nephrologist. Comprehensive clinical nephrology. Mosby; 2010.
[53] Köhler O, Belej K, Kaplan O. Lasers in urology. In: Lasers for medical applications diagnostics, therapy and surgery; 2013. p. 509–38. https://doi.org/10.1533/9780857097545.4.509.

[54] Mohamed S, Claes C, Tsang CW. Review of small gauge vitrectomy: progress and innovations. J Ophthalmol 2017;2017. https://doi.org/10.1155/2017/6285869.

[55] Sha X, Sun H, Zhao Y, Li W, Li WJ. A review on microscopic visual servoing for micromanipulation systems: applications in micromanufacturing, biological injection, and nanosensor assembly. Micromachines 2019;10(12). https://doi.org/10.3390/mi10120843.

[56] Hubschman JP, Wilson J, Tsao TC, Schwartz S. Robotic eye surgery. Ophthalmology 2010;117(4):857. https://doi.org/10.1016/j.ophtha.2009.11.010.

[57] Ergeneman O, et al. A magnetically controlled wireless optical oxygen sensor for intraocular measurements. Sensors (Peterborough, NH) 2008;8(1):29–37.

[58] Jacobson SA, Epstein AH. An informal survey of power MEMS. In: The international symposium on micro-mechanical engineering, vol. 12; 2003. p. 513–9. July. Available at: http://www.ttivanguard.com/montrealreconn/powerMEMS.pdf.

[59] Cavity T. The tympanic cavity. In: Middle ear mastoid microsurgery; 2017. https://doi.org/10.1055/b-0034-74917.

[60] Liem T. The organ of hearing and balance. Cranial Osteopat 1988;605–33. https://doi.org/10.1016/b978-044307499-8.50024-7.

[61] Nelson BJ, Kaliakatsos IK, Abbott JJ. Microrobots for minimally invasive medicine. Annu Rev Biomed Eng 2010;12(August):55–85. https://doi.org/10.1146/annurev-bioeng-010510-103409.

[62] Berris M, Shoham M. Febotics—a marriage of fetal surgery and robotics. Comput Aided Surg 2006;11(4):175–80. https://doi.org/10.3109/10929080600931579.

[63] Sheth KR, Van Batavia JP, Bowen DK, Koh CJ, Srinivasan AK. Minimally invasive surgery in pediatric urology. 2018;45(4):611–21.

[64] Mathis J, Raio L, Baud D. Fetal laser therapy: applications in the management of fetal pathologies. Prenat Diagn 2015;35(7):623–36. https://doi.org/10.1002/pd.4587.

[65] Sago H, Ishii K, Sugibayashi R, Ozawa K, Sumie M, Wada S. Fetoscopic laser photocoagulation for twin-twin transfusion syndrome. J Obstet Gynaecol Res 2018;44(5):831–9. https://doi.org/10.1111/jog.13600.

[66] Nott JP, Bonney EA, Pickering JD, Simpson NAB. The structure and function of the cervix during pregnancy. Transl Res Anat 2016;2:1–7. https://doi.org/10.1016/j.tria.2016.02.001.

[67] Anderson JG, Abrahamson K. Your health care may kill you: medical errors. In: Studies in health technology and informatics, 234; 2017. https://doi.org/10.3233/978-1-61499-742-9-13.

[68] Shademan A, Decker RS, Opfermann JD, Leonard S, Krieger A, Kim PCW. Supervised autonomous robotic soft tissue surgery. Sci Transl Med 2016;8:337. https://doi.org/10.1126/scitranslmed.aad9398.

[69] Fagogenis G, et al. Autonomous robotic intracardiac catheter navigation using haptic vision. Sci Robot 2019;4(29). https://doi.org/10.1126/scirobotics.aaw1977.

[70] Svoboda E. Your robot surgeon will see you now. Nature 2019;573(7775). https://doi.org/10.1038/d41586-019-02874-0.

[71] Deng C, Ji X, Rainey C, Zhang J, Lu W. Integrating machine learning with human knowledge. iScience 2020;23(11). https://doi.org/10.1016/j.isci.2020.101656.

[72] Lee J, Nam S, Park S. Energy-efficient control of mobile processors based on long short-term memory. IEEE Access 2019;7. https://doi.org/10.1109/ACCESS.2019.2923334.

[73] Yu L, Yu Y. Energy-efficient neural information processing in individual neurons and neuronal networks. J Neurosci Res 2017;95(11):2253–66. https://doi.org/10.1002/jnr.24131.

[74] Maass W. Energy-efficient neural network chips approach human recognition capabilities. Proc Natl Acad Sci U S A 2016;113(41):11387–9. https://doi.org/10.1073/pnas.1614109113.

[75] Chen W, et al. Wireless powered capsule endoscopy for colon diagnosis and treatment. Physiol Meas 2013;34(11):1545–61. https://doi.org/10.1088/0967-3334/34/11/1545.

[76] Basar MR, Ahmad MY, Cho J, Ibrahim F. Application of wireless power transmission systems in wireless capsule endoscopy: an overview. Sensors (Switzerland) 2014;14(6):10929–51. https://doi.org/10.3390/s140610929.

[77] Sun Y, Agostini NB, Dong S, Kaeli D. Summarizing CPU and GPU design trends with product data. arXiv; 2019.

[78] Hamilton KE, Schuman CD, Young SR, Bennink RS, Imam N, Humble TS. Accelerating scientific computing in the post-Moore's era. ACM Trans Parallel Comput 2020;7(1). https://doi.org/10.1145/3380940.

[79] Moore GM. Cramming more components onto integrated circuits with unit cost. Electronics 1965;38(8):114. Available at: https://newsroom.intel.com/wp-content/uploads/sites/11/2018/05/moores-law-electronics.pdf.

[80] Ladd TD, Jelezko F, Laflamme R, Nakamura Y, Monroe C, O'Brien JL. Quantum computers. Nature 2010;464(7285). https://doi.org/10.1038/nature08812.

[81] Huang HL, Wu D, Fan D, Zhu X. Superconducting quantum computing: a review. Sci China Inf Sci 2020;63(8). https://doi.org/10.1007/s11432-020-2881-9.

[82] Feizi N, Tavakoli M, Patel RV, Atashzar SF. Robotics and AI for teleoperation, tele-assessment, and tele-training for surgery in the Era of COVID-19: existing challenges, and future vision. Front Robot AI 2021;8. https://doi.org/10.3389/frobt.2021.610677.

[83] El-Atab N, et al. Soft actuators for soft robotic applications: a review. Adv Intell Syst 2020;2(10). https://doi.org/10.1002/aisy.202000128.

[84] Balch CM. Stress and burnout among surgeons. Arch Surg 2009;144(4):371. https://doi.org/10.1001/archsurg.2008.575.

[85] Klein M, Andersen LPH, Alamili M, Gögenur I, Rosenberg J. Psychological and physical stress in surgeons operating in a standard or modern operating room. Surg Laparosc Endosc Percutan Tech 2010;20(4). https://doi.org/10.1097/SLE.0b013e3181ed851d.

[86] Hurley AM, et al. SOS save our surgeons: stress levels reduced by robotic surgery. Gynecol Surg 2015;12(3):197–206. https://doi.org/10.1007/s10397-015-0891-7.

[87] Lemke M, et al. Colonoscopy trainers experience greater stress during insertion than withdrawal: implications for endoscopic curricula. J Can Assoc Gastroenterol 2019;(Xx):1–6. https://doi.org/10.1093/jcag/gwz031.

[88] Harvin G. Review of musculoskeletal injuries and prevention in the endoscopy practitioner. J Clin Gastroenterol 2014;48(7):590–4. https://doi.org/10.1097/MCG.0000000000000134.

[89] Villa E, Attar B, Trick W, Kotwal V. Endoscopy-related musculoskeletal injuries in gastroenterology fellows. Endosc Int Open 2019;07(06):E808–12. https://doi.org/10.1055/a-0811-5985.

[90] Austin K, Schoenberger H, Sesto M, Gaumnitz E, Teo Broman A, Saha S. Musculoskeletal injuries are commonly reported among gastroenterology trainees: results of a National Survey. Dig Dis Sci 2019;64(6):1439–47. https://doi.org/10.1007/s10620-019-5463-7.

[91] Lee SH, Park YK, Lee DJ, Kim KM. Colonoscopy procedural skills and training for new beginners. World J Gastroenterol 2014;20(45):16984–95. https://doi.org/10.3748/wjg.v20.i45.16984.

[92] Siau K, Hawkes ND, Dunckley P. Training in endoscopy. Curr Treat Options Gastroenterol 2018;16(3):345–61. https://doi.org/10.1007/s11938-018-0191-1.

[93] Chan B, Hussey A, Rubinger N, Hookey L. Patient comfort scores do not affect endoscopist behavior during colonoscopy, while trainee involvement has negative effects on patient comfort. Endosc Int Open 2017;05(12):E1259–67. https://doi.org/10.1055/s-0043-120828.

[94] Manfredi L, Natale G. New robotic technologies in cancer colon screening. Clin Cancer Drugs 2018;5(2):68–74. https://doi.org/10.2174/2212697x06666181220130020.
[95] Safi S, Thiessen T, Schmailzl KJG. Acceptance and resistance of new digital technologies in medicine: qualitative study. JMIR Res Protoc 2018;7(12). https://doi.org/10.2196/11072.
[96] Rogers EM, Singhal A, Quinlan MM. Diffusion of innovations. In: An integrated approach to communication theory and research. 3rd ed. Routledge; 2019.
[97] Porter M. What is value in health care? N Engl J Med 2010;363:2477–81.

Index

Note: Page numbers followed by "*f*" indicate figures, and "*t*" indicate tables.

I

J

K

L

M

W

X

Y

Z

FSC
www.fsc.org
MIX
Papier | Fördert gute Waldnutzung
FSC® C083411

Kolibri 360 GmbH
Ferdinand-Jühlke-Straße 7
99095 Erfurt, Deutschland
produktsicherheit@kolibri360.de